THE EFFECTS OF INDUCED HYDRAULIC FRACTURING ON THE ENVIRONMENT

Commercial Demands vs. Water, Wildlife, and Human Ecosystems

THE EFFECTS OF INDUCED HYDRAULIC FRACTURING ON THE ENVIRONMENT

Commercial Demands vs. Water, Wildlife, and Human Ecosystems

Edited by

Matthew McBroom, PhD

Apple Academic Press

TORONTO NEW JERSEY

Apple Academic Press Inc.
3333 Mistwell Crescent
Oakville, ON L6L 0A2
Canada

Apple Academic Press Inc.
9 Spinnaker Way
Waretown, NJ 08758
USA

Exclusive worldwide distribution by CRC Press, a member of Taylor & Francis Group

International Standard Book Number-13: 978-1-926895-83-3 (Hardcover)

Library of Congress Control Number: 2013950897

Library and Archives Canada Cataloguing in Publication

The effects of induced hydraulic fracturing on the environment: commercial demands vs. water, wildlife, and human ecosystems/edited by Matthew McBroom, PhD.

The chapters in this book were previously published in various places and in various formats.
Includes bibliographical references and index.
ISBN 978-1-926895-83-3
1. Hydraulic fracturing--Environmental aspects. 2. Water--Pollution. 3. Animals. 4. Biotic communities. 5. Shale gas industry--Environmental aspects. I. McBroom, Matthew W. (Matthew Wayne), 1972-, author, editor of compilation

TD195.G3E34 2013 333.8'2314 C2013-906623-3

Apple Academic Press also publishes its books in a variety of electronic formats. Some content that appears in print may not be available in electronic format. For information about Apple Academic Press products, visit our website at **www.appleacademicpress.com** and the CRC Press website at **www.crcpress.com**

ABOUT THE EDITOR

Matthew McBroom, PhD

Dr. Matthew McBroom is an associate professor at the Arthur Temple College of Foresrty and Agriculture at Stephen F. Austin State University in Nacogdoches, Texas, USA. His primary research interests are in the effects of land management on water resources, and how to design and implement land management activities to minimize potential impacts on water quality. The Texas Intensive Silviculture Study is one of his primary research projects, examining the effects of forest management activities such as clearcutting and site preparation with best management practices. He is also examining the dynamics of woody debris loading in the Sabine River, along with several other water resources studies.

CONTENTS

ACKNOWLEDGMENT AND HOW TO CITE

The chapters in this book were previously published in various places and in various formats. By bringing them together here in one place, we offer the reader a comprehensive perspective on recent investigations of hydraulic fracturing. Each chapter is enriched by being placed within the context of the larger investigative landscape.

We wish to thank the authors who made their research available for this book, whether by granting their permission individually or by releasing their research as open source articles. When citing information contained within this book, please do the authors the courtesy of attributing them by name, referring back to their original articles, using the credits provided at the end of each chapter.

LIST OF CONTRIBUTORS

Charles R. Anderson, Jr.
Colorado Division of Parks and Wildlife, Grand Junction, Colorado 81505 USA

Joseph R. Bidwell
Department of Zoology, Oklahoma State University, 430 Life Sciences West, Stillwater, OK 74078, USA

William Bourcier
Lawrence Livermore National Laboratory, 7000 East Avenue, Livermore, CA, 94550

R. Terry Bowyer
Department of Biological Sciences, Idaho State University, Pocatello, Idaho 83209 USA

Terence J. Centner
Department of Agricultural and Applied Economics, College of Agricultural and Environmental Sciences, The University of Georgia, Athens, GA 30602, USA

Naomi L. Cooper
Department of Zoology, Oklahoma State University, 430 Life Sciences West, Stillwater, OK 74078, USA

Sally Entrekin
Biology Department, University of Central Arkansas, Conway, AR

Michelle Evans-White
Department of Biological Sciences, University of Arkansas, Fayetteville, AR

Souheil Ezzedine
Lawrence Livermore National Laboratory, 7000 East Avenue, Livermore, CA, 94550

Wang Fengfan
Key Lab of Submarine Geosciences and Prospecting Techniques, Ministry of Education, Ocean University of China, Qingdao 266100, China

Madelon L. Finkel
Department of Public Health, Weill Cornell Medical College, New York, NY 10065, USA

Jonathan C. Fisher
Department of Zoology , Oklahoma State University, Stillwater, OK 74078

Matthew Fry
Department of Geography, University of North Texas, Denton, Texas 76203, United States

Li Gaolin
Key Lab of Submarine Geosciences and Prospecting Techniques, Ministry of Education, Ocean University of China, Qingdao 266100, China

Bernard D. Goldstein
Graduate School of Public Health, University of Pittsburgh, Pittsburgh, Pennsylvania, USA

Elizabeth Hagenbuch
Dynamac Corporation, contracted by the US EPA, Cincinnati, OH

Zhang Haiyan
Key Lab of Submarine Geosciences and Prospecting Techniques, Ministry of Education, Ocean University of China, Qingdao 266100, China

Heather Hatzenbuhler
Department of Agricultural and Applied Economics, College of Agricultural and Environmental Sciences, The University of Georgia, Athens, GA 30602, USA

Jake Hays
Physicians Scientists & Engineers for Healthy Energy (PSE), 452 West 57th Street Apt 3E, New York, NY 10019, USA

David J. Hoeinghaus
Department of Biology, University of North Texas, Denton, Texas 76203, United States

Liu Huaishan
Key Lab of Submarine Geosciences and Prospecting Techniques, Ministry of Education, Ocean University of China, Qingdao 266100, China

Jonathan Hunt
Lawrence Livermore National Laboratory, 7000 East Avenue, Livermore, CA, 94550

Brent Johnson
National Exposure Research Laboratory, US Environmental Protection Agency (EPA), Cincinnati, OH

John G. Kie
Department of Biological Sciences, Idaho State University, Pocatello, Idaho 83209 USA

Jill Kriesky
Graduate School of Public Health, University of Pittsburgh, Pittsburgh, Pennsylvania, USA

Thomas W. La Point
Department of Biology, University of North Texas, Denton, Texas 76203, United States

Adam Law
Weill Cornell Medical College, Cayuga Medical Center, Ithaca, NY 14850, USA

Patrick E. Lendrum
Department of Biological Sciences, Idaho State University, Pocatello, Idaho 83209 USA

Ryan A. Long
Department of Biological Sciences, Idaho State University, Pocatello, Idaho 83209 USA

Matthew McBroom
Arthur Temple College of Forestry and Agriculture, Stephen F. Austin State University, Nacogdoches, TX 75962, USA

Barbara Pavliakova
Graduate School of Public Health, University of Pittsburgh, Pittsburgh, Pennsylvania, USA

Alexandra G. Ponette-Gonzalez
Department of Geography, University of North Texas, Denton, Texas 76203, United States

Sarah Roberts
Lawrence Livermore National Laboratory, 7000 East Avenue, Livermore, CA, 94550

Tong Siyou
Key Lab of Submarine Geosciences and Prospecting Techniques, Ministry of Education, Ocean University of China, Qingdao 266100, China

Todd Thomas
Texas A&M Forest Service, Lufkin, TX 75901, USA

Ruthanne Thompson
Department of Biology, University of North Texas, Denton, Texas 76203, United States

Yanli Zhang
Arthur Temple College of Forestry and Agriculture, Stephen F. Austin State University, Nacogdoches, TX 75962, USA

LIST OF ABBREVIATIONS

ADEQ	Arkansas Department of Environmental Quality
ADQ	audit of data quality
API	American Petroleum Institute
ASTM	American Society for Testing and Materials
AU	assessment unit
BMP	best management practice
Br-DBP	brominated disinfection byproduct
BTEX	benzene, toluene, ethyl-benzene, and xylene
CASRN	Chemical Abstracts Service Registration Number
CBI	confidential business information
CBM	coalbed methane
CH_4	methane
cm	centimeter
CO_2	carbon dioxide
COGCC	Colorado Oil and Gas Conservation Commission
CERCLA	Comprehensive Environmental Response, Compensation, and Liability Act
CWA	Clean Water Act
CWT	centralized waste treatment facility
DBP	disinfection byproduct
DEP	Department of Environmental Protection
DFW	Dallas-Fort Worth
DOE	U.S. Department of Energy
DRO	diesel range organics
DSSTox	Distributed Structure-Searchable Toxicity Database Network
EDTA	ethylenediaminetetraacetic acid
EGS	enhanced geothermal systems
EPA	U.S. Environmental Protection Agency
Fe	iron

FORTRAN	formula translation
FRAC	Fracturing Responsibility and Awareness of Chemicals Act
g	gram
GIS	geographic information system
GRO	gasoline range organics
GWPC	Ground Water Protection Council
ha	hectare
HAA	haloacetic acid
HSPF	hydrologic simulation program
INJWELL	injection well
IRIS	integrated risk information system
J	joule
kg	kilogram
L	liter
LBNL	Lawrence Berkeley National Laboratory
LOAEL	lowest observed adverse effect levels
m	meter
MCL	maximum contaminant level
mg	milligram
MGD	million gallons per day
MHz	megahertz
mm	millimeter
MSDS	material safety data sheet
NAS	National Academy of Sciences
NDIC	North Dakota Industrial Commission
NEMS	National Energy Modeling System
NH_4^+	ammonia
NHD	national hydrography dataset
NOM	naturally occurring organic matter
NO_x	nitrogen oxide
NPDES	National Pollutant Discharge Elimination System
NRC	National Response Center
NYSDEC	New York State Department of Environmental Conservation
OSPER	Osage-Skiatook Petroleum Environmental Research

PADEP	Pennsylvania Department of Environmental Protection
PCA	principle component analyses
PO_4^+	phosphate
POTW	publicly owned treatment work
PPRTV	provisional peer-reviewed toxicity value
PWS	public water systems
QA	quality assurance
QAPP	quality assurance project plan
QC	quality control
RCRA	Resource Conservation and Recovery Act
rpm	rotations per minute
RRC	Railroad Commission of Texas
s	second
SDWA	Safe Drinking Water Act
SO_2	sulfur dioxide
SOP	standard operating procedure
SRB	Susquehanna River Basin
SRBC	Susquehanna River Basin Commission
SWAT	soil and water assessment tool
TDS	total dissolved solids
THM	trihalomethane
TIE	toxicity identification and evaluation
TOPKAT	toxicity prediction by komputer assisted technology
TOUGH	transport of unsaturated groundwater and heat
TSA	technical systems audit
TPH	total petroleum hydrocarbons
TSCA	Toxic Substances Control Act
TSS	total suspended solids
UCRB	Upper Colorado River Basin
UIC	underground injection control
US EIA	US Energy Information Administration
US FWS	US Fish and Wildlife Service
US GAO	US Government Accountability Office
US OMB	US Office of Management and Budget
USCB	US Census Bureau
USDA	US Department of Agriculture

USGS	U.S. Geological Survey
USHR	US House of Representatives
WET	whole effluent toxicity
WWTP	wastewater treatment plant
μm	micrometer

INTRODUCTION

Hydraulic fracturing, or "fracking" as it is commonly known, refers to the practice of using liquids at very high pressures to fragment oil and gas-bearing geologic strata, thereby allowing hydrocarbons to be harvested. This process, while it may increase energy resources, has myriad negative potential environmental impacts as well. This book looks not at the specifics of fracking as an industry, but instead at the environmental impacts. The first section looks at fracturing and the water supply; how hydraulic fracturing depletes finite water resources and contaminates water supplies. Section II looks at ecosystems and wildlife; fracking leads to habitat destruction and fragmentation, and various forms of wildlife species are either becoming endangered or are forced to find new areas to live. The final section examines the effect on human ecosystems and human health.

In the first chapter, Heather Hatzenbuhler and Terence J. Centner examine the regulation of water pollution at various horizontal drill sites. With the introduction of horizontal drilling, new commercial sources of energy have become available. Wells are drilled and injected with large quantities of water mixed with specially selected chemicals at high pressures that allow petroleum reserves to flow to the surface. While the increased economic activities and the outputs of domestic energy are welcomed, there is growing concern over negative environmental impacts from horizontal drilling in shale formations. The potential for water contamination, land destruction, air pollution, and geologic disruption has raised concerns about the merits of production activities used during extraction. The chapter looks at the impacts of horizontal drilling using hydraulic fracturing on water supplies and takes a comprehensive look at legislative and regulatory approaches to mitigate environmental risks in the Marcellus shale region. The overview identifies shortcomings associated with regulatory controls by local and state governments and offers two policy suggestions to better protect waters of the region.

The second chapter, by Matthew McBroom et al., looks at the impacts on surface water and, relatedly, soil erosion, rather than drilled wells. In their experiment, a 1.4 ha natural gas well pad was constructed in an intermittent stream channel at the Alto Experimental Watersheds in East Texas, USA (F1), while another 1.1 ha well pad was offset about 15 m from a nearby intermittent stream (F2). V-notch weirs were constructed downstream of these well pads and stream sedimentation and water quality was measured. For the 2009 water year, about 11.76 cm, or almost 222% more runoff resulted from F1 than F2. Sediment yield was significantly greater at F1, with 13,972 kg ha^{-1} yr^{-1} versus 714 kg ha^{-1} yr^{-1} at F2 on a per unit area disturbance basis for the 2009 water year. These losses were greater than was observed following forest clearcutting with best management practices (111–224 kg ha^{-1}). Significantly greater nitrogen and phosphorus losses were measured at F1 than F2. While oil and gas development can degrade surface water quality, appropriate conservation practices like retaining streamside buffers can mitigate these impacts.

Huaishan et al. examine the interactions between water supplies and the gas that is introduced during the fracturing process in chapter 3. They discover that in a certain frequency range, gas is an effective absorber and scatterer of sound, which changes the compressibility of water, and then changes the speed and frequency of sound. Gas continues rising, deforming, and dissolving. The same bubble of natural gas has different radii at different depths. By analyzing these changes, the resonance frequency of gas bubble, and its impacts on sound wave, characteristics of the influences of gas at different depths on the incident sound wave can be obtained. The main sound features of gas are relevant to the gas size, gas content, velocity, attenuation, resonance frequency, the scattering cross-section, and so forth. Sound models with hydrate and free gas in the water and sediment are established. Through the practical application to actual data, the sound characteristics yielded when the gas (or gas hydrate dissociation) escaped the water of seismic data are very clear.

In chapter 4, Bidwell et al. investigated potential biological effects of produced water contamination derived from occasional surface overflow and possible subsurface intrusion at an oil production site along the shore of Skiatook Lake, Oklahoma. They monitored basic chemistry and acute toxicity to a suite of standard aquatic test species in produced water and in samples taken from shallow groundwater wells on the site. Toxicity identification evaluations and ion toxicity modeling were used to identify toxic constituents in the samples. Lake sediment at the oil production site and at a reference site were also analyzed for brine intrusion chemically and by testing sediment toxicity using the benthic invertebrates, Chironomus dilutus, and Hyallela azteca. Sediment quality was also assessed with in situ survival and growth studies with *H. azteca* and the Asian clam, *Corbicula fluminea*, and by benthic macroinvertebrate community sampling. They found that the produced water was acutely toxic to the aquatic test organisms at concentrations ranging from 1% to 10% of the whole produced water sample. Toxicity identification evaluation and ion toxicity modeling indicated major ion salts and hydrocarbons were the primary mixture toxicants. The standardized test species used in the laboratory bioassays exhibited differences in sensitivity to these two general classes of contaminants, which underscores the importance of using multiple species when evaluating produced water toxicity. Toxicity of groundwater was greater in samples from wells near a produced water injection well and an evaporation pond. Principle component analyses (PCA) of chemical data derived from the groundwater wells indicated dilution by lake water and possible biogeochemical reactions as factors that ameliorated groundwater toxicity. Elevated concentrations of major ions were found in pore water from lake sediments, but toxicity from these ions was limited to sediment depths of 10 cm or greater, which is outside of the primary zone of biological activity. The study was able to demonstrate the utility of ion toxicity modeling to support data from toxicity identification evaluations aimed at identifying key toxic constituents in produced water. The study also demonstrated how geographic information systems, toxicity modeling, and toxicity assessment could be used to facilitate future site assessments.

Chapter 5 is a shorter, column-like piece by Fry et al. In it, they argue that the conflicts arising in the United States over water, as water grows scarce in some areas that have a high rate of fracking, foreshadow conflicts in other countries that also have cities situated over large natural-gas deposits. They use the example of Dallas-Fort Worth in Texas, USA to illustrate the challenges that are associated with balancing both energy needs and water sustainability in cities that already have a limited water supply.

Entrekin et al. argue a need for expanded research on the impact of natural gas extraction on the environment in chapter 6. Extraction of natural gas from hard-to-reach reservoirs has expanded around the world and poses multiple environmental threats to surface waters. Improved drilling and extraction technology used to access low permeability natural gas requires millions of liters of water and a suite of chemicals that may be toxic to aquatic biota. There is growing concern among the scientific community and the general public that rapid and extensive natural gas development in the US could lead to degradation of natural resources. Gas wells are often close to surface waters that could be impacted by elevated sediment runoff from pipelines and roads, alteration of streamflow as a result of water extraction, and contamination from introduced chemicals or the resulting wastewater. However, the data required to fully understand these potential threats are currently lacking. Scientists therefore need to study the changes in ecosystem structure and function caused by natural gas extraction and to use such data to inform sound environmental policy.

In chapter 7, Hunt et al. propose new ways of using silica gels in the fracturing process. Fractures and fracture networks are the principal pathways for migration of water and contaminants in groundwater systems, fluids in enhanced geothermal systems (EGS), oil and gas in petroleum reservoirs, carbon dioxide leakage from geological carbon sequestration, and radioactive and toxic industrial wastes from underground storage repositories. When dealing with EGS fracture networks, there are several major issues to consider, e.g., the minimization of hydraulic short circuits and losses of injected geothermal fluid to the surrounding formation, which in turn maximize heat extraction and economic production. Gel deployments to direct and control fluid flow have been extensively and successfully used in the oil industry for enhanced oil recovery. However, to the best of our knowledge, gels have not been applied to EGS to enhance heat extraction. Insitu gelling systems can either be organic or inorganic. Organic polymer gels are generally not thermostable to the typical temperatures of EGS systems. Inorganic gels, such as colloidal silica gels, however, may be ideal blocking agents for EGS systems if suitable gelation times can be achieved. In the current study, we explore colloidal silica gelation times and rheology as a function of SiO_2 concentration, pH, salt concentration, and temperature, with preliminary results in the two-phase field above 100 °C. Results at 25 °C show that it may be possible to choose formulations that will gel in a reasonable and predictable amount of time at the temperatures of EGS systems.

Chapter 8 provides an excerpt from a recent EPA study on the impacts of natural gas extraction. The first of the sections reproduced here, on scenario evaluations, shows how computer models are being used to identify conditions that may lead to impacts on drinking water resources from hydraulic fracturing. The EPA has identified hypothetical, but realistic, scenarios pertaining to the water acquisition, well injection, and wastewater treatment and waste disposal stages of the water cycle. Potential impacts to drinking water sources from withdrawing large volumes of water in semi-arid and humid river basins—the Upper Colorado River Basin in the west and the Susquehanna River Basin in the east—are being compared and assessed. Additionally, complex computer models are being used to explore the possibility of subsurface gas and fluid migration from deep shale formations to overlying aquifers in six different scenarios. These scenarios include poor well construction and hydraulic communication via fractures (natural and created) and nearby existing wells. As a first step, the subsurface migration simulations will examine realistic scenarios to assess the conditions necessary for hydraulic communication rather than the probability of migration occurring. The second section, on laboratory studies is largely focused on identifying potential impacts of inadequately treating hydraulic fracturing wastewater and discharging it to rivers. Experiments are being designed to test how well common wastewater treatment processes remove selected contaminants from hydraulic fracturing wastewater, including radium and other metals. Other experiments are assessing whether or not hydraulic fracturing wastewater may contribute to the formation of disinfection byproducts during common drinking water treatment processes, with particular focus on the formation of brominated disinfection byproducts, which have significant health concerns at high exposure levels. Samples of raw hydraulic fracturing wastewater, treated wastewater, and water from rivers receiving treated hydraulic fracturing wastewater have been collected for source apportionment studies. Results from laboratory analyses of these samples are being used to develop a method for determining if treated hydraulic fracturing wastewater is contributing to high chloride and bromide levels at downstream public water supplies. Finally, existing analytical methods for selected chemicals are being tested, modified, and verified for use in this study and by others, as needed. Methods are being modified in cases where standard methods do not exist for the low-level detection of chemicals of interest or for use in the complex matrices associated with hydraulic fracturing wastewater. Analytical methods are currently being tested and modified for several classes of chemicals, including glycols, acrylamides, ethoxylated alcohols, disinfection byproducts, radionuclides, and inorganic chemicals.

Chapter 9 moves into the section on wildlife, examining the effect of natural gas development on mule deer habitat. Lendrum et al. examine how the disruption of traditional migratory routes by anthropogenic disturbances has shifted patterns of resource selection by many species, and in some instances has caused populations to decline. Moreover, in recent decades populations of mule deer (*Odocoileus hemionus*) have declined throughout much of their historic range in the western United States. We used resource-selection functions to determine if the presence of natural-gas development altered patterns of resource selection by migrating mule deer. We compared spring migration routes of adult female mule deer fitted with GPS collars (n = 167) among four study areas that had varying degrees of natural-gas development from 2008 to 2010 in the Piceance Basin of northwest Colorado, USA. Mule deer migrating through the most developed area had longer step lengths (straight-line distance between successive GPS locations) compared with deer in less-developed areas. Additionally, deer migrating through the most developed study areas tended to select for habitat types that provided greater amounts of concealment cover, whereas deer from the least developed areas tended to select habitats that increased access to forage and cover. Deer selected habitats closer to well pads and avoided roads in all instances except along the most highly developed migratory routes, where road densities may have been too high for deer to avoid roads without deviating substantially from established migration routes. These results indicate that behavioral tendencies toward avoidance of anthropogenic disturbance can be overridden during migration by the strong fidelity ungulates demonstrate towards migration routes. If avoidance is feasible, then deer may select areas further from development, whereas in highly developed areas, deer may simply increase their rate of travel along established migration routes.

Chapter 10 again represents part of a longer study, this one by the New York State Department of Environmental Conservation. The section reproduced here examines the potential adverse impacts on ecosystems and wildlife from high-volume hydraulic fracturing operations. Four areas of concern related to high-volume hydraulic fracturing are: (1) fragmentation of habitat; (2) potential transfer of invasive species; (3) impacts to endangered and threatened species; and (4) use of state-owned lands. The dSGEIS concludes that high-volume hydraulic fracturing operations would have a significant impact on the environment because such operations have the potential to draw substantial development into New York, which would result in unavoidable impacts to habitats (fragmentation, loss of connectivity, degradation, etc.), species distributions and populations, and overall natural resource biodiversity. Habitat loss, conversion, and fragmentation (both short-term and long-term) would result from land grading and clearing, and the construction of well pads, roads, pipelines, and other infrastructure associated with gas drilling. The number of vehicle trips associated with high-volume hydraulic fracturing, particularly at multi-well sites, has been identified as an activity which presents the opportunity to transfer invasive terrestrial species. Surface water withdrawals also have the potential to transfer invasive aquatic species. The introduction of terrestrial and aquatic invasive species would have a significant adverse impact on the environment. State-owned lands play a unique role in New York's landscape because they are managed under public ownership to allow for sustainable use of natural resources, provide recreational opportunities for all New Yorkers, and provide important wildlife habitat and open space. Given the level of development expected for multi-pad horizontal drilling, the dSGEIS anticipates that there would be additional pressure for surface disturbance on State lands. Surface disturbance associated with gas extraction could have an impact on habitats on State lands, and recreational use of those lands, especially large contiguous forest patches that are valuable because they sustain wide-ranging forest species, and provide more habitat for forest interior species. The area underlain by the Marcellus Shale includes both terrestrial and aquatic habitat for 18 animal species listed as endangered or threatened in New York State that are protected under the State Endangered Species Law (ECL 11-0535) and associated regulations (6 NYCRR Part 182). Endangered and threatened wildlife may be adversely impacted through project actions such as clearing, grading and road building that occur within the habitats that they occupy. Certain species are unable to avoid direct impact due to their inherent poor mobility (e.g., Blanding's turtle, club shell mussel). Certain actions, such as clearing of vegetation or alteration of stream beds, can also result in the loss of nesting and spawning areas.

The final two chapters focus on the potential impact on human health. In chapter 11, Goldstein et al. examine the role of the environmental public health community in the case of the Marcellus Shale, a vast natural gas field underlying parts of Pennsylvania, New York, West Virginia, Virginia, and Maryland. Response to public concern about potential adverse environmental and health impacts has led to the formation of state and national advisory committees. Here, they review the extent to which advisory committees formed in 2011 by President Obama and governors of the states of Maryland and Pennsylvania contain individuals with expertise pertinent to human environmental public health. They also analyze the extent to which human health issues are of concern to the public by reviewing presentations at the public meeting of the Secretary of Energy Advisory Board (SEAB) Natural Gas Subcommittee formed by the U.S. President's directive. They find that at a public hearing held by the SEAB Natural Gas Subcommittee 62.7% of those not in favor of drilling mentioned health issues. Although public health is specified to be a concern in the executive orders forming these three advisory committees, we could identify no individuals with health expertise among the 52 members of the Pennsylvania Governor's Marcellus Shale Advisory Commission, the Maryland Marcellus Shale Safe Drilling Initiative Advisory Commission, or the SEAB Natural Gas Subcommittee. Despite recognition of the environmental public health concerns related to drilling in the Marcellus Shale, neither state nor national advisory committees selected to respond to these concerns contained recognizable environmental public health expertise.

The final chapter, by Finkel et al., makes an argument for the need for proactive health-related policies related to natural gas extraction. They state that high-volume horizontal hydraulic fracturing of shale formations has the potential to make natural gas a significant, economical energy source, but the potential for harm to human health is often dismissed by proponents of this method. While adverse health outcomes of medical conditions with long latency periods will not be evident for years and will depend on the exposure, duration of exposure, dose, and other factors, they argue that it would be prudent to begin to track and monitor trends in the incidence and prevalence of diseases that already have been shown to be influenced by environmental agents. The dirty downside of modern, unconventional natural gas development, as well as the potential for harm, is discussed.

PART I

FRACTURING AND WATER POLLUTION

CHAPTER 1

REGULATION OF WATER POLLUTION FROM HYDRAULIC FRACTURING IN HORIZONTALLY-DRILLED WELLS IN THE MARCELLUS SHALE REGION, USA

HEATHER HATZENBUHLER and TERENCE J. CENTNER

1.1 INTRODUCTION

In the last four years, horizontal drilling using many fractures along a horizontal wellbore has been used commercially to access the deepest shale gas (over 1800 m below the surface) in the United States [1,2]. Horizontal drilling employs turning a downward-plodding drill bit to continue drilling within a layer underneath the ground. Accompanying horizontal drilling is hydraulic fracturing, a well-stimulation technique that maximizes extraction of oil and natural gas in unconventional reservoirs such as shale, coalbeds and tight sands. Hydraulic fracturing involves injecting specially engineered fluids consisting of chemicals and granular material into the wells at incredible pressure to break up the fuel stores and stimulate the flow of natural gas or oil to the surface [1]. Once the well has been fractured, the pressure forces out some of the injection fluids containing chemicals, brines, metals, radionuclides and hydrocarbons [3]. For some wells, the toxic flowback fluids are removed and later injected into class II injection wells [4]. In other situations, the fluids are recycled or are transported to local wastewater treatment facilities. As a result of horizontal drilling, there has been a significant increase in the natural gas supply and a reduction in wholesale spot price of natural gas by nearly 50% [5].

The risks associated with all aspects of fracturing have been looked at from a variety of perspectives, but most concerns revolve around the use of water resources and their potential contamination [6]. Other risks are associated with surface spills [7,8]. The United States Environmental

Protection Agency (EPA) has been investigating drinking water contamination and is expected to complete an extensive study on all aspects of hydraulic fracturing in 2014 [9]. A conclusion that may be drawn from a review of recent scientific studies and incidences is that horizontal drilling accompanied by hydraulic fracturing poses threats to local environmental conditions and the health and safety of persons using land, water, and air resources.

1.2 FEDERAL AND REGIONAL POLICIES

Several federal and regional policies have been adopted to oversee potential risks related to hydraulic fracturing. However, amendments to the federal laws have limited the federal government's oversight of activities accompanying the development of shale gas resources. An overview of relevant legislation, summarized in Table 1, enumerates the role EPA and other agencies could play in minimizing negative impacts of natural gas production.

TABLE 1 Summary of federal and regional legislation

Legislation	Authority/Jurisdiction	Potential oversight for hydraulic fracturing
CERCLA–1980	None currently*/Clean-up of hazardous waste sites	Might hold companies responsible for clean-up and damages due to releases of hazardous materials at well sites and require reporting of toxic chemicals used in the fracturing process.
CWA–1972	EPA/Waters of the United States	NPDES stormwater permit required for discharges from well sites but could be extended to apply to temporary holding pits.
RCRA–1976	None currently*/ Hazardous wastes	Could require the listing of hazardous substances used in the injection fluids in addition to regulation of the resulting wastewater flowback.
SDWA–2005 amendment	None currently*/Drinking water of the United States	The UIC program could regulate subsurface emplacement fluids that would include injection for gas development and underground storage of waste fluids.
SRBC–1971 and DRBC–1961	Commissioners/Susquehanna and Delaware River Basins	Regulates deposits or withdrawals from the river basin so that fracturing operations need permits to withdraw water for injecting into wells or for depositing wastewaters back into the river system.

Note: * Exemptions exist that prohibit EPA from applying these standards to oil and gas extraction.

In 1972, the Clean Water Act (CWA) delineated the basic structure for regulating discharges of pollutants into waters and for establishing quality standards for surface waters under the authority of EPA [10]. Under the CWA's National Pollutant Discharge Elimination System program, stormwater permits were required for sediment runoff from construction sites and discharges of pollutants into surface waters [11]. The permitting system requires adoption of technology-based and water quality-based effluent limits [11,12]. Fracturing activities that inject liquid into the ground or store waters in temporary pits without any discharge are not regulated under the CWA. Thus, there is no federal oversight of fracturing activities until there is proof of fracturing contaminants in surface waters [13].

Congress acted to protect drinking water in the Safe Drinking Water Act of 1976 with protection through the implementation of an Underground Injection Control program regulating subsurface injections and storage of fluids. But, in the Energy Policy Act of 2005, Congress enacted an exclusion to this program.

> The term "underground injection"—(A) means the subsurface emplacement of fluids by well injection; and (B) excludes—(i) the underground injection of natural gas for purposes of storage; and (ii) the underground injection of fluids or propping agents (other than diesel fuels) pursuant to hydraulic fracturing operations related to oil, gas, or geothermal production activities [14].

While the Safe Drinking Water Act specifically excludes hydraulic fracturing from regulation, the use of diesel fuel in fracturing is regulated since it is defined as a hazardous contaminant [14].

Congress regulated hazardous waste from inception to disposal under the Resource Conservation and Recovery Act (RCRA) and EPA has developed a list of regulated substances [15]. However, RCRA does not regulate hazardous wastes involved in oil and gas extraction and production under RCRA Subtitle C. These materials are subject to state regulation under the less stringent RCRA Subtitle D solid waste regulations as well as other federal regulations, although states are also free to adopt more demanding provisions. In a publication regarding the exemption EPA says, "Although they are relieved from regulation as hazardous wastes, the exemption does not mean these wastes could not present a hazard to human health and the environment if improperly managed" [16].The absence of any federal requirement to disclosure hazardous chemicals used in fracturing is a major issue [17].

Hydraulic fracturing, like any deep drilling operation, is subject to the risk of leaks and spills that can cause areas to be contaminated by hazardous waste. In 1980, the Comprehensive Environmental Response, Compensation, and Liability Act (CERCLA) provided for the clean-up of abandoned hazardous waste and established liability to those who released the wastes to pay for clean-up [18]. Yet oil and gas exploration is exempt from clean-up of accidental spills, leaks, and problems from underground injection via the Energy Policy Act of 2005 [19]. Exploration and production companies cannot be held liable for damages under CERCLA, nor may they be sued by any entity for replacement of drinking water supplies or any health problems created as a result of their operations [20].

Applicable to fracturing regulation are two regional commissions that have jurisdiction over all water withdrawals from specific watersheds: the Delaware River Basin Commission and the Susquehanna River Basin Commission. Figure 1 illustrates the overlap of the Marcellus shale formation and several river basins. Because of regulations adopted by these commissions, all oil and gas production operations must obtain permits before they can pump millions of gallons of water to use in their wells. Therefore, these commissions play a critical role in the continuation of oil and gas development in the Marcellus shale region because hydraulic fracturing cannot occur without significant quantities of water.

The Delaware River Basin Commission is a regulatory body that was established in 1961 by a congressional compact. It includes a division engineer from the US Army Corps of Engineers and representatives from New York, Pennsylvania, New Jersey, and Delaware who are appointed individually by the executive office in each state [21]. Any decision of the Commission involves the approval of all members. The Commission has full water resource management authority, including water allocations and diversions. Any project that will withdraw or discharge water in or from the basin must be approved by a process that includes a public hearing. In 2009, the Delaware River Basin Commission banned new exploration and production of shale gas

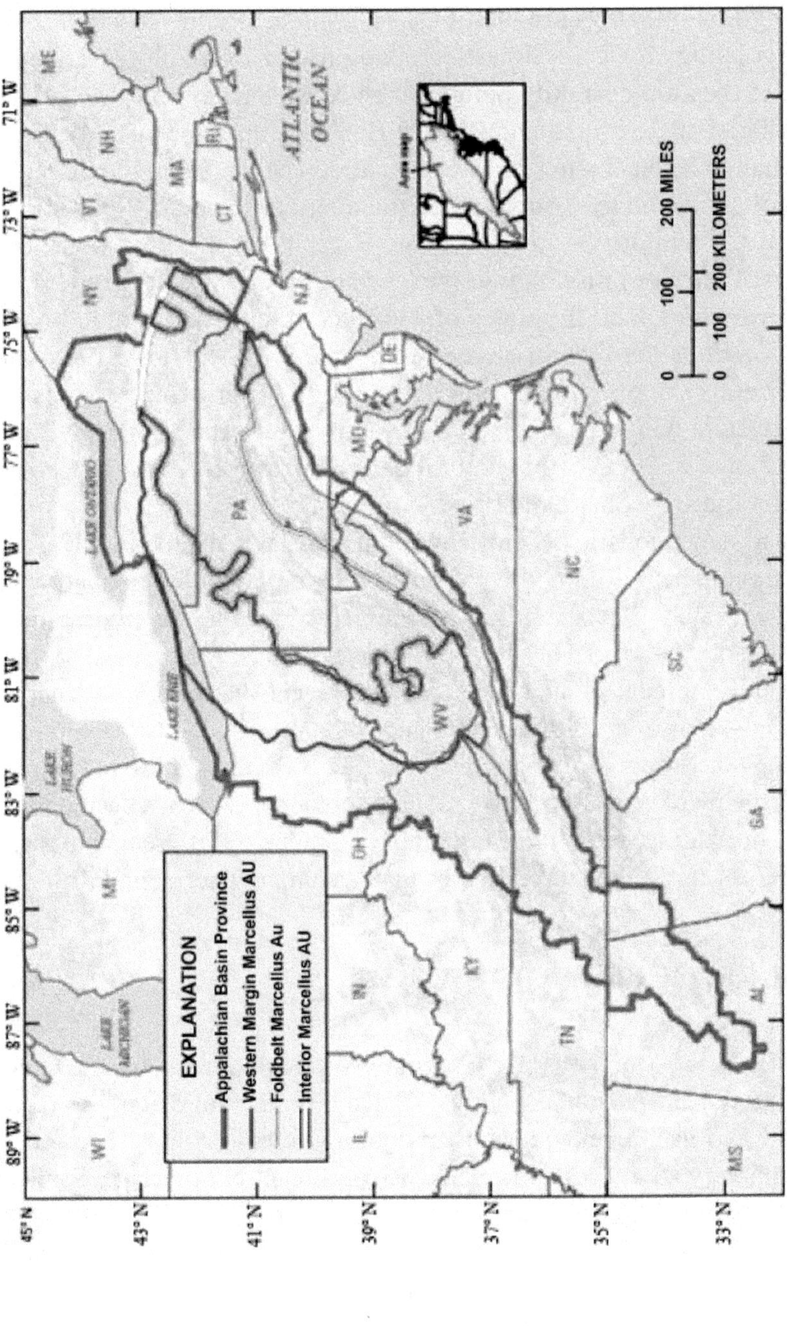

FIGURE 1: Map of the Marcellus shale assessment units (AU) which are located within the Appalachian Basin Province.

in the region until strong regulations are in place. Public comments on draft regulations for natural gas well pad projects were closed in April 2011, and they are currently being reviewed by the commission [22].

Similarly, the Susquehanna River Basin Compact established the Susquehanna River Basin Commission, another federal-interstate regulatory collaboration by Congress and the member states. It is parallel in structure and authority to the Delaware River Basin Commission. Any decision of the Commission involves the approval of all of the member parties, which include the states of Maryland, Pennsylvania, and New York, as well as the federal government. The Susquehanna River is in the Marcellus shale region so any hydraulic fracturing operation using surface waters will need a permit (see Figure 1). At the 15 March 2012 commission meeting, several natural gas drilling projects were rejected and many more reconsidered or tabled [23].

Due to exceptions to federal environmental laws detailed above, the federal government does not have a clear role to play in the regulation of hydraulic fracturing as a result of amendments made to the environmental laws detailed above. The exceptions have allowed for more liberal oil and gas development in areas not within the Delaware and Susquehanna river basins. However, EPA has announced that new federal standards for fracturing wastewater are being developed [24]. Even if these are implemented, the regulatory authority to address potential risks has been passed down to the states. States in the region have a range of different approaches to address environmental concerns that accompany horizontal drilling.

1.3 STATE POLICIES AND ACTIONS

Five states in the Marcellus Shale region, New York, Pennsylvania, Ohio, West Virginia, and Virginia (Figure 1), have different approaches to regulating oil and gas development involving horizontal drilling. These distinct regulatory structures, as well as the significant policy changes made by states in the last two years, illustrate a spectrum of possibilities and outcomes. These structures and outcomes can be used to guide future policy alternatives and decisions. A summary of the current policies and incidences can be found in Table 2.

TABLE 2: Summary of state regulations and outcomes

State	Regulatory Authority	Legislative Actions	Reported Incidents
New York	Department of Environmental Protection	• statewide moratorium • SPDES permit • disclosure of fracturing fluids • municipal zoning bans	• well water contamination from vertical wells
Ohio	Department of Natural Resources	• restrictions for impoundment pits • fees for wastewater disposal • electronic tracking	• earthquakes • well and surface water contamination
Pennsylvania	Department of Environmental Protection	• Act 13 of 2012 set stricter standards for oil and gas production and preempted most municipal regulations	• local water supply contamination
Virginia	Department of Mines Minerals and Energy	• the Gas and Oil Act allows the non-disclosure of chemicals and preempts municipal regulations	• noxious fumes • light pollution • well and surface water contamination • disruption from truck traffic
West Virginia	Department of Environmental Protection	• the Horizontal Well Act of 2011 preempted municipal regulations and exempts activities	• drinking water contamination and neurological disease • creek contamination with a massive fish kill

New York placed a moratorium on hydraulic fracturing in 2008 [25], and subsequently an executive order directed the state Department of Environmental Conservation to conduct a review and analysis of horizontal hydraulic fracturing [26]. The state has allowed hydraulic fracturing [27]; its horizontal hydraulic fracturing that is precluded. The state has developed a complex and comprehensive regulatory framework [28]. New York is also unique due to the local court battles between citizens and oil and gas companies concerning bans on hydraulic fracturing. Since 2008, 22 cities have rezoned to prohibit fracturing [29]. The city of Dryden is one of these local governments that banned horizontal hydraulic fracturing via a zoning law [30]. A natural gas production company filed suit against the city claiming that the municipality was overstepping its jurisdiction. In early 2012, a state superior judge ruled that the municipality was not preempted by state laws and had the right to tighten its land use regulations. Zoning bans by local governments across New York send a strong message about local disapproval of hydraulic fracturing and have established a precedent for other municipalities to limit the drilling rights of oil and gas companies. In New York, either a State Pollution Discharge Elimination System (SPDES) individual or general permit is required for fracturing activities that cause a discharge into surface waters [31]. For high-volume hydraulic fracturing activities, a special general permit has been proposed [32]. The special general permit addresses drilling operations from the construction phase through to the production phase, including well-site construction, soil disturbance, and potential contamination [32]. Hydraulic fracturing would be precluded in the New York City and Syracuse watersheds, on certain state lands, within 610 m of public drinking water supplies, and within 152 m of private wells. Furthermore, provisions require the identification and evaluation of fracturing fluid additives "to encourage the use of processes and substances that minimize the potential for environmental impacts" [32].

In Ohio, public concern about hydraulic fracturing came to a head on New Year's Eve 2011 when there was a 4.0 earthquake near the city of Youngstown. This seismic event followed several other earthquakes that began in March 2011, just three months after a 2804 m wastewater well was drilled in Youngstown for the storage of fracturing fluids. There is no record of seismic activity in this area during the previous 235 years [33]. Rather, the exponential growth in natural gas and storage well drilling in the area, jumping from an average of four new permanent-waste storage wells per year from 1990–2010 to 29 new wells in 2011, seems to have caused the seismic activity. Experts at the Ohio Department of Natural Resources concluded that the seismic disruptions were a result of brine injection related to hydraulic fracturing. Other research supports these claims, as it is widely understood that injecting fluid underground at high pressure can trigger earthquakes [34]. Any geologic disruption in an area where hazardous waste is permanently stored might result in contamination of ground and surface waters. In the last two years, Ohio has implemented further restrictions on impoundment pits located in urban areas, fees for disposal of wastewater via injections in wells, requirements for more comprehensive geologic data prior to permitting, and electronic tracking systems to identify the makeup of drilling wastewater fluids [35,36]. The

Ohio Department of Natural Resources has regulatory authority over hydraulic fracturing activities in the state.

In Pennsylvania, the Department of Environmental Protection has authority over hydraulic fracturing activities. In 2012, the state legislature passed Act 13 containing stronger and more detailed regulations including increased setback requirements for unconventional gas development, enhanced protection of water supplies, and strong, uniform, consistent statewide environmental standards [37]. This legislation included a uniformity provision that attempted to preempt "all local ordinances regulating oil and gas operations" [38]; however, a Pennsylvania court found this provision to be unconstitutional [39].

The legislature of Virginia decided to encourage the economical extraction of Virginia's coalbed methane [40]. Through the Gas and Oil Act, the state legislature preempted local regulations to give the Virginia Department of Mines, Minerals, and Energy the exclusive authority to regulate activities relating to oil and gas exploration and production [41]. The act establishes regulations and permitting requirements that govern mineral extraction. The act does not require the reporting of the chemical composition of fracturing fluids.

In West Virginia, a number of environmental problems have allegedly been caused by hydraulic fracturing activities, as documented by Earthjustice and mapped on their website of "fraccidents" [42]. The Department of Environmental Protection (DEP) issued a proposal in 2010 to rewrite regulations for drilling operations across the state. After months of talks with various stakeholder groups, the state legislature adopted the Horizontal Well Act in 2011 [43]. While the act delineates requirements that should help protect the environment, a number of provisions limit this protection [44]. For example, the act sets forth exceptions so that vertical and permitted wells escape further regulation [44]. Wells disturbing less than three acres or using less than 200,000 gallons of water in a 30-day period are not subject to the requirements of the act [43]. Turning to local regulations, the act specifically provides that the secretary of the West Virginia Department of Environmental Protection "has sole and exclusive authority to regulate" activities related to hydraulic fracturing so that municipal governments cannot interfere with drilling [43]. Furthermore, the secretary has "broad authority to waive certain minimum requirements" if deemed appropriate [43].

1.4 POLICY ALTERNATIVES

The absence of comprehensive controls and differences of regulatory approaches to horizontal drilling employing hydraulic fracturing between states do not provide adequate protection of local and regional water resources. The legal battles and state legislative revisions in the Marcellus shale region indicate significant public concern about the safety of horizontal well drilling. With the introduction of many fractures along a horizontal wellbore, there are new risks to be considered [45]. Furthermore, these fracturing activities pose risks to river systems and water quality that do not recognize manmade state and municipal boundaries. An individual state is unable to preclude pollutants from upstream states so that multistate or federal controls become important for the maximization of social, environmental, economic, and democratic outcomes for the Marcellus shale region [46]. In a similar manner, a local government may not be in a position to maximize outcomes for a region. Rather, by directing its focus on a small geographic area, a municipality may overlook broader, regional concerns.

The analysis of federal, regional and state regulatory controls over horizontal drilling identify two options for reducing risks accompanying hydraulic fracturing. The first option involves deleting the oil and gas production exemption set forth by the Energy Act of 2005 and requiring disclosure of hazardous chemicals employed in hydraulic fracturing. By deleting the exemption for oil and gas exploration and production, provisions of the Safe Drinking Water Act would offer additional oversight to fracturing activities involving chemicals being injected into the ground. In addition, requiring mandatory reporting of chemicals used in hydraulic fracturing would allow first responders to blowout accidents and other mishaps to have sufficient information for selecting appropriate responses. States often lack adequate controls and because the Marcellus shale formation spans multiple rivers and covers multiple states, a collective, standardized legal framework is needed to ensure equitable protection of the environment and to protect the economic interests of all parties involved.

In 2011, the Fracturing Responsibility and Awareness of Chemicals Act, which would repeal exemptions for hydraulic fracturing, was introduced in both houses of Congress [48]. However, the act remains in committees and, given the concern over rising energy prices, an initiative to restrict domestic energy production is an unpopular position for policy makers [49]. Moreover, the proposed FRAC Act does not require the public chemical disclosure requirements for fracturing fluids. Thus, the act fails to help identify sources of contamination that may occur from accidental releases and spills. States can enact requirements on the disclosure of chemicals, but most have chosen to include a provision for trade secret protection [50].

A second option to mitigate risks of water contamination by hydraulic fracturing is to strengthen safety controls for the disposal of flowback fluids. For drilling, damages from blowouts are a concern that can be addressed through better well construction standards and adequate construction monitoring and inspection [51]. For probabilistic events including unplanned accidents, the use of environmental impact assessments may reduce negative impacts as well as inspections [52]. Moreover, since hydraulic fracturing in the Marcellus shale region leads to increased concentrations of Ra226, Ra228, and Ba in flowback waters from Marcellus wells [53], more definitive and demanding treatment specifications for fracturing fluids discharged to publically owned treatment works may be needed to allay concerns that downstream water users are being harmed.

1.5 CONCLUSIONS

In the absence of consistent federal standards, individual states, driven by their short-term interests, are allowing actions that lead to long-term damage to common resources. This allows firms to avoid costs reflected in the negative externalities of production. With respect to horizontal drilling, a state's interest is economic gain through liberal gas production without full consideration of regional river basins and ground water supplies [54]. The examination of legal structures regulating hydraulic fracturing provides numerous examples of negative impacts on water quality as a result of poor management of drilling activities. To strengthen the protection of water sources in the Marcellus shale region, federal regulatory exemptions for oil and gas exploration should be deleted and additional resources should be allocated to the management of environmental risks accompanying hydraulic fracturing.

In a similar manner, local governmental actions addressing horizontal drilling may not be optimal. Tension exists between state and local governments over the regulation of hydraulic fracturing because local prohibitions on drilling can thwart state objectives. While the historic delegation of duties and responsibilities to municipal governments enable these governments to take actions on matters of local concern, state legislatures are having second thoughts about whether horizontal fracturing activities are local. Given changes in technology, communications, and transportation, issues relegated to local governments over past centuries may no longer be local. Interconnections of jobs, commerce, and social structures among local governments create externalities that cannot be meaningfully addressed by an individual municipality. Local governments may constitute archaic divisions that create impediments to the well-being of people and the economy of a state. Thus, in exercising their sovereignty, state legislatures are acting to preclude local decisions regarding fracturing that interfere with overriding state objectives.

REFERENCES

1. US Environmental Protection Agency. Plan to Study the Potential Impacts of Hydraulic Fracturing on Drinking Water Resources; EPA/600/R-11/122/November; US Environmental Protection Agency: Washington, DC, USA, 2011.
2. Weinhold, B. The future of fracking. Environ. Health Perspect. 2012, 120, A272–A279.
3. Finkel, M.L.; Law, A. The rush to drill for natural gas: A public health cautionary tale. Am. J. Public Health 2011, 101, 784–785.
4. Furlow, J.D.; Hays, J.R., Jr. Disclosure with protection of trade secrets comes to the hydraulic fracturing revolution. Tex. Oil Gas Energy Law 2011, 7, 289–355.
5. US Energy Information Administration. Short-Term Energy Outlook, Table 5b: U.S. Regional Natural Gas Prices; US Energy Information Administration: Washington, DC, USA, 2012. Available online: http://www.eia.gov/forecasts/steo/tables/?tableNumber=16#mstartcode=2007 (accessed on 25 September 2012).
6. Rahm, B.G.; Riha, S.J. Toward strategic management of shale gas development: Regional, collective impacts on water resources. Environ. Sci. Policy 2012, 17, 12–23.
7. Wiseman, H. Risk and Response in Fracturing Policy. University of Colorado Law Review, 2013, in press; FSU College of Law, Public Law Research Paper No. 594. Available online: http://ssrn.com/abstract=2017104 (accessed on 22 November 2012).
8. Wiseman, H. State Enforcement of Shale Gas Regulations, Including Hydraulic Fracturing. Energy Institute, University of Texas White Paper, 25 August 2011; FSU College

of Law, Public Law Research Paper No. 581. Available online: http://ssrn.com/abstract=1992064 (assessed on 15 November 2012).
9. DiCosmo, B. Jackson Downplays Concerns Over Broad EPA Oversight of Fracking Wells; Clean Energy Report; US Environmental Protection Agency: Washington, DC, USA, 2012.
10. US EPA. Summary of the Clean Water Act. US Environmental Protection Agency: Washington, DC, USA, 1972. Available online: http://www.epa.gov/lawsregs/laws/cwa.html (accessed 25 September 2012).
11. Navigation and Navigable Waters. United States Code, Sections 1311 and 1362, Title 33, Supplement 5, 2006.
12. Technology-Based Treatment Requirements in Permits. US Code of Federal Regulations, Section 125.3, Title 40, 2011.
13. Obold, J. Leading by example: The Fracturing Responsibility and Awareness of Chemicals Act of 2011 as a catalyst for international drilling reform. Colo. J. Int. Environ. Law Policy 2012, 23, 473–500.
14. Regulations for State Programs. United States Code, Section 300H, Title 42, Supplement 4, 2006.
15. US EPA. Summary of the Resource Conservation and Recovery Act; US Environmental Protection Agency: Washington, DC, USA, 2012. Available online: http://www.epa.gov/lawsregs/ laws/rcra.html (accessed on 25 September 2012).
16. US EPA. Exemption of Oil and Gas Exploration and Production Wastes from Federal Hazardous Waste Regulations; EPA530-K-01-004; US Environmental Protection Agency: Washington, DC, USA, 2002. Available online: http://www.epa.gov/osw/nonhaz/industrial/special/oil/oil-gas.pdf (accessed on 25 September 2012).
17. Wiseman, H. Trade secrets, disclosure, and dissent in a fracturing energy revolution. Columbia Law Rev. Sidebar 2011, 111, 1–13.
18. Comprehensive Environmental Response, Compensation, and Liability. United States Code, Sections 9601–9675, Title 42, Supplement 4, 2006.
19. Energy Policy Act of 2005. Public Law 109–58, Section 106, 2005.
20. McKay, L.K.; Johnson, R.H.; Salita, L.A. Science and the reasonable development of Marcellus shale natural gas resources in Pennsylvania and New York. Energy Law J. 2011, 32, 125–143.
21. Delaware River Basin Commission Home Page. Available online: http://www.state.nj.us/drbc/ (accessed on 25 September 2012).
22. Delaware River Basin Commission. Draft Natural Gas Development Regulations; Delaware River Basin Commission: West Trenton, NJ, USA, 2011. Available online: http://www.nj.gov/drbc/ programs/natural/draft-regulations.html (accessed on 25 September 2012).
23. Susquehanna River Basin Commission. SRBC met March 15: Reconsidered 22 and approved 20 additional projects; Denied 3 applications; Released proposed low flow policy for public comment. Available online: http://www.srbc.net/newsroom/NewsRelease.aspx?NewsReleaseID=81 (accessed on 25 September 2012).
24. US Environmental Protection Agency. EPA Announces Schedule to Develop Natural Gas Wastewater Standards/Announcement is part of administration's priority to ensure natural gas development continues safely and responsibly. Available online http://yo-

semite.epa.gov/opa/admpress.nsf/d0cf6618525a9efb85257359003fb69d/91e7fadb4b114c4a8525792f00542001!OpenDocument (assessed on 2 November 2012).
25. Applebome, P. Drilling Critics Face a Divide over the Goal of Their Fight. New York Times, 10 January 2012, p. A17.
26. Governor's Office, New York. Executive Order No. 41: Requiring further environmental review. Available online http://www.governor.ny.gov/archive/paterson/press/121110PatersonExecutiveOHydraulicFracturing.html (assessed on 2 November 2012).
27. Nolon, J.R.; Polidoro, V. Hydrofracting: Disturbances both geological and political: Who decides? Urban Lawyer 2011, 44, 507–532.
28. Mergen, A.C.; Aagaard, T.; Baillie, J.; Bender, P.; Beauduy, T.W.; Engelder, T.; Perry, S.; Ubinger, J.W., Jr.; Wiseman, H. "Shale" we drill? The legal and environmental impacts of extracting natural gas from Marcellus shale. Villanova Environ. Law J. 2011, 22, 189–224.
29. Calfee, C.; Weissman, E. Permission to transition: Zoning and the transition movement. Plan. Environ. Law 2012, 64, 3–10.
30. Anschutz Exploration Corporation v. Town of Dryden. 940 N.Y.S.2d 458. Supreme Court of New York, Tompkins County, 21 February 2012.
31. New York State Department of Environmental Conservation. SPDES General Permit for Stormwater Discharges from High Volume Hydraulic Fracturing Operations, Permit No. GP-0-XX-XXX, Fact Sheet; New York State Department of Environmental Conservation: Albany, NY, USA, 2011. Available online: http://www.dec.ny.gov/docs/water_pdf/hvhfgpfactsht.pdf (accessed on 27 September 2012).
32. New York State Department of Environmental Conservation. Proposed express terms 6 NYCRR parts 750.1 and 750.3. Available online: http://www.dec.ny.gov/regulations/77383.html (accessed on 27 September 2012).
33. Ohio Department of Natural Resources. Catalog and maps of Ohio earthquakes. Available online: http://www.dnr.state.oh.us/geosurvey/html/eqcatlog/tabid/8302/Default.aspx (accessed on 26 September 2012).
34. Raleigh C.B.; Healy J.H.; Bredehoeft J.D. An experiment in earthquake control at Rangley, Colorado. Science 1976, 191, 1230–1242.
35. Division of Mineral Resources Management—Oil and Gas. Ohio Administrative Code, Chapter 1501:9, 2011.
36. Ohio Department of Natural Resources. Preliminary Report on the Northstar 1 Class II Injection Well and the Seismic Events in the Youghstown, Ohio Area; Ohio Department of Natural Resources: Columbus, OH, USA, March 2012. Available online: http://ohiodnr.com/downloads/northstar/ UICreport.pdf (accessed 25 September 2012).
37. Pennsylvania Department of Environmental Protection. Act 13 of 2012. Available online: www.portal.state.pa.us/portal/server.pt/community/act_13/20789 (accessed on 25 September 2012).
38. Uniformity of Local Ordinances. Pennsylvania Consolidated Statutes, Section 3304, Title 58, 2012.
39. Robinson Township v. Commonwealth of Pennsylvania, 52 A.3d 463. Commonwealth Court of Pennsylvania, 26 July 2012.
40. Miller, T.; Kauffman, M.K. Is It a Deep Well or a Shallow Well and Who Cares? Energy & Mineral Law Institute: Lexington, Kentucky, USA, 2010; Chapter 12, Volume 31, pp. 404–431.

41. Exclusivity of Regulation and Enforcement. Code of Virginia, Section 45.1-361.5, Title 45.1, 2012.
42. Earthjustice. West Virginia and fracking. Available online: http://earthjustice.org/features/campaigns/ west-virginia-and-fracking (accessed on 25 September 2012).
43. Legislative Findings; Declaration of Public Policy. West Virginia Code, Section 22-61-2, Chapter 22, 2012.
44. Garvin, D.S.; Coordinator, L. Jr.; WV Environmental Coalition. Legislature Passes Weak Marcellus Shale Bill in Special Session; Sierra Club: San Francisco, CA, USA, 2012. Available online: http://westvirginia.sierraclub.org/newsletter/archives/2012/03/a_001.html (accessed on 26 September 2012).
45. He, H.; Dou, L.; Fan, J.; Du, T.; Sun, X. Deep-hole directional fracturing of thick hard roof for rockburst prevention. Tunn. Undergr. Space Technol. 2012, 32, 34–43.
46. 46. Vogel, E. Parceling out the watershed: The recurring consequences of organizing Columbia River Management within a basin-based territory. Water Alternat. 2012, 5, 161–190.
47. Wiseman, H.; Gradijan, F. Regulation of Shale Gas Development, Including Hydraulic Fracturing. Energy Institute, The University of Texas at Austin; 31 October 2011. University of Tulsa Legal Studies Research Paper No. 2011-11. Available online: http://papers.ssrn.com/ sol3/papers.cfm?abstract_id=1953547 (assessed 15 November 2012).
48. US House of Representations Bill No. 1084, Fracturing Responsibility and Awareness of Chemicals Act of 2011. Available online: http://www.govtrack.us/congress/bills/112/hr1084 (accessed 27 September 2012).
49. Bittle, S.; Rochkind, J.; Bosk, J. Confidence in U.S. Foreign Policy Index: Energy, Economy New Focal Points for Anxiety over U.S. Foreign Policy. Public Agenda: New York, NY, USA, 2008. Available online: http://www.publicagenda.org/files/pdf/foreign_policy_index_spring08.pdf (accessed on 25 September 2012).
50. Murrill, B.J.; Vann, A. Hydraulic Fracturing: Chemical Disclosure Requirements; CRS Report for Congress; Congressional Research Service: Washington, DC, USA, 2012.
51. Zhang, J. Pore pressure prediction from well logs: Methods, modifications, and new approaches. Earth. Sci. Rev. 2011, 108, 50–63.
52. Wiseman, H. Fracturing legislation applied. Duke Environ. Law Policy Forum 2012, 22, 361–384.
53. Haluszczak, L.O.; Rose, A.W.; Kump, L.R. Geochemical evaluation of flowback brine from Marcellus gas wells in Pennsylvania, USA. Appl. Geochem. 2012, in press.
54. Powers, E.C. Fracking and federalism: Support for an adaptive approach that avoids the tragedy of the regulatory commons. J. Law Policy 2011, 19, 913–971.

CHAPTER 2

SOIL EROSION AND SURFACE WATER QUALITY IMPACTS OF NATURAL GAS DEVELOPMENT IN EAST TEXAS, USA

MATTHEW MCBROOM, TODD THOMAS, and YANLI ZHANG

2.1 INTRODUCTION

Recent advances in drilling technology have resulted in a dramatic expansion in exploration for and development of oil and natural gas. Historically, single vertical wells were drilled into hydrocarbon traps in permeable rock formations where gas and oil had migrated to. Starting in the 1940s, water, sand, and other additives under high pressure were used to fracture low permeability hydrocarbon source rocks like shales. Due to the high cost of these operations relative to the value of the oil and gas recovered, this practice had only limited applicability. Recent advances in horizontal drilling technology coupled with higher prices for oil and natural gas have resulted in a significant increase in hydraulic fracturing or fracking. In addition, CO_2 emissions from natural gas combustion are 30%–40% lower than coal, NO_x emissions are 80% lower for natural gas, and emissions are almost 100% lower for SO_2, particulates, and mercury compared with coal [1]. Therefore, natural gas is seen as an acceptable bridge fuel until more sustainable energy sources become viable. This will likely result in greater development of natural gas resources in the future.

One area of very active drilling in the United States is East Texas, southwestern Arkansas, and western Louisiana. The Haynesville, Cotton Valley, Travis Peak, and other formations underlie this region and have been very productive, with a drilling success rate of over 99%. The

Haynesville shale has been the most productive formation and is between 3.1 and 4.3 km deep and about 91 m in thickness [2]. It is estimated to contain about 7 trillion m^3 of natural gas [3]. Drilling increased by over 300% in the Haynesville region from 2008 to 2012.

There are numerous concerns associated with oil and gas development and water resources. These include firstly, the large amount of water used in fracking. In the Barnett shale, fracking water use in 2010 was 308 Mm3, or about 9% of the total water used by the city of Dallas, Texas [4]. In addition, concerns exist about the possibility of fracking fluids contaminating aquifers. With regards to surface waters, leaking pipelines, reserve pits, and producer water spills are a significant hazard [5]. Finally, concerns exist about the erosion and sedimentation that can result from natural gas development. Sedimentation is among the greatest contributors to stream impairment in the United States [6].

In the Barnett shale region of north Texas, sediment yields from natural gas sites in Denton County were 54 t ha^{-1} yr^{-1}, much greater than the 1.1 t ha^{-1} yr^{-1} measured from undisturbed rangelands in this region [7]. The United States Environmental Protection Agency (USEPA) regulates small construction sites (0.4 ha or greater) for stormwater discharge and sediment movement. In the state of Texas, gas wells are not regulated by the state environmental agency as small construction sites and are not subject to the same regulations. In addition, little regulatory oversight is given to how the placement of well pads may impact surface water resources.

Best management practices (BMPs) to control stormwater discharge and nonpoint pollution for other industries like agriculture and forestry have been widely adopted in the USA. For example, over 95% of forestry operations in Texas employ these BMPs [8], and these BMPs have been proven to be very effective in reducing sedimentation from clearcutting and site-preparation [9]. Similarly, it is estimated that sedimentation from natural gas well sites could be reduced by as much as 93% by using BMPs [10].

The purpose of this study was to quantify the stormwater concentrations and losses of sediment, nutrients, and metals from a natural gas well site. Comparisons were made between a gas well site constructed in the stream channel and a site offset from the stream channel by 15 m to determine the extent to which well location may affect sediment loss and water quality. Comparisons were also made between these water quality impacts and impacts from other land uses in the watersheds.

2.2 MATERIALS AND METHODS

2.2.1 STUDY AREA

The study was conducted at the Alto Experimental Watersheds in the Neches River basin approximately 16 km west of the town of Alto in Cherokee County, Texas, USA (Figure 1). The study area is in the Gulf Coastal Plain and has a humid subtropical climate. Average summer temperatures are 27.2 °C and average winter temperatures are 9.5 °C, with a mean annual temperature of 18.7 °C. Annual rainfall in the region is 117 cm. The rain is distributed fairly evenly throughout the year with an average of 89 rain days a year, with April and May receiving the largest amount of rainfall [11].

The soils at the Alto Experimental Watersheds formed in Eocene sediments. The dominant surface formations are members of the Claiborne Group and are Sparta Sand and the Cook Mountain Formation [12]. These soils developed under mixed loblolly pine (Pinus taeda) and hardwood forests, have low inherent fertility and are most commonly classified as Alfisols and Ultisols. The most prevalent soil found in the watersheds is the Sacul Series (fine, mixed, active, thermic Aquic Hapludults) followed by the Tenaha Series (loamy, siliceous, semiactive, thermic Arenic Hapludults). Both soils are Ultisols with an argillic horizon and less than 35% base saturation. Teneha soils are well drained and runoff is negligible to medium with increasing slope [13]. Sacul soils are slowly permeable soils that formed in acidic, loamy and clayey marine sediments. They are moderately well drained with medium to very high runoff potential, and have a seasonally high water table that is within 61 to 122 cm of the soil surface in late winter and spring most years [13].

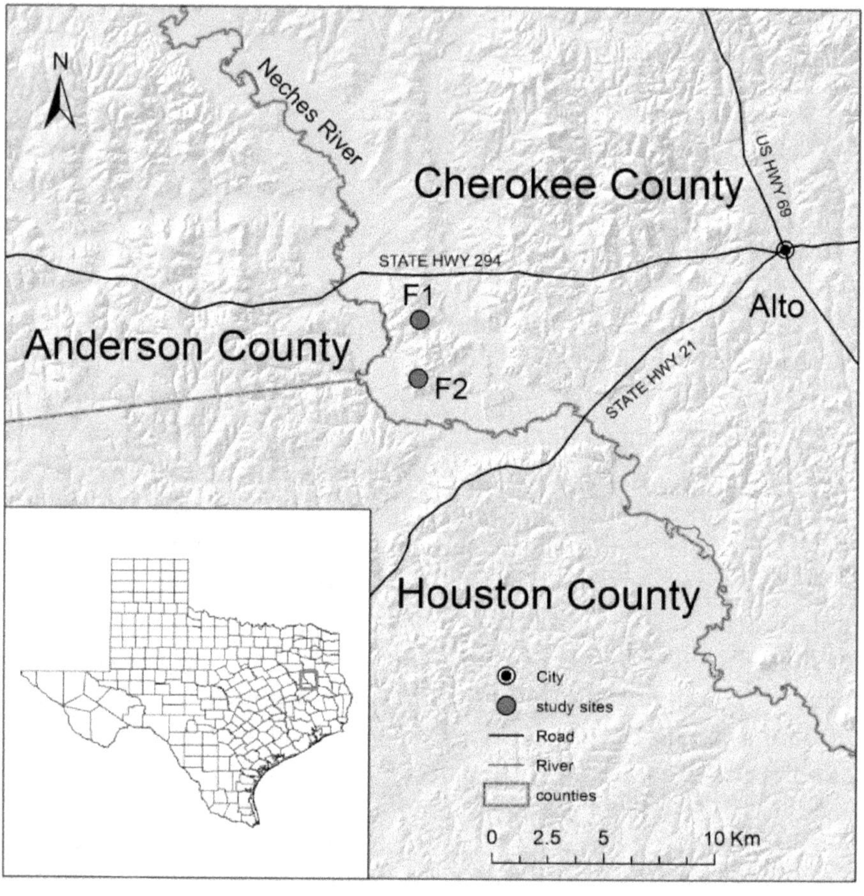

FIGURE 1: Location of study watersheds (F1 = no riparian buffer, F2 = 15 m riparian buffer) at the Alto Experimental Watersheds in Cherokee County, Texas, USA.

2.2.2 TREATMENTS

In the spring of 2008, two natural gas wells were drilled. At the first site (F1), the well pad was constructed directly in the channel of an intermittent stream and has a watershed area of 13.7 ha with the pad comprising 1.4 ha (Figure 2a). The stream was rechanneled around the north side of the pad following construction. At the second site (F2), the pad was offset from the creek channel by about 15 meters; this site has a watershed that consists of 4.5 ha with the well pad occupying 1.1 ha (Figure 2b).

In the process of constructing the well pad at F-1, fill material had to be brought in from an undisclosed location. The fill material consisted of 55.5% sand and 44.5% clay. Once this fill material had been brought in and the site leveled, iron ore gravel (16–150 mm diameter) was hauled in and spread over the majority of the pad with the exception of approximately one-quarter of the western end of the pad, which was used for a drilling fluid reserve pit. After drilling was completed, the reserve pit was filled with soil that was 40.2% sand, 14.1% silt, and 45.7% clay. This area was then seeded with ryegrass (*Lolium spp.*) While some of the seeds germinated, most did not grow or were carried away by surface runoff, resulting in bare soil.

The well pad at F2 required no fill material for pad construction due to the topography of the site. F2 was placed on the southern face of a large hill. Earth-moving equipment was used to modify the hill from a steep slope to a 1.1 hectare terrace suitable for operating large drilling equipment on. This soil was 65.1% sand, 9.5% silt, and 25.3% clay. After the terrace was constructed, iron ore gravel was spread similar to the method employed at F1. The back, southern portion used as a reserve pit for drilling fluids. The soil used to fill in the reserve pit was 21.7% sand, 32.1% silt, and 46.2% clay.

Both sub-watersheds where the gas well sites were constructed were dominated by loblolly pine. The northern portion of the F1 watershed was mixed hardwoods and pine; this area comprised approximately 3.5 hectares. The rest of the F1 watershed was 10–15 year old loblolly pine plantation. Approximately 2 hectares of the F2 watershed was 10–15 year old loblolly pine plantation while the rest was a mixed hardwood and pine stand. The portion of the watershed that was mixed hardwood and pine was composed of fairly large (≈50–100 cm) timber. These larger diameter trees consisted primarily of white and red oaks (*Quercus* spp.) and loblolly pine. This area of large mixed timber at both watersheds was the result of timber harvests in compliance with Texas BMPs, leaving the riparian forest as a contiguous buffer known as a streamside management zone (SMZ). The understory of both watersheds consists mostly of species such as dogwood (*Cornus florida*), sweetgum (*Liquidambar* styraciflua), vari-

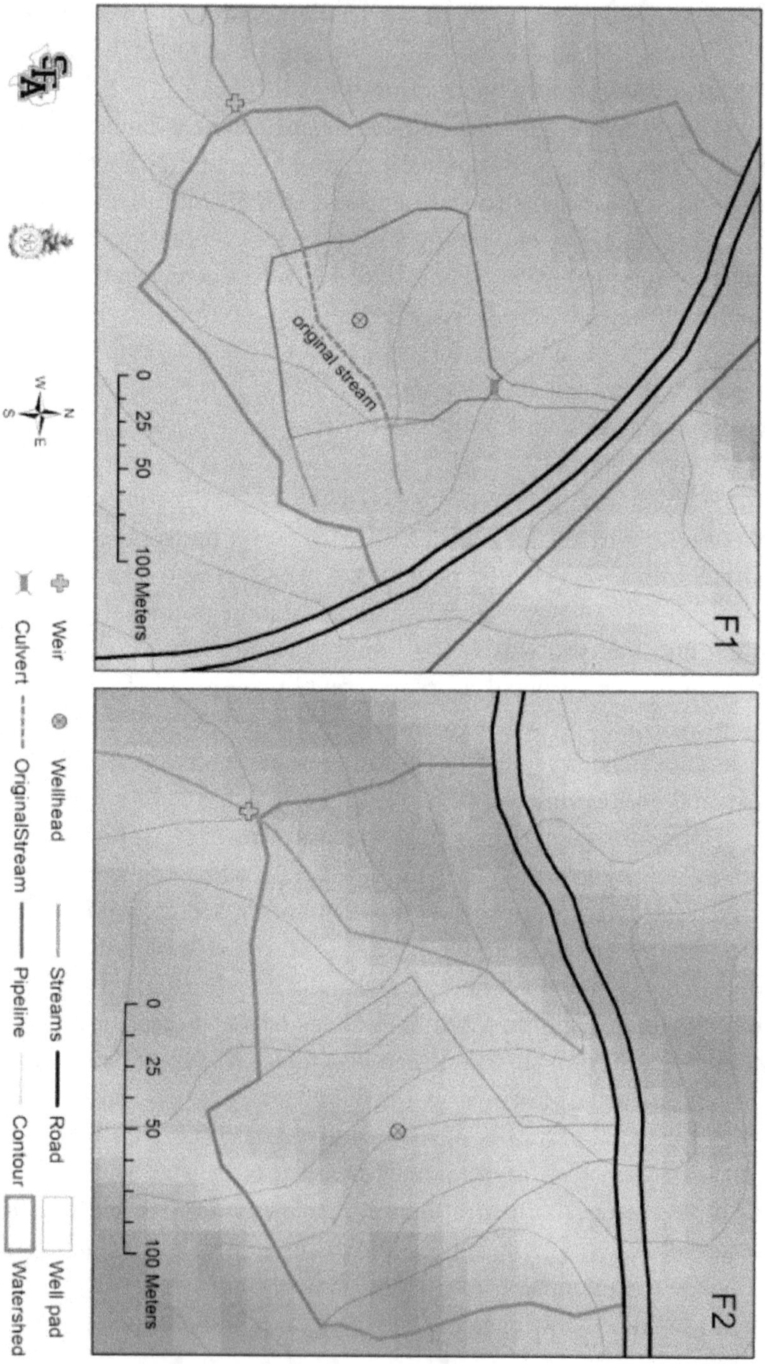

FIGURE 2. F1 and F2 natural gas well pad layout at the Alto Experimental Watersheds, Texas, USA

ous magnolias (*Magnolia* spp.), various hickories (*Carya* spp.), yaupon (*Ilex vomitoria*), sassafras (*Sassafras albidum*), and American beautyberry (*Calicarpa americana*).

2.2.3 WATER QUANTITY AND QUALITY

In both streams, a v-notch weir was constructed approximately 80 m downstream from the pad (Figure 3).

In each weir, an AquaRod® water level monitor was installed in the mouth of the flume. Unfortunately, stage data obtained from the AquaRods® were unreliable due the unexpectedly high sediment loads deposited in the weirs burying the capacitance rods. Streamflow was therefore estimated using the ArcAPEX model from precipitation measured at the sites [14]. ArcAPEX was calibrated and validated for these watersheds in earlier studies [15]. Rain gauges were located throughout the watershed and after each storm event precipitation data were collected.

As a result of the streamflow being ponded by the front plate of the weir, the coarse sediments were deposited in the drop box section on the floor of the weir. After each rain event this sediment was removed and weighed to determine the amount of sedimentation occurring in the stream channel (Figure 4). Dry mass was determined from a sub-sample of this sediment. The amount of sediment deposited in the drop box was later added to the amount of suspended sediment losses in stormflow. These losses were quantified using the flow estimated by ArcAPEX multiplied times the total suspended sediment (TSS) values that were obtained from stormwater samples. Sampling occurred from September 2008 to March 2010.

Water samples were collected from each weir using one of two techniques. The first technique utilized a Nalgene® Storm Water Sampler (Figure 3). Within 24 h of each storm runoff event the sample bottle was removed and a clean, acid rinsed bottle was placed in the cylinder. These samplers were frequently buried by the large volumes of sediment. When this occurred, the second method was used, the grab sample method, in which a 1 L sample bottle was placed in the flow of the stream and a water sample was taken. Grab samples typically represented the recession phase of the hydrograph. Once the samples were collected from the field they were brought to the laboratory for analysis. The samples in the lab were analyzed using a Hach® DR/890 Datalogging Colorimeter and a Hach® sensION 156 Portable pH/Conductivity Meter according to approved United States Environmental Protection Agency (USEPA) methods [16]. Parameters analyzed included total suspended solids (TSS), total dissolved solids (TDS), pH, conductivity (EC), total nitrogen (TN), ammonia (NH 4+), nitrate nitrogen (NO^{3-}), nitrite nitrogen (NO^{2-}), total phosphorus (TP), ortho-phosphate (PO^{4+}) sulfate (SO^{4+}), iron (Fe), turbidity, color, salinity, calcium hardness and magnesium hardness. A paired T-test was employed to determine if mean water quality values were different by site at = 0.05.

(a) (b)

FIGURE 3: In-channel instrumentation for measuring total runoff (V-notch weir), stream level (Aquarod®), water quality (Nalgene® Stormwater Sampler), and sediment (drop box) on the F2 sub-watershed before a storm event (a) and after a 6.3 cm rain event in April, 2009 at F1 (b); at the Alto Experimental Watersheds in Texas, USA.

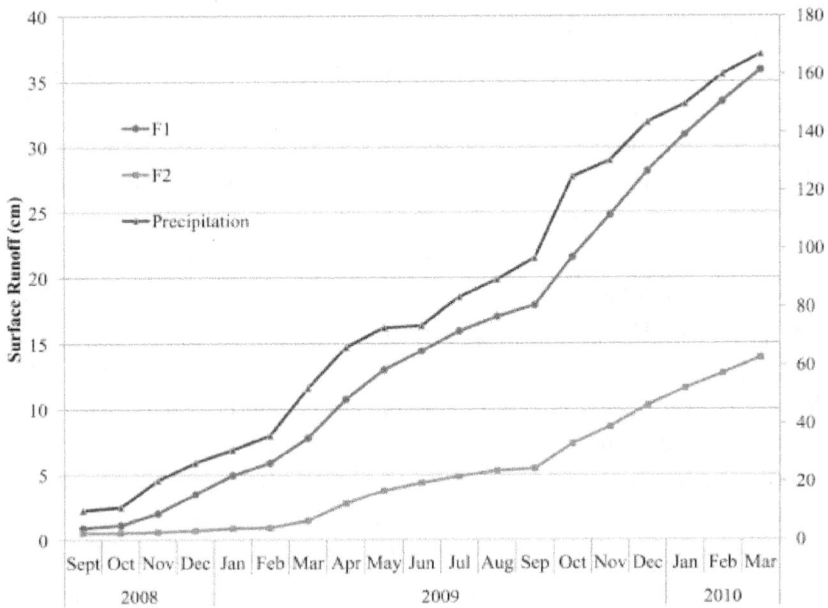

FIGURE 4: Cumulative ArcApex simulated water yield and rainfall for two natural gas well locations, one placed directly in the stream channel (F1) and the other offset from the channel by a 15 m buffer (F2) at the Alto Experimental Watersheds, Texas, USA.

2.3 RESULTS

In the small forested watersheds of East Texas, stream flow in headwater streams is typically intermittent and is mostly a product of storm runoff. The simulated water yield at F1 was significantly greater ($p < 0.0001$) than the water yield at F2 (Figure 4). In the first month of data collection (September 2008) the water yield at F1 was 0.915 cm and 0.545 cm at F2. Due to lower than average precipitation in the month of October, there was a decrease in storm runoff, but this decrease was most pronounced at F2, with 0.216 cm and 0.001 cm at F1 and F2 respectively. This trend continued throughout the study period, regardless of season. Percent runoff efficiency (runoff divided by precipitation) was different for two watersheds, 33.0% at F1 and 12.3% at F2.

Soil compaction of the well pad was much greater than in the rest of the watershed. The mean bulk density of the well pad at F1 was 2.04 g cm^{-3}. Mean bulk density measurements taken in the surrounding watershed were 1.3, 1.19, and 0.99 g cm^{-3} for logging sets, skid trails, and undisturbed forest floor respectively.

Sediment yield was also significantly greater ($p < 0.001$) from F1 that F2 (Figure 5). Starting in September 2008, the sediment yield was 83 kg ha^{-1} at F1 versus 10 kg ha^{-1} at F2. Continuing through the winter of 2009, the total yield continued to increase at F1 over F2. The total sediment yield for the 2009 water year (September 2008–August 2009) was 19,561 kg versus 785 kg at the F1 and F2 watersheds, respectively. However, this does not take into account the differences in the percent of the watershed that was actually disturbed by the well site. The well site occupied about 24% of the total watershed area at F2 versus about 10% at F1. Therefore, it is also useful to compare the sediment yields per unit area disturbed by natural gas development in order to make meaningful comparisons with the clearcut watersheds. On this basis, the equivalent sediment losses for F1 and F2 were 13,972.1 and 714 kg $ha^{-1}yr^{-1}$ for the 2009 water year respectively, or 16,896 and 1,087 kg $ha^{-1}yr^{-1}$ for F1 and F2, respectively, annualized for the entire 19 month (September 2008–March 2010) study period. About 56% of the sediment loss recorded at F1 was deposited in the flume, with less than 44% moving in the suspended form. However, at F2, 98% of the sediment moved in the suspended form over the study period, with only 2% being deposited in the flume. Since sediment filled the flume on F1 for several runoff events, it is possible that these loss values underestimate the amount of coarse sediments actually eroded from the pad.

In terms of concentrations of other water quality parameters, differences between F1 and F2 were less pronounced (Table 1). For nutrients, only PO^{4+} was significant, with the mean value being significantly greater at F2 than at F1. At F1, pH was also significantly greater, though these values were well below the Texas water quality standard minimum value of 6.0. Color was significantly greater at F1 than F2, probably associated

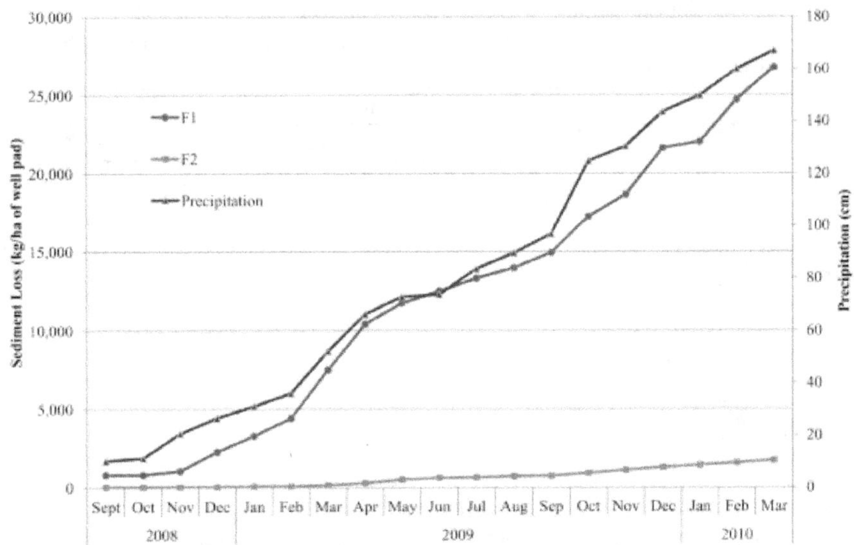

FIGURE 5: Cumulative sediment yield and rainfall for two natural gas well locations, one placed directly in the stream channel (F1) and the other offset from the channel by a 15 m buffer (F2) at the Alto Experimental Watersheds, Texas, USA.

with the higher amounts of sediment eroded from the pad at F1. However, there were no significant differences in either TSS or TDS. Salinity was significantly greater at F1 than F2, and this could have been attributed to an accidental spill of saline producer water that occurred in October 2008, but more sampling would have been required to establish this. The volume and chemical properties of this salt water was spilled was not tested. However, this spill did result in the death of several loblolly pine trees and understory vegetation down gradient of the well pad (Figure 6).

When nutrient and metal concentrations were converted to mass losses per hectare, all of the losses were greater from F1 than F2, with TDS, TN, NO^{3-}, PO^{4+}, SO^{4+}, and Fe being significantly greater ($\alpha < 0.05$) using the T-test (Table 2). Since streamflow was significantly greater at F1 throughout the study period (Figure 4), it would be expected that mass losses would also be greater.

TABLE 1: Mean concentrations for water quality parameters measured below two natural gas well sites (F1 and F2) from October 2008–March 2010 at the Alto Experimental watersheds in East Texas, USA.

Water quality parameter	Mean		*T*-test *p*-value
	F1	F2	
Total Nitrogen (TN, mg L^{-1})	2.78	2.50	0.26
Ammonia (NH_4^+, mg L^{-1})	1.55	0.57	0.27
Nitrate (NO_3^-, mg L^{-1})	2.78	0.74	0.15
Nitrite (NO_2^-, mg L^{-1})	0.02	0.03	0.50
Total Phosphorus (TP, mg L^{-1})	0.57	0.72	0.59
Ortho-Phosphate (PO_4^+, mg L^{-1})	0.16	**0.30**	0.01
Total Suspended Solids (TSS, mg L^{-1})	335.72	288.33	0.40
Total Dissolved Solids (TSD, mg L^{-1})	281.43	415.44	0.13
pH	**4.90**	4.53	0.04
Conductivity (μS cm^{-1})	461.06	554.65	0.30
Color (CU)	**1231.28**	576.58	0.04
Calcium Hardness (mg L^{-1})	1.23	0.75	0.23
Magnesium Hardness (mg L^{-1})	2.81	2.95	0.87
Iron (Fe, mg L^{-1})	5.55	4.36	0.18
Salinity (mg L^{-1})	0.24	**0.41**	0.02
Sulfate (SO_4^+, mg L^{-1})	6.43	5.30	0.23

Note: Bold values were significantly greater based on the paired *t*-test at α=0.05

TABLE 2: Total values for mass losses (kg ha−1) for water quality parameters measured below two natural gas well sites (F1 and F2) from October 2008–March 2010 at the Alto Experimental Watersheds in East Texas, USA.

Water quality parameter	Mean		*T*-test *p*-value
	F1	F2	
Total Nitrogen (TN)	**10.84**	3.08	0.00
Ammonia (NH_4^+)	4.55	0.67	0.112
Nitrate (NO_3^-)	**11.84**	0.84	0.035
Nitrite (NO_2^-)	0.10	0.05	0.217
Total Phosphorus (TP)	2.53	1.42	0.059
Ortho-Phosphate (PO_4^+)	**0.72**	0.47	0.042
Total Suspended Solids (TSS)	**1,196**	418	0.000
Total Dissolved Solids (TSD)	**969**	559	0.032
Iron (Fe, mg L^{-1})	**19.2**	5.54	0.001
Sulfate (SO_4^+, mg L^{-1})	**21.53**	6.43	0.000

Note: Bold values were significantly greater based on the paired *t*-test at α=0.05

FIGURE 6: Mortality of loblolly pine overstory trees (red/brown needles) and understory vegetation at F2 at the edge of the streamside buffer strip following an accidental spill in October 2008 of saline water produced during natural gas extraction at the Alto Experimental Watersheds in Texas, USA.

2.3.1 WATER QUANTITY AND QUALITY

In both streams, a v-notch weir was constructed approximately 80 m downstream from the pad (Figure 3).

In each weir, an AquaRod® water level monitor was installed in the mouth of the flume. Unfortunately, stage data obtained from the AquaRods® were unreliable due the unexpectedly high sediment loads deposited in the weirs burying the capacitance rods. Streamflow was therefore estimated using the ArcAPEX model from precipitation measured at the sites [14]. ArcAPEX was calibrated and validated for these watersheds in earlier studies [15]. Rain gauges were located throughout the watershed and after each storm event precipitation data were collected.

As a result of the streamflow being ponded by the front plate of the weir, the coarse sediments were deposited in the drop box section on the floor of the weir. After each rain event this sediment was removed and weighed to determine the amount of sedimentation occurring in the stream channel (Figure 4). Dry mass was determined from a sub-sample of this sediment. The amount of sediment deposited in the drop box was later added to the amount of suspended sediment losses in stormflow. These losses were quantified using the flow estimated by ArcAPEX multiplied times the total suspended sediment (TSS) values that were obtained from stormwater samples. Sampling occurred from September 2008 to March 2010.

Water samples were collected from each weir using one of two techniques. The first technique utilized a Nalgene® Storm Water Sampler (Figure 3). Within 24 h of each storm runoff event the sample bottle was removed and a clean, acid rinsed bottle was placed in the cylinder. These samplers were frequently buried by the large volumesof sediment. When this occurred, the second method was used, the grab sample method, in which a 1 L sample bottle was placed in the flow of the stream and a water sample was taken. Grab samples typically represented the recession phase of the hydrograph. Once the samples were collected from the field they were brought to the laboratory for analysis. The samples in the lab were analyzed using a Hach® DR/890 Datalogging Colorimeter and a Hach® sensION 156 Portable pH/Conductivity Meter according to approved United States Environmental Protection Agency (USEPA) methods [16]. Parameters analyzed included total suspended solids (TSS), total dissolved solids (TDS), pH, conductivity (EC), total nitrogen (TN), ammonia (NH 4+), nitrate nitrogen (NO 3 −), nitrite nitrogen (NO2 −), total phosphorus (TP), ortho-phosphate (PO4+) sulfate (SO 4+), iron (Fe), turbidity, color, salinity, calcium hardness and magnesium hardness. A paired T-test was employed to determine if mean water quality values were different by site at = 0.05.2.4 Discussion

2.4.1 STORM RUNOFF

Total runoff from these two natural gas well locations was much greater than would be expected from undisturbed areas in this region. In the undisturbed forested areas, direct surface runoff is uncommon. However, due to the significant increase in bare, compacted soils surface runoff was much more frequent. In addition, the significantly higher bulk density on the well locations resulted in less infiltration. McBroom *et al.* found that for nearby undisturbed forests, annual runoff ranged between 0.64 and 10.32 cm, depending on rainfall. Following clearcutting of the watersheds reported by McBroom *et al.* [17], annual runoff ranged between 7.82 and 9.79 cm. This was comparable to runoff measured at F2 in the 2009 water year of 9.58 cm. However, the clearcut reported by McBroom *et al.* covered an average of 75% of the total watershed area, where the well location at F2 only occupied about 10% of the total watershed area. Even when the gas well pad was offset by 15 m from the stream, it still had a proportionally greater impact on runoff than forest management. For the well pad directly in the stream channel, the effects on runoff were much greater, with 24.67 cm of runoff in the 2009 water year. In addition, runoff efficiency following clearcutting on adjacent watersheds increased from 1% pre-harvest to 9% post harvest, compared with 33% and 12% on F1 and F2, respectively.

2.4.2 SEDIMENT LOSSES

In terms of total sediment yield, results from this study are much greater than reported from proximate watershed studies, indicating the greater relative impact of natural gas development. For undisturbed forestlands, sediment yield averaged about 42 kg ha−1 [17]. Following clearcut harvesting and site preparation in 2003, losses increased from 111 to 224 kg ha−1 yr−1, though these differences were not found to be statistically significant [17]. In that study, a streamside management zone (SMZ) with a minimum total width of 30 m was retained around all stream channels. In 1981 these same watersheds were clearcut harvested and no SMZ was retained, and the following site preparation, sediment losses averaged 2917 kg ha−1 first year after harvest [18]. Losses returned to levels measured in undisturbed forests by the second year after harvest in both 1981 and in 2003 [15]. While large sediment plumes were observed to have eroded from both gas well locations, at F2 lobes of coarser sediments were trapped by the riparian vegetation and surface cover before reaching the stream channel. On F1, the 13,972 kg ha−1 of disturbance for 2009 largely resulted from sediment moving from the fill slope on the back side of the pad directly into the stream channel (Figure 7).

Construction of a natural gas well location in Denton, Texas resulted in 54,000 kg ha^{-1} yr^{-1} of erosion [7]. This represents sediment that eroded from the pad, but may not have necessarily entered the stream channel. Using the RUSLE 2.0 model, Waschal et al. concluded that good sediment control practices and BMPs can reduce sediment yields from natural gas well pads by 52%–93%. Similarly in the current study, the 94% difference in sediment between F1 and F2 can be attributed in part to the 15 m riparian buffer on F2 and better stormwater management.

One area of continued concern on F1 is that no efforts at site stabilization or revegetation were attempted following the initial failed attempt at seeding with rye grass. Significant rill and small gully erosion resulted from storm runoff flowing off the compacted pad area and down onto the sloping fill material where the reserve pit had been. Unlike results reported by Williams et al [7], after four years, the F1 well pad continued to erode with little evidence of natural stabilization, and natural vegetation remained sparse due to the poor condition of this fill material as a plant growing medium.

Similar to what was found with natural gas wells in the Fernow Experimental Forest in West Virginia, silt fences were inadequate at stopping these large sediment volumes [19]. Silt fences were installed down-gradient of the well location during construction, but they were installed about

0.25 m above the old stream channel on F1 and were overwhelmed by the large sediment loads, making silt fence ineffective at controlling these large volumes of sediment (Figure 8). Like with the wells constructed in the Fernow [19], improper installation resulted in the ineffectiveness of silt fence as a stormwater BMP. Silt fences functioned as intended on F2 due to proper installation and a lower overall sediment load that did not overwhelm their design capacity.

Beyond the continued erosion of the well pad, another significant concern that exists is that this deposited sediment will have long term consequences for the aquatic ecosystem. The original stream channel below the well on F1 was buried by about 0.5 m of sediment and the original pool, riffle, and glide aquatic habitats were obliterated. Sediment loading of this magnitude can have dramatic effects on lotic food webs [20]. For streams in the southeastern United States, hundreds of years if not millennia may be required to naturally purge large volumes of sediments out of regional stream and river networks [21]. This represents a localized legacy sediment issue comparable to what occurred in this region due to poor agricultural practices in the 19th Century. Effective and systematic implementation of soil conservation practices is needed to ensure that significant land use alterations do not impair surface waters in regions with extensive natural gas development.

2.4.3 WATER QUALITY

The effects of natural gas development on water quality parameters were less significant than with sedimentation. Concentrations of most parameters were not significantly different between F2 an F1. However, overall runoff volumes were greater at F1 than F2, so when concentrations were converted to mass losses, the most of the water quality parameters were significantly greater at F1 than F2. The larger runoff volumes from F1 may have diluted the concentrations, but the overall mass export was significantly greater. This indicates that reducing the export of nutrients and metals from natural gas well pads is dependent on effective stormwater management. At F1, there was no buffering between the well pad and the stream, meaning that direct contributions of contaminants occurred without the benefits of filtration provided by riparian buffer strips.

FIGURE 7: Sediment plume below the F2 natural gas well location trapped by riparian forest vegetation before entering the stream channel at the Alto Experimental Watersheds in Texas, USA.

As noted in the Results section, a producer water spill at F2 did result in the death of several trees along the stream channel, and this may account for the significantly higher salinity values at F2 (Figure 6). Differences in water quality were not observed with other parameters. Ground water was pumped into the stream channel immediately after the spill for several days in order to dilute the effects of the spill. While the water quality of the spilled water was not characterized, this remediation measure may have been adequate to reduce the impacts on water quality parameters that could be directly measured. However, the death of the riparian trees immediately in the flow area of the spill indicates that the direct ecological effects may require different remediation strategies.

2.5 SUMMARY AND CONCLUSIONS

Natural gas development is important for maintain economic prosperity and for providing a necessary energy source until renewable energy sources become more viable [22]. However, significant impacts on surface water resources were measured in this study when a gas well pad was constructed with little attention given to surface drainage patterns. Unfortunately, this was not an isolated incident on this lease area, with a pad being constructed in a perennial stream a few km north from F1 and another pad platted and surveyed over another intermittent stream nearby. Erosion rates that result from this practice are orders of magnitude greater than other land uses in this region. The 13,972 kg ha^{-1} yr^{-1} per unit dis-

FIGURE 8: Silt fence installed below the natural gas well pad at F1 illustrating the ineffectiveness of this sediment control technique due to poor installation and large sediment volumes eroded at the Alto Experimental Watersheds in Texas, USA.

turbance area recorded at F1 for the 2009 water year compared with the 714 kg ha−1 yr−1 recorded at F2 indicates that natural gas wells can be constructed without significant water quality degradation when necessary erosion control measures are implemented. However, once stream channels are filled in and obliterated, remediative BMPs like silt fence and revegetation are unlikely to have a significant effect in reducing erosion and minimizing aquatic habitat degradation. The stormwater generated by even relatively small rain events washed pollutants directly off the pad into the stream, with no opportunity for deposition and filtration.

Since construction of gas well pads in the state of Texas is not currently regulated like other construction sites, the responsibility for ending the practice of stream channel obliteration for gas well pad construction falls on the industry to self-regulate this practice. There is a precedent for effective industrial self-regulation in Texas, where forest practices like clearcutting along intermittent and perennial streams are not regulated by state or federal environmental agencies. After research demonstrated that clearcutting could have significant impacts on water resources [23], voluntary BMPs that restrict forest harvesting along streams were adopted by the forest industry in Texas by the mid-1980s. After an extensive education and outreach campaign, 98% of forestry activities in Texas voluntarily retained streamside buffers by 2011 [8]. Like the production of wood and fiber, development of natural gas resources is necessary for society. However, this must be conducted with effective and systematic implementation of soil and water conservation practices that ensure these land use changes will not impair surface waters in regions where extensive natural gas development will occur.

REFERENCES

1. EIA (Energy Information Administration). *Annual Energy Outlook with Projections to 2035*; DOE/EIA-0383(2012); U.S. Energy Information Administration: Washington, DC, USA, 2012.
2. Grant, J.; Parker, L.; Bar-Ilan, A.; Kemball-Cook, S.; Yarwood, G. *Development of Emissions Inventories for Natural Gas Exploration and Production Activity in the Haynesville Shale*; East Texas Council of Governments: Kilgore, TX, USA, 2009.Available online: http://www.netac.org/ UserFiles/File/NETAC/9_29_09/ Enclosure_2b.pdf (accessed on 12 November 2012).

3. Louisiana Department of Natural Resources (LDNR). *Louisiana Hydraulic Fracturing State Review*; Louisiana Department of Natural Resources: Baton Rouge, LA, USA, 2011.
4. Nicot, J.P.; Scanlon, B.R. Water use for shale-gas production in Texas, USA. *Environ. Sci. Technol.* **2012**, *46*, 3580–3586.
5. Entrekin, S.; Evans-White, M.; Johnson, B.; Hagenbuch, E. Rapid expansion of natural gas development poses a threat to surface waters. *Front. Ecol. Environ.* **2011**, *9*, 503–511.
6. United States Environmental Protection Agency (USEPA). *National Water Quality Inventory: Report to Congress*; EPA 841-R-08-001; United States Environmental Protection Agency Office of Water: Washingdon, DC, USA, **2009**; p. 43.
7. Williams, H.F.L.; Havens, D.; Banks, K.; Wachal, D. Field-based monitoring of sediment runoff from natural gas well sites in Denton County, Texas, USA. *Environ. Geol.* **2008**, *55*, 1463–1471.
8. Simpson, H.; Coup, C.; Duncan, C. *Voluntary Implementation of Forestry Best Management Practices in East Texas*; Texas Forest Service: Lufkin, TX, USA, 2011; p. 54.
9. McBroom, M.W.; Beasley, R.S.; Chang, M.; Ice, G.G. Water quality effects of clearcut harvesting and forest fertilization with best management practices. J. Environ. Qual. 2008, 37, 114–124.
10. Wachal, D.J.; Banks, K.E.; Hudak, P.F.; Harmel, R.D. Modeling erosion and sediment control practices with rusle 2.0: A management approach for natural gas well sites in Denton County, TX, USA. Environ. Geol. 2009, 56, 1615–1627.
11. Chang, M.; Clendenon, L.D.; Reeves, H.C. Characteristics of a Humid Climate, Nacogdoches, Texas; College of Foestry, Stephen F. Austin State University: Nacogdoches, TX, USA, **1996**; p. 211.
12. University of Texas at Austin Bureau of Economic Geology. Geologic Atlas of Texas, Palestine Sheet; University of Texas at Austin Bureau of Economic Geology: Austin, TX, USA, **1968**.
13. Mowery, I.C. Soil Survey Cherokee County Texas; United States Department of Agriculture Soil Conservation Service: Washington, DC, USA, **1959**; p. 68.
14. Tuppad, P.; Winchell, M.; Wang, X.; Srinivasan, R.; Williams, J. ArcAPEX: ArcGIS interface for agricultural policy environmental extender (APEX) hydrology/water quality model. *Int. Agric. Eng. J.* **2009**, 18, 59–71.
15. Wang, X.; Saleh, A.; McBroom, M.W.; Williams, J.R.; Yin, L. Test of APEX for nine forested watersheds in East Texas. *J. Environ. Qual.* **2007,** 36, 983–995.
16. Hach Company. Dr/890 Colorimeter Procedures Manual, 9th ed.; Hach Company: Loveland, CO, USA, **2009**; Volume 48470-22; p. 614.
17. McBroom, M.W.; Beasley, R.S.; Chang, M.; Ice, G.G. Storm runoff and sediment losses from forest clearcutting and stand reestablishment with best management practices in the southeastern United States. *Hydrol. Process*. **2008**, 22, 1509–1522.
18. Blackburn, W.; Wood, J.; DeHaven, M. Storm flow and sediment losses from site-prepared forestland in east Texas. *Water Resour. Res.* **1986,** 22, 776–784.
19. Adams, M.B. Effects of Development of a Natural Gas Well and Associated Pipeline on the Natural and Scientific Resources of the Fernow Experimental Forest; U.S. Department of Agriculture, Forest Service, Northern Research Station: Newtown Square, PA, USA, 2011.

20. Henley, W.; Patterson, M.; Neves, R.; Lemly, A.D. Effects of sedimentation and turbidity on lotic food webs: A concise review for natural resource managers. *Rev. Fish. Sci.* **2000**, 8, 125–139.
21. Jackson, C.; Martin, J.; Leigh, D.; West, L. A southeastern piedmont watershed sediment budget: Evidence for a multi-millennial agricultural legacy. *J. Soil Water Conserv.* **2005**, 60, 298–310.
22. Kerr, R.A. Natural gas from shale bursts onto the scene. *Science* **2010**, 328, 1624–1626.
23. Ice, G.G. History of innovative best management practice development and its role in addressing water quality limited waterbodies. *J. Environ. Eng.* **2004**, 130, 684–689.

McBroom, M., Thomas, T., and Zhang, Y. Soil Erosion and Surface Water Quality Impacts of Natural Gas Development in East Texas, USA. Water 2012, 4, 944-958. doi:10.3390/w4040944. Reprinted with permission from the author.

CHAPTER 3

ANALYSIS OF WATER FEATURES IN GAS LEAKAGE AREA

LIU HUAISHAN, WANG FENGFAN, TONG SIYOU, LI GAOLIN, and ZHANG HAIYAN

3.1 INTRODUCTION

There are various forms of marine gas, they may be generated by ocean waves, fish's swim bladder, submarine wake, shallow gas leak, hydrate decomposition, and so forth, they have a wide range of effects on sound signal used in marine seismic exploration, such as changing the speed of sound and attenuating signal energy. Various phenomena caused by natural gas have been found in the sea shallow gas and gas hydrates area, for example, speckled reflection "flame-" shaped reflection, and so forth. These phenomena do not exist in an area without shallow gas and gas hydrates.

Carstensen and Foldy proceeded with theoretical study of bubble scattering effects on sound propagation ignoring the relevance of the spatial distribution of bubbles. They obtained the bubble's radius by measuring the frequency of the sound generated by bubble, combined with the relationship between resonant frequency and bubble radius, and also obtained experimental data of the transmission and reflection coefficient of sound in the mixture. Experiments show that when the same bubble radius, and the incident wave frequency and the bubble resonance frequency are the same, the sound attenuates most, which is consistent with the theoretical value. Fox et al. and others got the sound phase velocity and the sound absorption coefficient when the frequency is between $10 \sim 390\,\text{kHz}$ through incident continuous wave on the bubbly water. Keller and Kolodner incident sound with frequency 15 ~ 100 kHz into the mixed-phase medium with the bubble (the radius of the 0.16–0.5 mm), the result confirmed the Foldy theory once again. Silberman measured the velocity and the attenuation coefficient of bubbly liquid in the static wave tube, the bubble radius is 0.08 ~ 0.26 inches, and bubble volume fraction of 0.03 to 1.0%, sound frequency is 60 Hz ~ 20 kHz. Xiaoliang et al. first proposed the concept of mixed media, obtained the sound wave equation in bubbly mixed media [4]. The researchers also found that the vibration characteristics of the bubble should be included in dynamic equations of mixed body. Prosperetti linearized the vibration of air bubbles under the action of sound, studied heat conduction effects on the attenuation of sound, and found that the polytropic index and thermal conductivity were related to sound frequency. Kameda and Matsumoto gave the mathematical model of nonlinear vibration of the bubble, and gave accurate description of the scope and characteristics of the effective equation. From his study we can find that Foldy theory and Van Wijingaarden equations are applicable in the case of lower volume fraction of bubbles that the interaction between bubbles can be ignored. Commamder [7, 8] and Prosperitti developed this theory based on Porsperetti's and others' accurate description of the dynamics on the bubble, and discussed the linear pressure wave propagation in bubbly water.

Li et al. discussed the influences of the relevance of the spatial distribution of bubbles to sound propagation in bubbly water with consideration of Twersky's theory and sound absorption of bubble. The results show that when considering the relevance of the spatial distribution of bubbles, the attenuation coefficient of the peak moves backward and the peak is lower and flatter than ignoring the relevance. With the increasing of bubbles' volume fraction, the relevance of their spatial distribution increases, the resonant frequency of the bubble increases, and its scattering amplitude becomes smaller when resonating. Yonghui used ultrasonic testing method to measure sound transmission loss and velocity ratio of the gas-water miscible media, the used sound frequencies were 0.5 MHz, 1.25 MHz, 2 MHz, and 2.5 MHz. Xiaoliang et al. and others developed a series of analytical formulas that calculate the sound propagation characteristics in bubbly water and the nonlinear parameters of media for the fact that the actual bubble size always exists distribution. They studied the sound wave propagation features such as velocity and attenuation in bubbly water.

All the studies above are based on analysis of the impact extent of bubbles on the water, there are no studies on the sound characteristics of sediments and water containing hydrate and free gas in the gas leakage area. This paper studies the influence of gas on the effective sound absorber and scatter in a certain frequency range, natural gas radius at different depths when the gas continues to rise, deformation, or dissolved in the sea water. Then this paper analyzes that when these processes occur, how the resonance frequency of natural gas and the sound characteristics of it change. This paper established sound models with hydrate and free gas in the water and sediment through the analysis of gas leakage area (including hydrate and free gas) and real seismic data.

3.2 THE RISING VELOCITY AND TRACK OF GAS

The rising of natural gas in water is a very complex process, and its rising velocity and track are related to the diameter and shape of natural gas. When calculating the rising process of natural gas, in order to simplify the calculation process, in theory, the diameter of natural gas is considered as the volumetric spherical to calculate, ignoring the internal activity and driven from setting velocity of solid particle in water.

The resultant force of natural gas in water equals to buoyancy force minus resistance force;

$$F_1 - F_2 = \frac{\pi}{6} d^3(\rho_w - \rho_a)g - \frac{\pi}{2} C_D \rho_w \frac{v_0^2 d^2}{4} \tag{1}$$

In (1), d is the diameter of natural gas, ρ_w is the density of water, ρ_a is the density of air, g is the acceleration of gravity, C_d is resistance coefficient, and $v_0^2d^2/4$ is the projected area of natural gas in vertical direction.

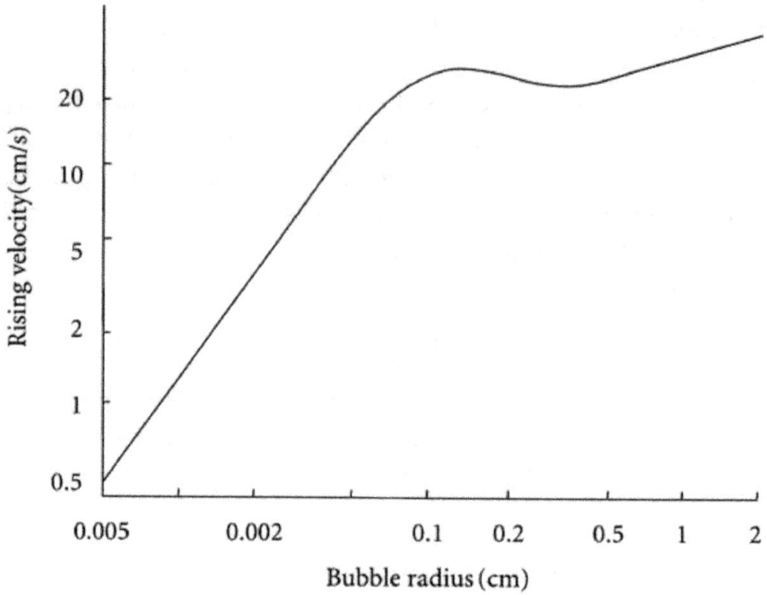

FIGURE 1: Curve between the rising velocity and diameter of natural gas.

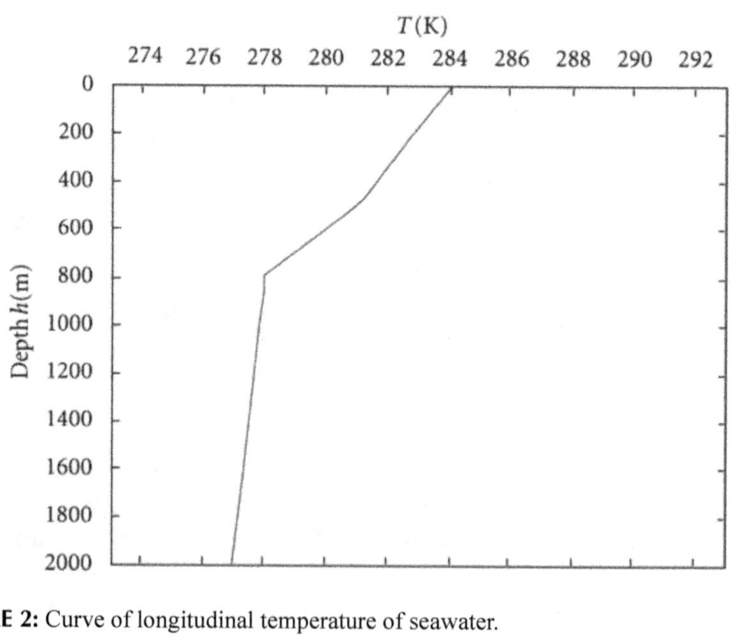

FIGURE 2: Curve of longitudinal temperature of seawater.

According to Newton's second law, we can know that $F = ma$, $m = \pi d^3 \rho_a / 6$, $a = dv_a/dt$, v_a is the rising velocity of natural gas, if we put it into (1), we can get that

$$\frac{dv_a}{dt} = \frac{(\rho_w - \rho_a)g}{\rho_a} - \frac{3C_D \rho_w v_a^2}{4\rho_a d} \tag{2}$$

When $dv_a/dt = 0$, v_a is the ultimate velocity;

$$v_0^2 = \frac{4gd(\rho_w - \rho_a)}{3\rho_w C_D} \tag{3}$$

From hydromechanics knowledge, we can know that resistance coefficient C_D can be determined by Reynolds number R_e;

$$R_e = \frac{v_a d}{v} = \frac{\rho_w v_a d}{\mu} \tag{4}$$

In (4), v is motion viscosity coefficient and μ is dynamic viscosity coefficient, respectively. According to the difference of Reynolds number, the value of can be divided into three areas: laminar flow, transition, and turbulence. The velocity formula calculated from this is also different.

According to hypothetical conditions, (3) has a small range of application and usually is suitable to bubble whose diameter is less than 150 μm in static water.

While in the rising process of moderate natural gas, there are two forms of motions [5]: (1) Rolling motion of rigid body, zigzag or spiral motion. (2) Changes in shapes and expanding, which is oscillation of natural gas. These forms of motion often exist at the same time, and movement of moderate natural gas is very complicated. But we can simplify its model to a perfect ellipsoidal and then consider its additional motions.

While with regard to big natural gas, its shape can only be acquired from observation and the expression of its velocity is considerably complex.

Therefore, the rising velocity of natural gas in static water is closely related to its linear density. According to huge amounts of experimental

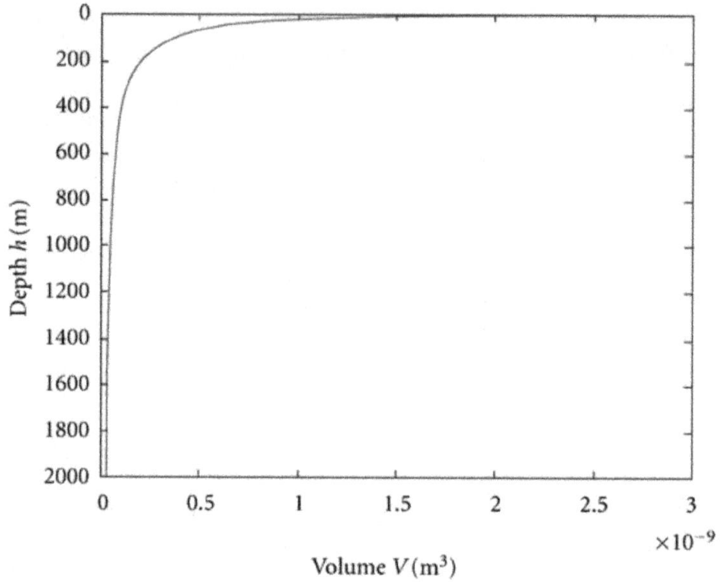

FIGURE 3: The relationship between the balance volume of natural gas and depth.

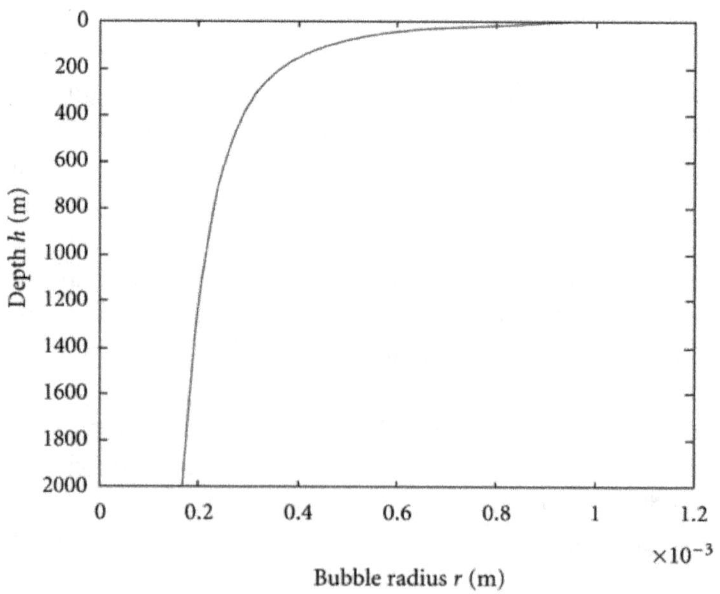

FIGURE 4: Curve of the balance radius of natural gas with depth.

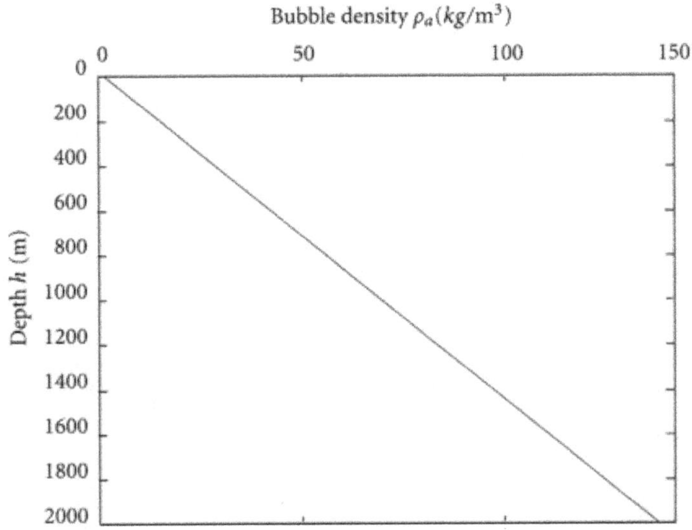

FIGURE 5: Curve of the density of natural gas with depth.

data and theoretical analysis, we can get the relation curve between the rising velocity and diameter of natural gas which is shown in Figure 1. It can be seen from Figure 1 that the velocity of natural gas reaches a maximum value when the diameter is about 0.1 cm, and then changes very little with the increasing of diameter while the shape of natural gas changes from spherical to ellipsoidal. With regard to small natural gas, its characteristics of motion are linear. And for big natural gas, its characteristics of motion is approximate to linear, but its shape usually becomes distorted and not spherical any more. When the diameter is greater than 1000 µm, the motion of natural gas is not stable anymore and is not the function of diameter, so it is hard to calculate the velocity of natural gas. In experimental observation, its rising velocity is approximate to 30 cm/s.

In fact, in the rising process of natural gas, the volume will increase with the decrease of pressure and its shape will also change. The formation of water vapor in its interior changes the density of natural gas. And because of the existence of surface tension, the gas in natural gas will dissolve in the water. Experiments show that the natural gas whose diameter is 0.1 cm can completely dissolve in water within about 20 min.

From the above analysis, we can see that the big natural gas in sea water will rise to the sea surface quickly, and tiny natural gas mainly

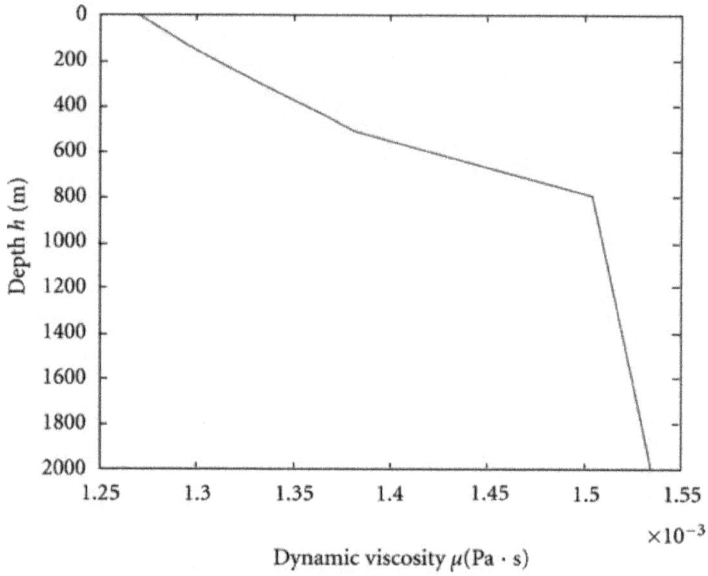

FIGURE 6: Curve of seawater viscosity coefficient with depth.

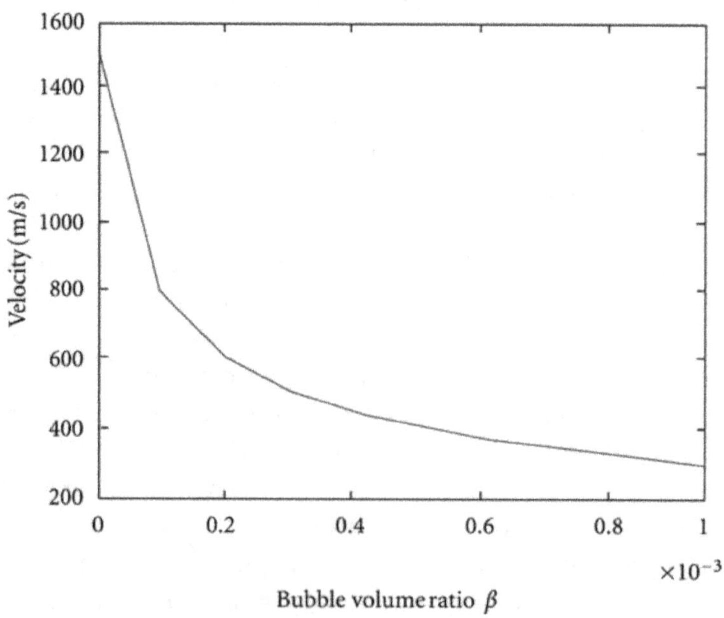

FIGURE 7: The curve of relation between sound velocity and the ratio of gas volume.

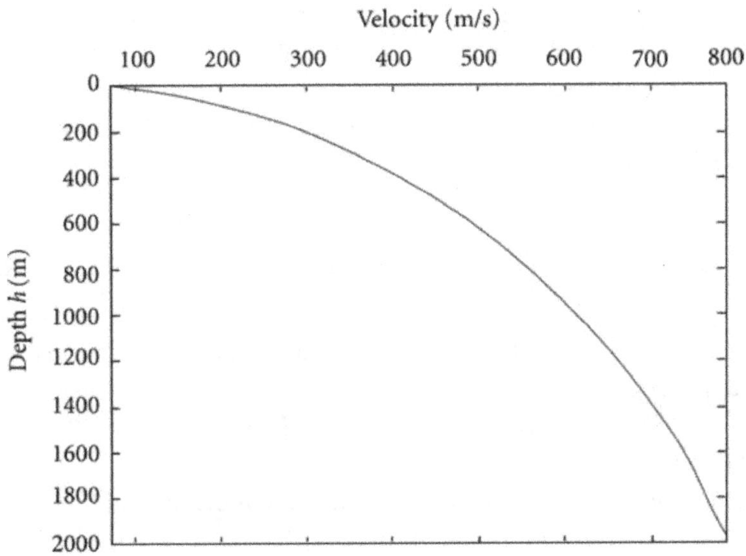

FIGURE 8: Sound velocity changing with depth in the gas layer.

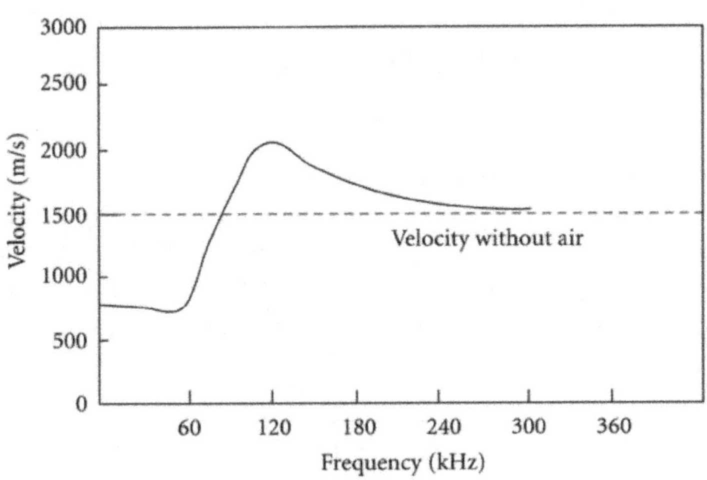

FIGURE 9: Curve of sound velocity measured in the gas of same size (diameter is 0.011 cm) under wide frequency range [2].

exists in sea water. The minimum stable diameter in sea water is about 1 μm. Although this natural gas is still subjected to the influence of buoyancy force and gas diffusion to dissolve and rise constantly, amounts of measuring results in sea show that the densities of tiny natural gases in sea water will keep a dynamic balance state and diameters of natural gases are all less than 500 μm [11].

3.2.1 THE RISING FEATURE OF NATURAL GAS IN WATER

Assuming that there is some natural gas in marine, its shape always remained spherical in the process of rising in static water, the radius in sea level is 1 mm, and the volume is $V_0 = 4.1888 \times 10^{-9}\ m^2$, which is full of methane gas.

The pressure in the sea level is one standard atmospheric pressure, and the seawater pressure in the depth of h is $P_1 = P_0 + \rho_w gh$. Suppose the seawater is incompressible fluid, $\rho_w = 1026$ kg/m³ is the average density of seawater, and the gravity acceleration is $g = 9.81$ m/s². The temperature in sea surface is 284 K. The change of temperature is shown in Figure 2.

The molar mass of methane gas is 16 g/mol, according to the equation of state of ideal gas:

$$\frac{P_0 V_0}{T_0} = \frac{P_1 V_1}{T_1} = nR, \tag{5}$$

$R = 8.31\ \text{Jmol}^{-1}\text{K}^{-1} = 0/082 \times 10^{-3}\ \text{m}^3\ \text{atm K}^{-1}\text{mol}^{-1}$

We can obtain that the amount of substance of this natural gas is $n = 1.7987 \times 10^{-7}$ mol, its quality is $m = n \times M = 2.8879 \times 10^{9}$ kg, and the density in the surface is $\rho_a = m/V_{0-} = 0.68705$ kg/m³, then the relationship between volume of gas and depth can be obtained:

$$V_1 = \frac{P_0 V_0}{P_0 + \rho_w gh} \frac{T_1}{T_0} \tag{6}$$

In the equation, T_0 is the temperature in the sea surface, and T_1 is the temperature in the depth of h.

Assume that the internal gas composition and quality of the natural gas do not change in different depths of ocean water, then according (6) and the given data, we can obtain the change curve of gas volume with depth (Figure 3).

At the same time, the change of radius r with depth h can be found by formula $r = (3V/4\pi)^{1/3}$ (Figure 4). As shown in the figure, at the beginning, the change of volume and radius of the bubble in the water depth of several hundred meters is very rapid, but along with the increase of depth, the change rates of volume and radius become smaller and smaller. According to the relationship between the quality of natural gas and volume , the change curve of the density of natural gas can be worked out (Figure 5).

Without considering the composition change of natural gas, there is a linear relationship between the density of natural gas and depth:

$$\rho_a = \frac{mT_0}{P_0V_0T_1}(\rho_w gh + P_0) \tag{7}$$

We can calculate that the volume of natural gas is $V = 2.05 \times 10^{-2}$ mm^3 and the radius is $r = 0.16966$ mm in the depth of 2000 m.

Thus we can assume that there is a methane gas with balance radius of 0.16966 mm in the depth of 2000 m below the sea surface, which rises freely in the case of static seawater.

Because the maximal radius of this natural gas is only 1 mm, according to the actual observation data and calculation results, its maximal Reynolds number in seawater is far less than 1000.

As known from the fluid mechanics, the object moves relative to the fluid in the fluid whose surface has a "boundary layer." The microregiment near objects in the layer is static relative to the object, and the fluid micro regiment on the outside of the layer has the same speed as fluid. So there are velocity grads and viscous force in the boundary layer, which can serve as resistance to the object.

According to the given conditions, the dynamic viscosity coefficients μ of different depths are calculated using water and steam properties (Figure 6).

Smaller objects move slowly in the fluid with large stickiness, which means, in the case of small Reynolds numbers, the resistance, called viscous resistance, is the main factor. The famous Stokes formula describes the viscous resistance of spherical objects: $F = 6\pi\mu vr$, r is the sphere radius, v is the velocity of sphere, and μ is viscosity. This formula is correct as the condition that Reynolds number is much smaller than 1. That means when $Re \leq 1$, it fits the conditions of Stokes resistance formula, $C_D = 24/Re$, combined (4):

$$\frac{dv_a}{dt} = \frac{(\rho_w - \rho_a)g}{\rho_a} - \frac{18\mu v_a}{\rho_a d^2} \tag{8}$$

While $1 \leq Re \leq 1000$, resistance coefficient C_D is calculated by corrected White experience formula:

$$C_D = \frac{24}{Re} + \frac{6}{\sqrt{Re}} + 0.4 \tag{9}$$

Take formula (9) into formula (8):

$$\frac{dv_a}{dt} = \frac{(\rho_w - \rho_a)g}{\rho_a} - \frac{3\rho_w v_a^2}{4\rho_a d}\left(\frac{24\mu}{\rho_w v_a d} + 6\sqrt{\frac{\mu}{\rho_w v_a d}} + 0.4\right) \tag{10}$$

Using formula (10) can calculate the transient acceleration of bubble in the process of rising, and making $dv_a/dt = 0$ can obtain the final speed when the force is balance.

3.2.2 THE VELOCITY OF NATURAL GAS IN WATER

It is measured that the air dissolved in fluid (liquid) has no influence on the velocity of sound, even the air in the liquid is saturated. But if there is small gas saturated suspended in the water, even very little, it may decrease the sound velocity. The sound velocity is related to the proportion of the gas in liquid and the frequency. Because the gas has a better compression than water, the compressibility of the water will change as the increasing of the proportion of the air.

When the air's volume is smaller than the resonance bubbles, that means the frequency of sound is less than the resonance frequency, according to the theory of mixed liquid, we can get the sound velocity through the resonance of the water and air and the density. We can suppose k, k_a, k_w to represent the compressibility of mixed liquid, air, and water separately, and presume ρ, ρ_a, ρ_w are the corresponding density, when β (the proportion of the air) in the water is very little, because the $k_a >> k_w$, $\rho_a << \rho_w$, and $\beta << 1$, we can get the velocity:

$$v = \left(\frac{1}{\rho k}\right)^{\frac{1}{2}}$$

$$= \left\{\frac{1}{[\beta\rho_a + (1-\beta)\rho_w][\beta k_a + (1-\beta)k_w]}\right\}^{\frac{1}{2}}$$

$$[5\,pt] = \left[\frac{1}{\rho_w k_w(1+\beta k_a/k_w)}\right]^{1/2} = v_0\left[\frac{1}{1+\beta k_a/k_w}\right]^{1/2}$$

$$= v_0\left(\frac{1}{1+2.5\times10^4\beta}\right)^{1/2} \tag{11}$$

In (11), v_0 is the sound velocity of water without air. We can get that when the ratio of the gas in the mixed liquid is 0.01, the velocity is about the 53.5% of that without air. Supposing the velocity is v_0 = 1500 m/s when there is no air, the curve is shown in Figure 7.

Conversely, when the frequency is higher than the resonance frequency of the mixed liquid with least air, which means that the frequency is higher than all the resonance frequency of the air, the influence of suspension gas can be ignored.

If the single bubbles distributing inside an area changed to be uniform, supposing the starting volume ratio is 0.01%, the sound velocity varies following the depth, as in Figure 8. Through calculating, near the sea level, the velocity of sound in the layer is only 5% of that without gas.

When the frequency of sound approximately equals the air frequency, the velocity of sound wave has relations with not only the gas content but also the frequency of sound wave. If the frequency is approaching the resonance frequency, the speed of sound has a big change. Figure 9 is the smoothing speed curve of a group of homogeneous gas drawn by Fox et al. under the conditions as these, the average diameter is 0.011 cm, and the density $V_a = 2 \times 10^{-4}$. When the frequency is resonant frequency, the speed of sound is the same as the velocity without air.

From Figure 9 we can see that when the frequency of sound wave is less than the resonant frequency of gas, the sound speed decreases because of the air; conversely, when the frequency of sound wave is higher than the resonant frequency of gas, the gas has little infection on the sound velocity; when the frequency of sound wave is approaching to the resonant frequency of gas, the sound velocity changes largely as the change of frequency.

When the frequency of the sound wave is almost the same as the resonant frequency of gas, and the diameter of sphere bubble distributed evenly far less than the wave length of sound, the accurate speed and reduction under the effects of gas resonance is expressed as

$$\left(\frac{c_0}{c}\right)^2 = \frac{1+AX}{2}\left\{1 \pm \left[1+\left(\frac{AY}{1+AX}\right)^2\right]^{1/2}\right\}, \alpha = \frac{c\pi f AY}{c_0^2} \quad (12)$$

In the equation, $A = \rho c_0^2/p_0 \gamma$, γ is the adiabatic exponent, c_0 is the speed of water without bubble, p_0 is the pressure of static water, ρ depends the density of gas and water mixture, and the parameters X and Y are presented as

$$X = \frac{\beta(1-f_*^2)}{(1-f_*^2)^2+\delta_*^2}, Y = \frac{\beta\delta_*}{(1-f_*^2)^2+\delta_*^2} \quad (13)$$

Among (13), $f_* = f/f_0$, $\delta_* = \delta f_*^2$, f is the frequency of the sound wave, f_0 is the resonant frequency of gas bubble, and the δ is the damping constant.

3.2.3 REFLECTION AND TRANSMISSION OF SOUND WAVE

When the distance between the gases is less than the sound wavelength, the gas medium can be regarded as homogeneous medium, because the gas in the sea water changes its compressibility, making the average density and velocity changed, the interface between gas-bearing water and seawater is the wave impedance interface. Therefore, the gas reservoir can be seen as an intermediate of the sea water medium at this time. If the wavelength is much longer than the radius of the gas ($r << \lambda/2\pi$), suppose that a plane wave normally incident to a gas reservoir with the thickness of H, and the gas in it has the uniform size and uniform distributed (Figure 10), then the incident sound pressure is $P_i = A_i \exp(i(wt - k_w z))$, where A_i is amplitude of the incident sound pressure; $P_r = A_r \exp(i(wt - k_w z))$ is the reflection sound pressure, where A_r is amplitude of the refection sound pressure; the transmission wave sound pressure is $P_T = A_T \exp(i(wt - k_w z)$ where A_T is amplitude of the sound pressure; the layer transmission wave sound pressure is $P_1 = A_1 \exp(i(wt - k_w z)$, where A_1 is amplitude of the sound pressure; the reflection sound pressure is , $P_2 = A_2 \exp(i(wt - k_w z)$ where w is the sound harmonic vibration circular frequency, and ρ_w and ρ_m are the density of sea water and gasbearing, and c_w and c_m are the sound speed of sea water and gas-bearing, and $k_w = w/c_w$ and $k_m = w/c_m$ are the sound wave number of sea water and gas-bearing. The plane wave particle velocity is $v = v_0 \exp(wt - kx)$, and the velocity amplitude is $v_0 = p_0/c_p$. Let , $Z = cp$, then Z_w and Z_m are wave impedances of sea water and gas-bearing.

The velocity and pressure on interface between sea water and gasbearing are continuous, so the boundary conditions are as follows

(1) when $z = 0$, pressure: $P_i + P_r = P_1 + P_2$, that is,

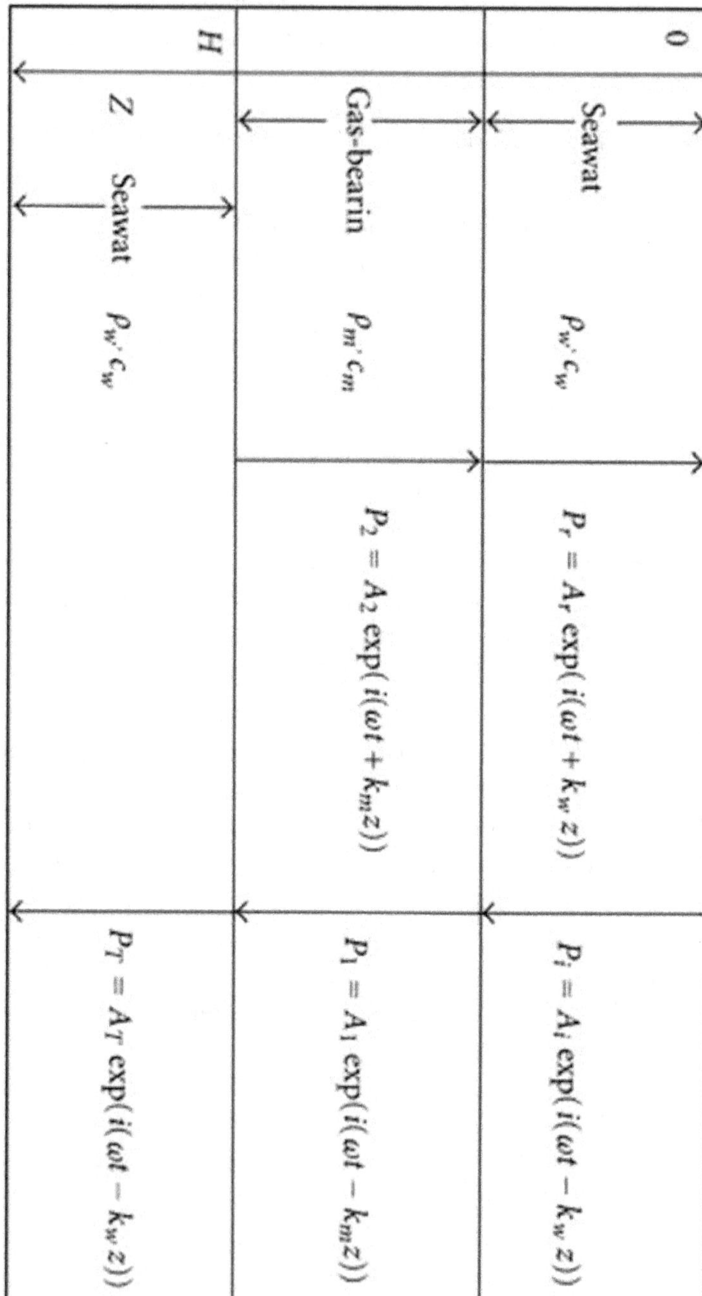

FIGURE 10: Reflection and transmission of sound wave in gas reservoir.

$$A_i + A_r = A_1 + A_2, \tag{14}$$

velocity: $(P_i - P_r)/Z_w = (P_1 - P_2)/Z_m$, that is,

$$\frac{A_i - A_r}{Z_w} = \frac{A_1 - A_2}{Z_m} \tag{15}$$

(2) when $z = H$, pressure: $P_1 + P_2 = P_T$, that is,

$$A_1 \exp(-ik_m H) + A_2(ik_m H) = A_T \exp(-ik_m H), \tag{16}$$

velocity: $(P_1 - P_2)/Z_m = P_{T1}/Z_W$, that is,

$$\frac{A_1 exp(-ik_m H) - A_2 exp(ik_m H)}{Z_m} = \frac{A_T exp(-ik_w H)}{Z_W} \tag{17}$$

Simultaneous equations (14), (15), (16), and (17), and making $R = A_r/A_i$, $T = A_T/A_i$, $B_1 = A_1/A_i$, $B_2 = A_2/A_i$, then

$$R = \frac{(Z_W^2 - Z_m^2)[1 - \exp(i2k_m H)]}{(Z_m + Z_w)^2 exp(i2k_m H) - (Z_w - Z_m)^2}$$

$$T = \frac{4Z_m Z_w \exp[i(k_m + k_w)H]}{(Z_m + Z_w)^2 exp(i2k_m H) - (Z_w - Z_m)^2}$$

$$B_1 = \frac{2Z_m(Z_m + Z_w) exp(i2k_m H)}{(Z_m + Z_w)^2 exp(i2k_m H) - (Z_w - Z_m)^2}$$

$$B_2 = \frac{2Z_m(Z_w - Z_m)}{(Z_m + Z_w)^2 exp(i2k_m H) - (Z_w - Z_m)^2} \tag{18}$$

The particle displacement is $\xi = (v_0/iw)\exp(i(wt - kx))$ for plane wave, and the displacement amplitude is $\xi_0 = v_0/iw$, at the point of $x = 0$, $R = A_r/A_i = v_{r0}/v_{i0} = \xi_{r0}/\xi_{i0}$ is the ratio of the displacement amplitude of incident and reflected waves on the interface. Similarly, T is the ratio of the amplitude of incident and wave transmission through the gas, B_1 is the ratio of the amplitude of incident and wave transmission of the above interface, B_2 is the ratio of the amplitude of incident and wave transmission of the following interface.

Assuming a floating gas layer, in which the size of the gas is uniform distribution, with the thickness of $H = 0.01$ m remaining unchanged, the size of the gas at the depth of 2000 m is the size of gas assumed in this paper. The volume ratio is $\beta = 0.01\%$. When a plane wave with the fre-

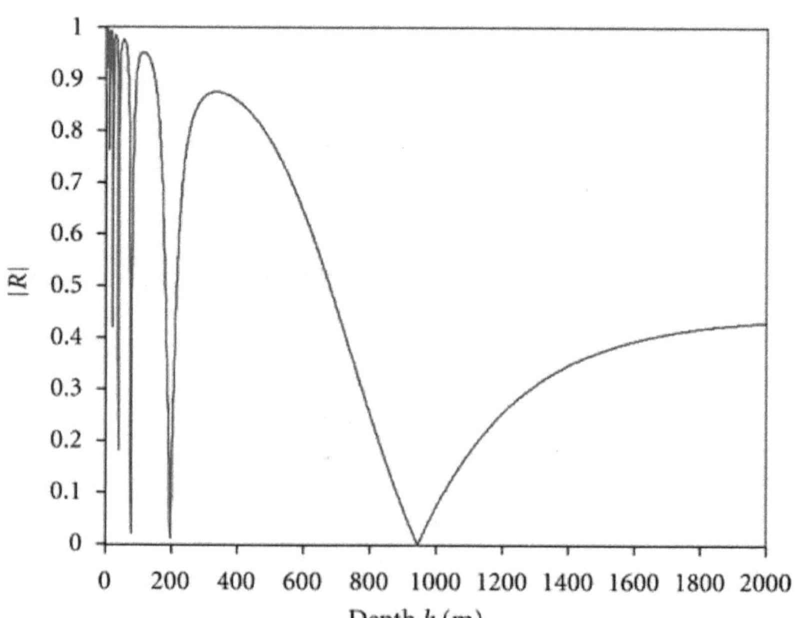

FIGURE 11: Curve of |R| changed with depth h.

quency of 30 kHz normal incident, the amplitude ratio R of incident and reflection is shown as Figure 11.

As can be seen from Figure 11, |R| has certain cyclical changes. When $k_m H = n\pi$ $(n = 0, 1, 2, ...)$, the thickness of the bubble layer is an integer multiple of half wavelength, then |R| is minimum, for example, at the point of $h = 947$ m, the wave is almost the transmission sound. When $k_m H = (2n - 1)\pi/2$,, the thickness of the gas layer is an odd multiple of the quarter wavelength, then is maximum, for example, at the point of $h = 338$ m, almost all the sound wave reflects back.

3.2.4 SCATTERING EFFECT OF NATURAL GAS

The attenuation yielded by sound absorption needs to be considered in the actual gas. While the gas is forced to vibrate, some sound energy is burned because of the system's internal friction and distributed to the surrounding water in the form of heat, but the large vibration attenuation prevented the temperature balance in every moment. When the gas is compressed, the heat is diffused along with the increasing temperature, so the bubble began to cool prior to the expansion. When the volume of gas is compressed to a minimum, the temperature is slightly reduced because of the heat shed to the water. The same is true when the expansion. Therefore, the process carried out in the gas is the procession between adiabatic process and isothermal process. Taking all these circumstances into account, the scattering cross-section expression of real gas is

$$\sigma_s = \frac{4\pi r^2}{\left(\frac{f_0^2}{f^2} - 1\right)^2 + \delta^2} \tag{19}$$

in (19), δ is total damping value, of which expression is

$$\delta = \frac{f_0^2}{f^2}\beta + \eta + \frac{C_1}{c\rho\eta} \tag{20}$$

in (20), β is normal number; $\beta << 1$, C_1 is constant, to represent friction effect.

From (19), can obtain the ration of scattering cross-section and geometric cross section:

$$\frac{\sigma_s}{\pi r^2} = \frac{4}{\left(\frac{f_0^2}{f^2} - 1\right)^2 + \delta^2} \tag{21}$$

When the sound frequency is closed to resonance frequency, $\sigma_{s0}/\pi r^2 = 4/\delta_0^2$, δ_0 is called the total damping constant.

For the real gas, the absorption of sound energy must be considered. The sum of scattering energy and absorption energy is called dissipation energy while the dissipation section is

$$\sigma_e = \sigma_s + \sigma_a = \frac{4\pi r^2(\delta/\eta)}{\left(\frac{f_0^2}{f^2} - 1\right)^2 + \delta^2} \tag{22}$$

So the absorption cross-section can be calculated as

$$\sigma_a = \frac{4\pi r^2 \left(\frac{\delta}{\eta} - 1\right)}{\left(\frac{f_0^2}{f^2} - 1\right)^2 + \delta^2} \tag{23}$$

It can be seen that the relationship of dissipation section and scattering cross-section is $\sigma_e/\sigma_s = \delta/\eta$.

3.3 EFFECT OF ACTUAL DATA

Deep -ea high-resolution multichannel seismograph system developed by Ocean University of China was used in March 2009. It is an all new digital

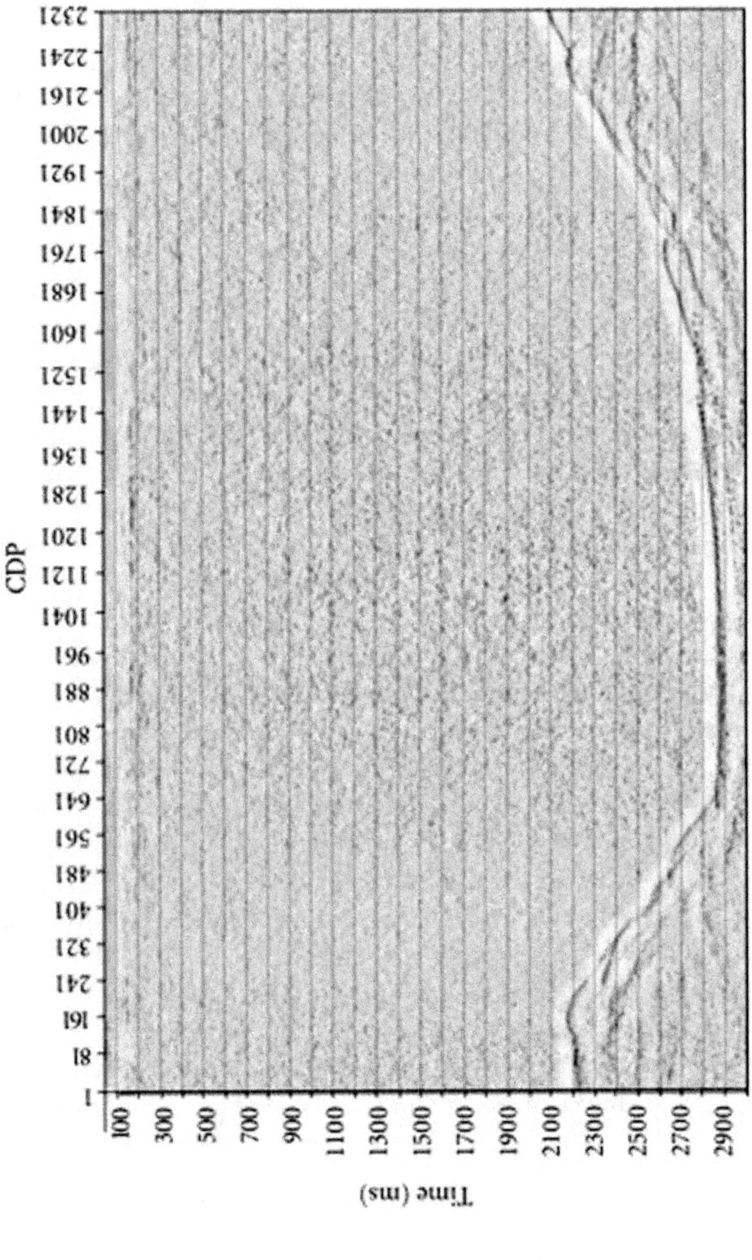

FIGURE 12: Characteristics of water bodies which appear in the high-resolution seismic profile of gas hydrate leakage area in the South China Sea.

marine seismic hydrophone and the most advanced new marine seismic hydrophone all-digital instrument system in the world. Compared with the current analog seismic hydrophone system, its domestic applicability is stronger, and its resolution and SNR are higher. Each hydrophone has 24 channels and can be combined into a system of channels. The diameter of hydrophone is only 38 mm and the wall thickness is 1/8 inch. That is easy for transport and delivery and is durable. Using 6.25 m group interval and ms broadband technology, it can be applied in oil exploration, marine engineering, and stratigraphic section measurement to obtain high-resolution seismic data. The high-resolution seismic data acquisition was firstly used in gas hydrates area near Dongsha Islands in the South China Sea. Record length of survey line is 3 s, the sampling interval is 1 ms, and the number of coverage is 6 times.

For feature of study areas that have a wide range of leakage area (including hydrate and free gas) in favor gas hydrate [12, 13], various gas hydrate leakage areas (including gas hydrate and free gas) distributions have been found through the geological survey studies. With the recent increase in understanding and investigating instrument performance improvement, we can further understand the distribution of gas hydrate leakage area (including gas hydrate and free gas) and seismic character (Figure 12), through the study on gas hydrate leakage area (including hydrate and free gas) according to seismic survey data.

For further research on the seismic character in shallow gas leakage area (including hydrate and free gas), according to the research and experimental observation of water body features (Figure 13) on high-resolution seismic profile in the Bohai Sea shallow gas leakage area, the following sound properties of sediments and water that contains gas have been found. (1) The nature of the sound parameters of the gas. (2) The sound characteristics in the water contains natural gas (gas in shallow water, and the vibration process, resonance frequency, sound velocity, attenuation, reflection, and transmission). (3) The sound characteristics of sediments contain natural gas (resonant frequency, bubble damping, sound velocity, and attenuation).

In recent years seismic reflection characteristics of gas hydrates in deep water (Figure 12) and shallow gas in shallow water (Figure 13) leakage area (including hydrate and free gas) were summarized. Thus the seismic characteristics of the natural gas leakage area (containing gas hydrate and free gas) is divided into three categories according to spatial location. (1) The seismic characteristics in the formation: sound blanket, sound curtains, sound disturbances, irregular strong reflecting surface interface, and phase-down on both sides. (2) Features in the seabed surface: submarine pockmark and large collapse pit. (3) Features in water layer: sound plumes, cloud-like move, and dash-dot reflection. Combination of theory and case analysis shows that (1) there exists gas escaped from buried natural gas area in the sound plumes, cloud-like disturbance, and dash-dot reflection areas in seawater; (2) the region of sound blanket space represents areas of high gas concentration, a variety of sound disturbances represents areas with low gas concentration, and the gas concentration in sound curtains between the front two conditions.

3.4 CONCLUSIONS

The following conclusions can be obtained by the study of features of bubbly water in gas leakage area (including hydrate and free gas) through the study of characteristics of elastic parameters of deposition in the gas leakage area (including hydrate and free gas), and the use of sound theory of water and sediment containing hydrate and free gas.

(1) We get the changes of the volume and radius when the gas rises at different depths, and also get the expression of acceleration in the static water. Besides, since the dynamic viscosity becomes smaller when the gas rises, the force gas suffered and the velocity keeps increasing until it arrives at the surface. Radius and volume of gas changes with depth with different velocities, in the context of 400 m below sea level, the velocity changes very fast, and as the depth increases the changes become smaller.

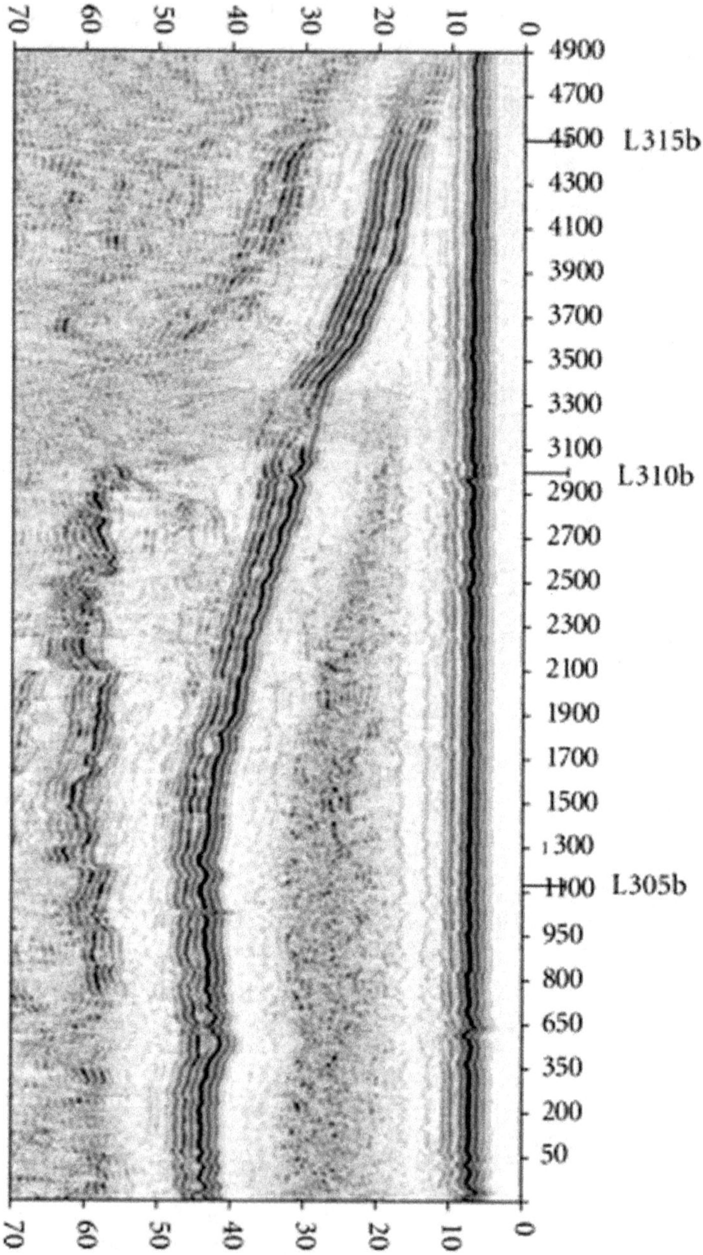

FIGURE 13: Characteristics of water bodies which appear in the high-resolution seismic profile of shallow gas hydrate leakage area in the Bohai Sea.

(2) We also give the expression of reflection coefficient of gas, the size of the reflection coefficient not only related to impedance of both sides of the interface, but also related to the product of wave number and thickness of the gas, the reflection coefficient shows a certain periodicity. The gas becomes bigger when rising, volume ratio grows and sound is attenuated faster, Theoretically, there are only 5% of no gas medium reaching sea level.
(3) The size of the cross-section represents the gas range of the sound scattering, When the pressure is constant, scattering cross-section reaches the maximum near the resonance frequency, these conditions are changing as the gas rises, and the maximum of the scattering cross-section is only one-tenth of its geometric cross-section. Sound energy endure losses through internal friction and thermal conduction when the actual gas vibrates. This loss of performance for the scattering and absorption of air bubbles, absorption cross-section is less than the scattering cross section.
(4) The main sound features of gas are relevant to gas size, gas content, velocity, attenuation, resonance frequency and the scattering cross-section, and other conditions. We can clearly see the sound characteristics of seismic data that yielded when the gas (or gas hydrate dissociation) escaped the water.

REFERENCES

1. E. L. Carstensen and L. L. Foldy, "Propagation of sound through a liquid containing bubbles," Journal of Acoustical Society of America, vol. 19, no. 3, pp. 481–501, 1947.
2. F. E. Fox, S. R. Curley, and G. S. Larson, "Phase velocity and absorption measurements in water containing bubbles," Journal of the Soundal Society of America, vol. 27, no. 3, p. 534, 1955.
3. E. Silberman, "Sound velocity and attenuation in bubbly mixtures measured in standing wave tubes," Journal of Acoustical Society of America, vol. 29, no. 8, pp. 925–933, 1957.
4. Z. Xiaoliang, Z. Zhemin, and Z. Lin, "Analytical description of sound wave propagation in a bubbly liquid and its strong nonlinearity," Applied Sounds, vol. 18, no. 6, pp. 18–23, 1999.

5. A. Prosperetti, "Thermal effects and damping mechanisms in the forced radial oscillations of gas bubbles in liquids," Journal of Acoustical Society of America, vol. 61, no. 1, pp. 17–27, 1977.
6. M. Kameda and Y. Matsumoto, "Nonlinear oscillation of a spherical gas bubble in acoustic fields," Journal of the Acoustical Society of America, vol. 106, no. 6, pp. 3156–3166, 1999.
7. K. W. Commander and A. Prosperetti, "Linear pressure waves in bubbly liquids: comparison between theory experiments," Journal of the Acoustical Society of America, vol. 85, no. 2, pp. 732–746, 1989.
8. A. Prosperetti, L. A. Crum, and K. W. Commander, "Nonlinear bubble dynamics," Journal of the Acoustical Society of America, vol. 83, no. 2, pp. 502–514, 1988.
9. F. Li, J. Sun, and J. Huang, "Study on propagation of sound wave in a bubbly liquid," Journal of Northwestern Polytechnical University, vol. 16, no. 2, pp. 241–245, 1998. View at Scopus
10. G. Yonghui, "Concentration of entrained air in air-water mixture as measured by sound transmission loss," Applied Sounds, vol. 18, no. 1, pp. 32–35, 1999.
11. S. Vagle and M. F. David, "The measurement of bubble-size distribution by soundal backscatter," Journal of Atmos-Pheric and Oceanic Technology, vol. 9, pp. 630–644, 1992.
12. L. Huaishan and Z. Zhengyun, "Seismic data processing approaches for the study of gas hydrates in the East China Sea," Journal of Ocean University of Qingdao, vol. 1, no. 1, pp. 87–92, 2002.
13. H. Liu, G. Huang, Y. He, S. Tong, S. Cui, and J. Zhang, "AVO character research of natural gas hydrates in the East China Sea," Journal of Ocean University of Qingdao, vol. 8, no. 3, pp. 270–276, 2009.

CHAPTER 4

FIELD VALIDATION OF TOXICITY TESTS TO EVALUATE THE POTENTIAL FOR BENEFICIAL USE OF PRODUCED WATER

JOSEPH R. BIDWELL, JONATHAN C. FISHER, and NAOMI L. COOPER

4.1 INTRODUCTION

A major issue facing oil and gas producers in both offshore and inland facilities is the disposal of produced water that is pumped to the surface along with the fossil fuels. In addition to high levels of dissolved salts, produced water may contain dissolved and free-phase hydrocarbons, trace metals, and radionuclides, all of which could have a significant impact on surface and groundwater quality (Boelter et al. 1992). Generally, the volume of produced water increases as a well ages and may be as great as 98% of the total volume of extracted fluids (Stephenson, 1992). On average, produced water accounts for approximately 80% of the residual waste associated with oil and gas production that requires some type of management or disposal, and concerns regarding the environmental effects and disposal of this waste have been reported from as early as the 1920's (Wiebe et al., 1924; Clemens and Jones, 1954). The volume of produced water that must be dealt with is far from trivial, with an estimated 18 billion barrels generated by U.S. onshore wells in 1995 alone (Veil, 2003).

Disposal of produced water in inland areas is accomplished through reinjection, release into holding ponds or surface waters, commercial hauling and treatment, or reuse. Cost estimates for deep well injection of produced water range from \$0.50 to \$1.75/barrel, while commercial hauling and treatment costs may exceed \$5.00/barrel (Jackson and Meyers 2002). Beneficial reuse options include livestock and wildlife watering, aquaculture, aquifer recharge and wetland remediation (for selected reviews, see GWPC 2002). The costs of produced water reuse are largely associated with the degree of treatment required to remove potential contaminants, with the extent of treatment depending on the overall quality of the water (ALL Consulting 2003).

Numerous technologies exist either to prevent produced water production through downhole separation (separates oil and water at bottom of a production well before water is brought to the surface, Veil et al. 1999) or to remove organic contaminants via photodegradation (Moraes et al. 2004) or microorganism induced degradation (Woolard and Irvine 1995; Dalmacija et al. 1996). Even if hydrocarbons are removed at the surface, the brine can still negatively affect receiving water communities (Kefford 2000). Desalinization of brines is a viable, but potentially expensive solution for beneficial reuse of produced water. Evaporation allows reclamation of salts from produced water, which have been used by local highway departments in New York for deicing (Atkinson et al. 1992). Produced water was found to be unsuitable for irrigation in Wyoming due to a build-up of salts in the root zone (Ganjegunte et al. 2005). Beneficial reuse of produced waters is likely limited by the multiple contaminants present and the presence of certain contaminants (such as nonessential or heavy metals or radionucleotides) that would completely preclude most potential reuses.

For onshore locations, release of produced water for any beneficial purpose is largely regulated at the state level, and quality assessment is based on the degree to which the water meets single chemical water quality standards or passes prescribed toxicity assessments in accordance with the National Pollutant Discharge Elimination System (NPDES) permitting program (Mancini and Stilwell, 1992; ALL Consulting 2003, see also US Code of Federal Regulations, 40 CFR Part 435). Approaches that assess the quality of produced water therefore have important implications for the particular treatment regime, associated treatment costs, and ultimate potential for beneficial use.

Toxicity assessments for both freshwater and marine systems involve the use of selected "standard" test species that are exposed to dilutions of the material being evaluated (e.g. produced water), with the response of the organisms determined through the use of either lethal (mortality) and/

or sublethal (growth/reproduction) endpoints. While these standard laboratory tests represent the most common method currently used to evaluate the potential toxicity of aqueous wastes, the extent to which test results represent actual conditions in the field has been questioned, and the need for field validation of tests has been identified (Cairns et al. 1996).

In the case of toxicity evaluations of produced water, constituents that may negatively affect an organism in the relatively sterile environment of a laboratory test beaker may be effectively detoxified in a field setting. For example, dissolved solids may be diluted by water flow in receiving systems, while organic constituents may adsorb to suspended material in the water column and be rendered unavailable for uptake by aquatic organisms. To date, toxicity assessment of produced water and associated field validation of laboratory test results has largely focused on offshore production in marine systems (e.g. Mendelssohn et al. 1990; Henderson et al. 1999; Washburn et al. 1999). While laboratory evaluations of marine produced water toxicity have indicated the potential for detrimental effects of untreated water (Krause 1995), a number of field validations have suggested that effects are significantly minimized when the produced water is subject to the diluting effects of the environment. For example, Mendelsshon et al. (1990) observed no effects of produced water discharges on marsh vegetation in freshwater and brackish wetlands in Louisiana, an effect they attributed to dilution and the behavior of the more dense produced water in the receiving system. Reed et al. (1994) observed that in-field effects of produced water on giant kelp were limited to sites within 5 m of the outfall even though laboratory test results indicated likely effects at sites farther afield. Douglas and Veil (1996) stated that produced water toxicity tests that focus on marine environments must be interpreted in light of the potential modification of the waste once it is discharged to the receiving system.

Unfortunately, relatively few studies have evaluated how well toxicity tests predict produced water effects in freshwater environments. This is significant, since the majority of the 37 U.S. states that are engaged in oil and gas production are landlocked (Produced Waters Database, 2002), meaning that 1) the potential for beneficial reuse will be greatest in these inland areas and 2) operators who wish to release produced water for beneficial reuse would be subject to discharge permit requirements that may include toxicity assessments with freshwater species.

Of those studies that have assessed the effects of produced water on freshwater organisms, elevated ion concentrations, organics and ammonia have been identified as the primary constituents causing observed toxicity (Ho and Caudle 1997; Sauer et al. 1997), and each of these components could be significantly modified upon entering a receiving system. However, field studies that validate these laboratory-based effects are practically nonexistent. In one combined laboratory and field investigation of produced water generated from an oil and gas field in Wyoming, the majority of field sites that were influenced by produced water releases were able to support functional biological communities, indicating the beneficial use of the produced water was appropriate (Mancini and Stilwell, 1992; Ramirez, 1993).

Given the growing interest in reuse of produced water and the associated increase in toxicity assessments that will accompany its release, it is imperative to generate field data that will evaluate how well laboratory bioassays of produced water represent the true potential for environmental effects and whether existing discharge standards are appropriate. The present study sought to first use standard laboratory bioassays to evaluate the aquatic toxicity of produced water from an oil field production site located along the shoreline of a freshwater reservoir in Oklahoma, including the identification of key constituents in the produced water that was associated with any observed toxicity. Since previous studies by the United States Geological Survey (USGS) indicated produced water was entering the reservoir littoral zone through both occasional surface flow and potential subsurface infiltration, the second major objective was to determine if there were any apparent effects on organisms in the field, and if these effects were consistent with the results of laboratory bioassays with the produced water itself.

4.1.2 STUDY OBJECTIVES

Characterize the zone of produced water infiltration through chemical analyses on the sediment from the littoral and sublittoral zone of the lake.

Evaluate the potential for effects and temporal changes in produced water quality using laboratory bioassays along with toxicity identification and evaluation (TIE) to indicate toxic constituents of produced water.

Determine the effects of produced water intrusion on benthic macroinvertebrate communities using field sampling and *in situ* experiments.

Evaluate how representative laboratory toxicity tests are of field conditions.

4.2 EXPERIMENTAL METHODS

4.2.1 FIELD SITE

The field site for this study was located at Skiatook Lake, a 10,500-acre impoundment in Osage County, Oklahoma. Osage County ranks among the top oil and gas producing areas in the state, with some 38,000 oil wells. Of these, about 13,000 wells lie within the Skiatook Lake watershed (about 820 km^2). In many instances, wells occur within a few tens of meters of the lake shoreline or the banks of streams that feed into the system. The produced waters derived from these wells are normally reinjected into underlying formations for disposal. However, inadvertent spills, leaks, and accidents, have lead to both historic and present release of produced water in the riparian zone and into the lake.

The field site itself was one of the focal points of the U.S. Geological Survey's Osage-Skiatook Petroleum Environmental Research (OSPER) Project and was designated "OSPER Site B" (see project summary as described on the USGS Toxics Hydrology Program Homepage, http://ok.water.usgs.gov/skiatook/).

For the purposes of this study, the OSPER site will be referred to as the "oil production site" or "production site". Both the oil production site and the associated field reference site were located along the southern side of the lake on land contained within the grounds of the Zink Ranch. Oil production at the production site began in 1938, which predates lake impoundment in 1984.

Previous USGS studies at the site have indicated salt scarring from surface flow of produced water and subsurface transport of produced water salts and hydrocarbons from at least three locations; an old tank battery, an injection well, tank, and pit, and a large active tank battery and associated pit. The USGS data also indicate that highly saline produced water has been seeping from the active pit and flows toward the lake in the 1.2-1.5 meter thick layer of subsurface colluvium and alluvium and penetrates less than 1 m into the underlying impermeable shale. Modeling suggests that the produced water takes about 2-4 years to reach the lake through the mixed colluvium and alluvium. During the course of the study, the USGS had a tacit agreement with United States Environmental Protection Agency Region 6 to not take substantial enforcement action at this site while the research project was ongoing.

The water chemistry of the lake as a whole suggests that the system is not substantially contaminated by produced water from oil production sites. However, a reconnaissance survey of a number of tributary streams in the watershed under low-flow conditions conducted by the Osage Nation Environmental and Natural Resources Department shows that several streams draining into the lake have elevated salinity derived from oil production activities. The OSPER Site B was considered an ideal site to undertake the field component of this study because the sources of produced water were particularly close to the lake, the cove affected is narrow and restricted, and the preliminary data derived from the USGS studies clearly illustrate that produced water is entering the aquatic system.

4.2.2 PHASE 1 RESEARCH ACTIVITIES

As part of the OSPER project, the USGS drilled 29 shallow (one to three meters deep) groundwater wells at the production site that allow for the characterization of shallow groundwater and assessment of produced water intrusion (Figure 2, but also see http://ok.water.usgs.gov/skiatook/). Six 33-m transects were established at the production and reference site. Three transects running west to east (off of wells BE07 and BE62, and approximately 8 meters north of the BE07 transect) and one running north to south (of groundwater well BE62) were established at the production site, with two transects running north to south at the reference site.

4.2.2.1 TASK 1.0 – DEFINITIVE CHARACTERIZATION OF PRODUCED WATER INFILTRATION INTO THE RECEIVING SYSTEM.

Produced water was collected from onsite storage tanks in one-liter amber bottles with Teflon coated lids, filled from the bottom with a hose to reduce volatile chemical loss and minimize headspace, and held at 4°C. Bimonthly monitoring of produced water (September 2005 to May 2007) from onsite storage tanks included basic water quality measures (ammonia, pH, conductivity, alkalinity, and hardness). Hydrocarbons were extracted and analyzed (US EPA method 8015B, Accurate Labs Inc., Stillwater, OK) for BTEX (benzene, toluene, ethyl-benzene, and xylene), gasoline range organics (GRO) and diesel range organics (DRO). These categories were meant to quantify the amount and general molecule size range of the hydrocarbons that may be present in water samples. Finally, produced water samples were sent to the Oklahoma Cooperative Extension Service's Soil, Water, and Forage Analytical Laboratory at Oklahoma State University. This laboratory measured concentrations of sodium, calcium, magnesium, potassium, nitrate, chloride, sulfate, boron, and bicarbonate.

To test for produced water intrusion, one well volume (the volume of water contained within and extractable from a well) of water from all accessible groundwater wells was collected using a hand pump concurrently with produced water sampling and filtered using Whatman GF/A glass fiber filters to remove sediment. Basic physicochemical properties (pH -Fisher Scientific (Pittsburgh, PA) Accumet AP62 portable pH meter, conductivity -Fisher Scientific Accumet AP65 portable conductivity meter, alkalinity -US EPA method 310.1, and hardness -US EPA method D1126-96) were measured on groundwater samples. During February and May of 2007, well water samples were sent to the Oklahoma Cooperative Extension Service's Soil, Water, and Forage Analytical Laboratory at Oklahoma State University to characterize the same ion concentrations as those measured for produced water. Further analytical characterization of groundwater samples was not feasible because maximum sample volume typically did not exceed 500 mL, and was commonly 300 mL or less.

Three aquatic sediment cores (30cm long by 1 cm wide -Wildco (Buffalo, NY) hand corer with polyethylene sample sleeves) were taken along each transect at the study sites during field sampling to a depth of 30 cm. Though greater numbers of core samples were originally proposed, the rocky substrate of the lake bottom required multiple attempts to drive each core sample into the substrate. The sampling accomplished represents one field day each season devoted to diving and attempting to collect core samples. Pore water was extracted from 5-cm sections of the cores in 50 ml tubes via centrifugation at 4100 rpm for 30 minutes at 23°C. Extracted pore water was then analyzed for brine contamination using specific conductance as a surrogate. Based on guidance from the analytical laboratory, samples were sent to the Oklahoma Cooperative Extension Service's Soil, Water, and Forage Analytical Laboratory at Oklahoma State University for major ions analysis, as was done with the produced water samples. The analytical laboratory measured as many ions as feasible (depending on sample volume and ion content) starting with chloride, then measuring cations, and then anions. Ion concentrations were used to predict potential pore water toxicity to the standardized bioassay organism *D. magna* using an Ion Toxicity model (Tietge et al. 1997). The pore waters collected were of insufficient volume to allow quantification of bicarbonate concentrations and these concentrations were assumed to be either zero or the maximum value detected (938mg/L) during toxicity modeling. When the concentrations were assumed to be zero if bicarbonate concentrations were greater than this value then predicted toxicity would increase, whereas if bicarbonate concentrations were lower then predicted toxicity would decrease. This resulted in both conservative estimates (worst-case scenario) and best-case scenario estimates of potential toxicity with which comparisons could be made.

During the Fall of 2005 water along the sediment-water interface in the lake was sampled at each oil production site transect to test for produced water flow into Skiatook Lake. Water samples were collected in 20 ml syringes every 0.5 m along each transect and tested for conductivity. These results were then compared with conductivity observed in the lake during the course of this study.

4.2.3 PHASE 2 RESEARCH ACTIVITIES

4.2.3.1 TASK 1.0 – ROUTINE LABORATORY TOXICITY TESTING OF PRODUCED WATER AND GROUNDWATER SAMPLES

Static renewal acute (48hr) and chronic (7d) testing of the produced water samples collected for water analyses were run within 48 hours of collection and in addition to chemical analyses during bimonthly sampling events. All water samples were held on ice in the field and at 4°C once in storage at the laboratory. Acute tests utilized the test organisms *D. pulex* and *D. magna* (US EPA 2002a), while chronic tests used *Ceriodaphnia dubia* and the fathead minnow *Pimephales promelas* (US EPA 2002b).

4.2.3.2 TASK 2.0 – TOXICITY IDENTIFICATION EVALUATION OF PRODUCED WATER SAMPLES

Toxicity Identification Evaluation (TIE) is a US EPA (1992) standardized series of water manipulations and toxicity tests meant to identify toxic constituents of a complex mixture or effluent. Manipulations performed and tested on *D. pulex* included unadjusted effluent; chelation with EDTA (to test for metal toxicity); oxidation with sodium thiosulfate; adjustment to pH 6, 7, and 8 (to test for pH-dependent toxicants); and aeration, filtration (through a glass fiber filter to remove particulates), and C18 solid phase extraction (SPE, removes nonpolar organic chemicals) at initial pH, pH 3 and pH 11. Evaluations were conducted in February 2006, March 2007, and September 2007.

Mount et al. (1992, 1997) derived an Ion Toxicity Model that uses empirically derived statistical models to predict major or essential ion (Na, K, Ca, Mg, Cl, SO_4, and HCO_3) toxicity to three standardized bioassay organisms (*Pimephales promelas, Ceriodaphnia dubia,* and *Daphnia magna*) in response to concerns about increased discharges of oil-field produced waters. This model uses major ion concentrations as predictors of organism survival in a multiple logistic regression. Major ion salts were a significant component of produced water toxicity and this model has previously been used to assess toxic constituents of produced waters (Tietge et al. 1997). Evaluations were supplemented with comparisons of produced water toxicity as determined during routine laboratory toxicity testing with *D. magna, C. dubia*, and fathead minnows with values predicted using an Ion Toxicity Model (Gulley et al. 1992). This model accounts for toxicity due to major ion salts, in each round of toxicity monitoring without extensive toxicity identification evaluation (TIE) procedures. Major ion salts are an expected component of produced water toxicity and this model definitively demonstrates the contribution of these salts to overall mixture toxicity.

Further experiments were conducted to test for the effects of hydrocarbons on test organisms. Full toxicity tests with *D. magna* were conducted on produced water samples in June 2007. Tests were performed on raw produced water, produced water with hydrocarbons removed (using C18 solid phase extraction), and in laboratory water spiked with the material retained on the C18 SPE columns (following US EPA 1991). These data, coupled with predictions of toxicity based on major ion salt concentrations were then used calculate the effect hydrocarbons had on observed produced water toxicity.

4.2.4 PHASE 3 RESEARCH ACTIVITIES

Water quality monitoring and static renewal acute toxicity testing (48hr) with *Daphnia pulex* (US EPA 2002a) on selected groundwater wells coincided with bimonthly monitoring of produced water quality. Wells were selected to capture gradients in groundwater quality from potential produced water sources (evaporation pond and injection well) to the lakeshore. *D. pulex* was chosen as the test organism because Oklahoma Department of Environmental Quality regulations (252:690-3-29) specify this organism for acute whole effluent toxicity (WET) testing and the volume of sampled water from the groundwater wells was insufficient to conduct a battery of tests on multiple test species as was done for produced water samples.

In February and May 2007 groundwater samples were sent to the Oklahoma Cooperative Extension Service's Soil, Water, and Forage Analytical Laboratory to generate major ion profiles for each sampling well. These ion profiles were then incorporated into an Ion Toxicity Model and predictions from this model were used to create a Geographic Information System (GIS) surface to model predicted groundwater toxicity across the entire study site. Briefly, ArcMap 9.1 used the predicted groundwater toxicity of each well to predict groundwater toxicity for groundwater not directly sampled and based these predictions on the values for the nearest groundwater wells and the distance to those wells. The specific interpolation method used here is called ordinary kriging. The resulting surfaces show predicted toxicity of groundwater across the site.

4.2.4.1 TASK 1.0 – ASSESSMENT OF BENTHIC COMMUNITY CONDITION

Multi-plate samplers were employed to assess benthic community response to field conditions. Sediment cores were originally proposed and attempted, but the rocky lake substrate prevented reliable and replicated use. Multi-plate samplers may better sample local diversity, particularly that associated with hard substrates, and they also may reduce sample variability, though artificial substrates select for organisms that colonize them (Rosenberg and Resh 1982). Four multi-plate samplers were deployed at one location along each transect plus four more in the submerged brine pit for six weeks, twice a year, during historic high and low water levels (Cover and Harrel 1978), though sampling depth along each transect were within 1.5 meters of each other. Macroinvertebrate samples were sorted and identified to the lowest practical taxon (Merrit and Cummins 1996; Epler 2001; Smith 2001). Macroinvertebrate community composition and Shannon-Wiener Diversity were compared between sites.

4.2.4.2 TASK 2.0 – IN-SITU BIOASSAYS

Tests with the midge *Chironomus dilutus*, the amphipod *Hyallela azteca* (US EPA 2000), and the Asian Clam *Corbicula fluminea* coincided with field sampling. Midges and amphipods were collected from existing laboratory cultures (US EPA 2000), while clams were collected from a local infested stream (tail waters of Canton Lake, Blain County, Oklahoma). Midge and amphipod experiments utilized cylindrical, clear plastic, vented exposure chambers (12.5 cm long by 7 cm wide, following Burton et al. 2005), whereas clams were exposed in orange mesh bags. Both tests utilized ten organisms per experimental unit (an exposure chamber or mesh bag) and occurred at three locations: reference site, submerged brine pit, and along transect two, which is in line with the major salt scar. There were also three position treatments of amphipod enclosures with four enclosures per treatment at each location: containing surficial sediment, laying on the sediment, and suspended in the water column. Midge and amphipod enclosures were harvested after a ten-day incubation when survival and mean growth per chamber endpoints were measured. Asian Clams were individually marked (8 x 4 mm type FPN shellfish tags, Hallprint Pty Ltd., Victor Harbor, South Australia) such that each clam from each experimental unit was measured before and after incubation so that there were ten replicate organisms per experimental unit. Marked clams of approximately two centimeters in length measured using a digital caliper (to the nearest 0.005 mm) along the longest shell dimension (anterior to posterior) were placed into orange mesh bags (experimental unit) and then secured on the lake benthos with marking flags. At least two mesh bag experimental units were placed at each location and incubated for thirty days after which growth and survival endpoints were measured (Soucek et al. 2001). These experiments coincided with benthic macroinvertebrate community sampling in May (high water) and September (low water) each year. Amphipod experiments were conducted on August 14, 2007, September 9, 2007, and October 6, 2007. Asian clam experiments were conducted on August 17, 2005, May 16, 2006, October 5, 2006, and September 8, 2007.

4.2.4.3 TASK 3.0 – COMPARISON OF LABORATORY AND FIELD DATA

Laboratory sediment toxicity tests with amphipods were conducted in addition to *in situ* bioassays (US EPA 2000). Sediment was collected from the top 5cm of the lake bottom using the plastic sample containers used for storage and placed on ice for transport back to the laboratory where they were held at 4°C until testing. Samples were from the same locations as the field experiments and allowed powerful comparisons between laboratory and field experiments. Sediment tests occurred within eight weeks of sample collection (US EPA 2000). The numerous methods employed here also allowed comparison of the various assessment techniques between field and laboratory conditions. Statistical Analyses

Most statistical analyses were conducted using SAS 9.1 (PROC GLIMMIX, The SAS Institute, Cary, NC). Asian clam data was analyzed using analysis of covariance with initial organism size as the covariate. Macroinvertebrate data was natural log transformed and analyzed using a multivariate analysis of variance to explore differences in sample location and sample season. Toxicological endpoints such as LC50's and confidence limits were performed using the Comprehensive Environmental Toxicity Information System (CETIS) version 1.1.2 (Tidepool Scientific Software, McKinleyville, CA).

4.3 RESULTS AND DISCUSSION

4.3.1 *PHASE 1*

4.3.1.1 TASK 1.0 – DEFINITIVE CHARACTERIZATION OF PRODUCED WATER INFILTRATION INTO THE RECEIVING SYSTEM.

Produced water is a complex mixture containing many potential toxicants. These constituents include major ions (Na, Ca, K, Mg, Cl, SO_4), ammonia, hydrogen sulfide, petroleum hydrocarbons, BTEX (benzene, toluene, ethyl-benzene, and xylenes), phenols, naphthalenes, zinc, and other heavy metals (Fucik 1992; Schiff et al. 1992; Stromgren et al. 1995; Smith et al.

1998; Kharaka et al. 2005). These constituents vary between and within different geologic basins depending on geology and hydrology (Daly and Mesing 1995; Collins 1985).

In the present study, produced water major ion profiles and basic water quality parameters were collected from December 2005 to September 2007. Measured parameters ranged as follows: pH (3.7 to 6.9), conductivity (84,800 to 186,000 μS/cm), alkalinity (0 to 68 mg/l CaCO3), hardness (8,615 to 186,000 mg/l CaCO3), sodium (21,399 to 48,677 mg/l), calcium (2,351 to 8,642 mg/l), magnesium (668 to 1,689 mg/l), potassium (75 to 535 mg/l), nitrate (0 to 3 mg/l), chloride (36,395 to 93,989 mg/l), sulfate (83 to 353 mg/l), bicarbonate (0 to 83 mg/l), ammonia (27.7 to 63.6 mg/l), and boron (0.77 to 4.26 mg/l). Hydrocarbon concentrations were measured starting in March 2006 through September 2007 and ranged from 0.42 to 6.53 mg/l for BTEX, 1.0 to 10.0 mg/l GRO, and 2.6 to 9.6 mg/l DRO. Sodium chloride salts dominated the major ions and hydrocarbons were evenly split between benzene, GRO, and DRO.

Some of the field component of the project was negatively affected by extreme weather conditions during the 2006 and 2007 field seasons. During the summer of 2006, much of the state of Oklahoma was in the midst of a severe drought which lead to a significant drop in lake levels such that most of the underwater sampling sites were exposed. Then in 2007, heavy rain through the spring and early summer caused lake levels to rise so that the majority of groundwater wells became inaccessible after the May sampling period and the underwater transects could not be accessed due to extremely poor visibility and extensive amount of debris in the water. These weather conditions precluded sampling for chemical analyses and initiation of any in situ studies.

In addition, beginning in December 2006 and through March 2007, oil production was shut down at the site and eventually resulted in a change in site operations. These changes initially resulted in diminishing quantities of produced water that could be collected from the site and changes in operations after March changed how produced water was managed at the site. This change meant that rather than being available from the on-site tank battery, produced water could only be collected from an injection well on the northernmost side of the site when that well was in use, resulting in inconsistencies in produced water availability. Produced water quality also changed after this site management change, evidenced by increased hydrocarbon concentrations and greater variability in major ion concentrations (both the minimum and maximum measured concentrations for all ions were measured after this period). Before this change the southern injection well (INJWELL14) was the predominantly-used well, and produced water could be easily collected from the tank battery located along the evaporation pond.

Groundwater wells were sampled from September 2005 through May 2007. Flooding during the spring of 2007 rendered the majority of groundwater wells inaccessible to sampling after the May sampling period (the tops of the wells were below the lake surface). Water quality parameters were measured during each sampling period and ranged from 3.90 to 8.24 for pH, from 1,374 to 38,100 μS/cm for conductivity, from 0 to 1,178 mg/l CaCO3 for alkalinity, and from 169 to 20,000 mg/l $CaCO_3$ for hardness. In general, pH and alkalinity were lowest while conductivity and hardness were greatest in wells located near the brine injection well.

During February 2007, major ion profiles were measured on each sample well. These data were then analyzed using a principle components analysis (PCA) to explore possible trends in groundwater ion composition. Principle components analysis creates a linear combination of all measured variables for each sample. These linear combinations are then used to create new axes on which the samples can be plotted. A symmetric biplot (which scales variables and samples equally to produce an even appearing graphic) of these profiles for each sample well was created using the PCA Wells were plotted along the first two PCA axes and these axes explained 89.5% of the variation in well ion profiles. The first (horizontal) PCA axis potentially represents biogeochemical processing of intruded produced water (such as bicarbonate buffering, chloride exchange for sulfate via hematite, and bacterial removal of nitrate) or dilution from other groundwater sources. The second (vertical) PCA axis represents the dilution gradient from raw produced water to lake water.

Pore water samples collected from replicate sediment cores along each field transect yielded different results at the reference site compared with the oil production site. Reference core samples exhibited more constant pH, conductivity, and ion concentrations with depth. Conductivity and ion concentrations tended to increase with sample depth for the oil production site samples. pH dropped dramatically with depth on the first transect whereas pH was relatively constant on transect three and increased on transect two and in the submerged brine pit. Major ion profiles of each sediment core section tended to reflect conductivity readings with increased conductivity corresponding to greater concentration of all ions. Potential toxicity of these pore waters to the test organism *Daphnia magna* was predicted using an ion toxicity model and indicated that some pore waters may cause mortality if organisms were exposed to these waters. Potential toxicity is reported in toxicity units (TU's) which is the predicted 48-hr LC50 to *D. magna* divided by the original concentration. Toxicity units greater than one indicate samples where an LC50 would be observed if a bioassay had been possible.

Overall, raw produced water from the oil production site was characterized primarily by decreased pH, excesses in major ion salts (sodium, calcium, and chloride), and other contaminants (ammonia, boron, and hydrocarbons). Tietge et al. (1997) examined produced waters from around the United States and found sodium chloride and sodium bicarbonate salts as the primary toxicants, and additional toxicity resulted from nonpolar organic compounds or other unidentifiable toxicants.

Oil production operations have been shown to contaminate nearby groundwater with excess metals. For example, non-essential metals, including arsenic, cadmium and copper, were reported in groundwater samples adjacent to a petrochemical complex in Saudi Arabia (Sadiq and Alam 1997). The most commonly used disposal method, and the method currently in use at the site used for the present study, is water flooding where produced water is reinjected back into the petroleum seam to improve production. This process is associated with fewer environmental effects compared with other options, though groundwater contamination is possible. For example, a freshwater aquifer in Turkey was contaminated from a brine reinjection operation up to

18.7 km away from the injection source (Okandan et al. 2001). Hudak and Blanchard (1997) suggested that oil field brine contamination from oil and gas wells in northern Texas increased groundwater concentrations of chloride and bromide.

Brine contamination of the near-surface aquifer at both OSPER sites has been reported previously (Kharaka et al. 2002), and groundwater modeling at the OSPER-B site (oil production site) demonstrated that contaminating brine travels between two and four years in the near-surface aquifer before entering Skiatook Lake (Herkelrath and Kharaka 2002). Sediment cores taken throughout the oil production site, especially those south of transect one, exhibited conductivity and ion concentrations consistent with produced water contamination. Contamination increased with depth, suggesting dilution or leaching of contaminated pore water by lake water. All groundwater wells exhibited pH, conductivity, alkalinity and hardness at levels intermediate between produced water and lake water. Groundwater samples tended to have more bicarbonate and sulfate compared to produced water samples, although wells located near the injection well were more similar to raw produced water samples, supporting the previous findings of brine contamination at the site (Kharaka et al. 2002).

Water samples taken from the sediment-water interface during the fall of 2005 indicated no conductivity abnormalities that were beyond the normal variation in conductivity observed in Skiatook Lake during the study (Figure 9). As such, this conductivity survey was unable to locate evidence of contaminated groundwater flow into Skiatook Lake. The use of mini-piezometers to indicate upwelling of groundwater in the sub-littoral zone of the lake was also unsuccessful due to the rocky nature of the lake substrate near the shoreline at the oil production site.

These results indicate that produced water appears to be a significant contaminant of groundwater at the oil production site, particularly near the injection well, and also may occur in pore water of lake sediments. Toxicity modeling indicates that pore waters would probably not affect biota at sediment depths less than 10cm at the most contaminated sampling locations. Based on these results, the potential plume of contaminated groundwater underneath Skiatook Lake may extend beyond the study area, but likely does not significantly infiltrate surficial sediments or the water column.

4.3.2 PHASE 2

4.3.2.1 TASK 1.0 – ROUTINE LABORATORY TOXICITY TESTING OF PRODUCED WATER AND GROUNDWATER SAMPLES

Acute and chronic toxicity tests with the four standard test organisms were conducted from January 2006 through March 2007 (Figure 10). Acute 48-hr LC50's ranged from 0.94% to 4.13% for D. pulex, 2.68% to 5.36% for D. magna, 2.06% to 2.74% for *C. dubia*, and 7.44% to 11.19% for fathead minnows. Acute 48-hr endpoints for *C. dubia* and fathead minnows were generated from daily monitoring of the chronic 7-d tests and these organisms were fed during the acute period, unlike *D. pulex* and *D. magna* tests. Chronic 7-d LC50's for *C. dubia* ranged from 1.78% to 2.74% and from 2.96% to 5.80% for fathead minnows. Chronic 7-d EC50's for *C. dubia* reproduction ranged from 1.00% to 1.66% and from 2.20% to 6.51% for fathead minnow growth.

In previous studies examining the effects of produced water on aquatic test species, the freshwater cladoceran, *Ceriodaphnia dubia*, was found to be more sensitive to produced water than the fathead minnow, *Pimephales promelas* (Boelter et al. 1992; Fucik 1992). Boelter et al. (1992) tested water from streams receiving produced water in Wyoming and found that *C. dubia* was sensitive to major inorganic ions present in the discharge, while fathead minnows did not exhibit a response. Laboratory toxicity tests were shown to be valid measures of stream quality when compared with benthic macroinvertebrate samples in Alabama streams receiving produced water (Mount et al. 1992). Despite these findings, laboratory toxicity testing may overestimate produced water effects because dilutions used for such tests may represent effluent concentrations only present at the immediate discharge point (Douglas and Veil 1996). Evaluating the environmental effects of produced water discharges should incorporate multiple lines of investigation including chemical quality of the effluent and receiving system, toxicity testing, biomonitoring, and specific continuous monitoring of conductivity and chloride as indicators of brine water (O'Neil et al. 1992b).

Bimonthly monitoring of groundwater acute (48 hr) toxicity to *D. pulex* occurred from September 2005 through May 2007. Toxicity was greatest near a brine injection well and lowest in a well adjacent to the evaporation pond in a vegetated area. Toxicity (expressed as percent of the groundwater sample) at the first transect (northernmost), running along a salt scar from the injection well to the lake, ranged from 2.12% to 31.5% in BE 53 and 13.45% to 36.75% in BE18. Toxicity along the transect going from the evaporation pond to the lake along a salt scar ranged from 5.76% to 35.35% in BE11, 12.04% to 35.35% in BE08, and 2.99% to 30.82% in BE07. Observed toxicity in the vegetated transect running from the evaporation pond to the lake ranged from 48.07% to no observed effects in BE59, 11.71% to 56.81% in BE61, and 9.25% to 65.56% in BE16.

Major ion profiles for groundwater wells sampled in the spring of 2007 were used to generate surfaces in GIS. Briefly, ArcMap 9.1 used the predicted groundwater toxicity of each well to predict groundwater toxicity for groundwater not directly sampled and based these predictions on the values for the nearest groundwater wells and the distance to those wells. The specific interpolation method used here is called ordinary kriging. The resulting surfaces show predicted toxicity of groundwater across the site. The graphical representations of predicted toxicity support the previous finding that toxicity is greatest near the injection well on the northwestern side of the site. Acute 48-hr toxicity tests on selected groundwater wells using *D. magna* were compared with predictions from an Ion toxicity model to validate the toxicity predictions used in the GIS model. The predictions fell relatively close for all wells tested, except BE53, which is the closest groundwater test well to the brine injection well. In samples from this site, the observed toxicity to *D. magna* was greater than model predictions. These results suggest that for all groundwater wells, except for those closest to produced water sources, toxicity results from elevated major ions. Further, these data support the use of the GIS model surfaces for other produced water contaminated sites, with some limitations. This modeling assumes that all toxicity is derived from major ion excesses. Toxicity testing with the test organism modeled, as was done here, is needed to detect the presence of other toxic groundwater constituents and validate predictions.

4.3.2.2 TASK 2.0 – TOXICITY IDENTIFICATION EVALUATION OF PRODUCED WATER SAMPLES

Toxicity identification evaluation (TIE) of produced water resulted in reduced toxicity due to chelation by ethylenediaminetetraacetic acid (EDTA, indicating metal toxicity), oxidant reduction with sodium thiosulfate (indicating chloride or other oxidant toxicity) filtration at pH 11 (potential hydrocarbon toxicity), and aeration at pH 10 (volatile chemicals) when compared to the baseline toxicity of unmodified produced water. These results suggest that on the February 2006 and September 2007 runs, produced water toxicity came from metals (including major ion salts), possibly chloride, hydrocarbons removed from solution by high pH, and volatile chemicals.

Comparisons between observed 48 h LC50's and those predicted from major ion concentrations differed depending on the test organism. Fathead minnow lethal concentrations were most similar to predicted values, indicating that major ions were the dominant, if not only source of toxicity to these organisms. Observed 48-hr LC50's for C. dubia and *D. magna* indicated greater toxicity than predicted values and were likely affected by toxicants other than major ions. The TIE's run using D. pulex support the finding of toxicity greater than can be predicted from major ions alone and that toxicants other than major ions are responsible for mixture toxicity. When Tietge et al. (1997) compared toxicity test results of six produced waters from across the United States, two samples were found to exhibit toxicity beyond what was predicted by ion toxicity modeling.

Removal of hydrocarbons from produced water using C18 solid phase extraction did not significantly reduce 48-hr mixture toxicity to D. magna ($P = 0.8181$), however the retained non-polar organics did elicit a toxic response from test organisms when eluted into a clean solution. Overall, produced water toxicity contained 22.91 toxicity units (TU), of which 7.77 TU's can be explained by major ion toxicity and 8.48 resulted from hydrocarbons and other non-polar organics. The remaining 6.66 toxicity units fell well within the 95% confidence interval (13 to 34 toxicity units) of the original mixture 48-hr LC50. Despite the demonstration of toxicity derived from hydrocarbons, plots of mixture TU's, with TU's accounted for by ion toxicity removed (total TU's minus major ion TU's, from the Ion toxicity model), versus total petroleum hydrocarbons (TPH) result in no demonstrable relationship for any of the test organisms (Figure 17). Produced water caused test organism mortality at concentrations below 10%, and for some organisms (*D. pulex*) below 1%. Fathead minnows were least sensitive to produced water with 48hr LC50's averaging 8.42%, followed by *D. magna* at 3.69%, then C. dubia at 3.12%, and with D. pulex at 2.36% as the most sensitive. These results are similar to those found for produced water from around the United States by Tietge et al. (1997), but are more toxic than the majority of samples examined by Fucik (1992). Fucik (1992) suggested that hydrocarbons and semi-volatile compounds were responsible for test organism mortality, in addition to dissolved solids, in produced waters where cladoceran toxicity was greater than that found for fathead minnows. While produced water toxicity did vary over time, no discernable pattern emerged from the data.

Toxicity identification evaluations of produced waters from multiple locations in the United States have consistently identified elevated salinity as a source of toxicity (Tietge et al. 1997). Other sources of toxicity include major ion imbalances, hydrocarbons, ammonia, and hydrogen sulfide (Boelter et al. 1992; Sauer et al. 1997; Elias-Samlalsingh et al. 2004). Gulley et al (1992) and Mount et al. (1992) found that toxicity resulting from major ion salinity could be accurately predicted and accounted for using statistically-derived ion toxicity models. Due to concerns regarding increasing discharges of oil field produced waters, statistical models of major ion toxicity to *Pimephales promelas*, *Daphnia magna*, and *Ceriodaphnia dubia* have been developed to predict toxicity resulting from exposure to saline waters (Gulley et al. 1992). Major ion concentrations and toxicity tests checked against predictions from the ion toxicity model could accurately determine whether further phase I TIE was necessary by accounting for all toxic units observed. Produced-water contaminated groundwater from sites along Skiatook Lake have shown elevated concentrations of major ions, strontium, iron, dissolved organic carbon, and acetate (Kharaka et al. 2005). Ion toxicity modeling incorporated into TIE procedures determined which produced water mixture constituents contribute to overall toxicity (Tietge et al. 1997).

Previous produced water investigations using the ion toxicity model coupled with TIE procedures (Tietge et al. 1997), the demonstration of toxic levels of hydrocarbons present in site produced water, and TIE's performed in this study, indicate hydrocarbons are also a potential component of overall toxicity. Salts have already been shown as an important source of toxicity in groundwater and lake sediment pore waters. Hydrocarbons may degrade quickly or adsorb to organic matter present in the groundwater matrix and result in little measured effect. Other potential toxicants, such as ammonia, may be masked by other mixture components present at more toxic concentrations. For example, produced water samples from this study contained at least eight toxic units of major ion salt toxicity, whereas total ammonia at pH 6.5 exhibits a 96h LC50 to fathead minnows at 254 mg/l (Thurston et al. 1981), over four times that found in site produced waters.

4.3.3 PHASE 3

4.3.3.1 TASK 1.0 – ASSESSMENT OF BENTHIC COMMUNITY CONDITION

Several methods can be used to assess benthic macroinvertebrate community structure including *in-vitro* and *in-situ* sediment toxicity assessments, and macroinvertebrate community sampling. *In situ* studies incorporate experimental controls into field conditions to better represent site conditions while retaining some of the advantages of laboratory experiments. Unfortunately, *in situ* studies do not always mirror field or laboratory study results, confounding applicability. The most environmentally-relevant method to measure disturbance effects on resident communities is to compare the community structure of potentially impacted communities with unaffected communities, though this approach is time consuming and subject to environmental variability that may overwhelm treatment effects.

Sediment-dwelling or benthic macroinvertebrates are ideal study organisms because they are ubiquitous, diverse, relatively immobile (compared with freshwater vertebrates), and have lifecycles that, depending on species, can last from one month to multiple years (Rosenberg and Resh 1993). Currently, 49 of 50 US states employ benthic macroinvertebrate surveys as a part of routine biomonitoring activities (Carter et al. 2006). Benthic macroinvertebrate monitoring has been used to assess the impact of short-term saline inputs on stream communities (Marshall and Bailey 2004) and to determine the effects of metals (cadmium, copper, lead, and zinc) along salinity gradients (Peeters et al. 2000).

In the present study, benthic macroinvertebrates exhibited few differences between sample sites. Of the nineteen organisms found in more than 10% of samples, only the aquatic earthworm, *Lumbriculus* spp. (tolerance value of eight, higher numbers correspond to greater tolerance of stressful environmental conditions, Carter et al. 2006), was found in greater abundance at the reference site, whereas the mayflies *Stenonema femoratum* (tolerance value of four) and *Paracloedes* spp. (tolerance value of four) were both found in greater abundances at the oil production site. Shannon-Wiener diversity did not differ between locations at the oil production and reference sites, though richness and diversity tended to be greater at the oil production site compared with the reference.

Previous freshwater studies of the instream biological effects of produced water discharges summarized by O'Neil (1992a), found few or no significant effects on resident biota. However, some chronic brine discharges from oil operations have been found to significantly change benthic communities. For example, Olive et al. (1992) reported a lower percent Trichoptera (Caddisflies) and decreased invertebrate density in a stream impacted by brine discharges. Biologic Integrity of invertebrate communities in two Ohio brine-receiving streams were unchanged compared to reference sites, though greater percentages of salt-tolerant diatoms were present and chloride concentrations did not exceed 74 mg/L (Olive et al. 1992). In a drainage system chronically exposed to high levels of salts from oilfield operations, Short et al. (1991) found that while fishes and Dipteran larvae were tolerant of salinities as high as 10 ‰, mayfly larvae (Ephemeroptera) were absent when salinities exceeded 2 ‰. O'Neil et al. (1992a,b) observed that more benthic macroinvertebrate taxa were found at a site in the Warrior Basin, Alabama where there was less produced water-associated sulfate. Most previous studies of the environmental effects of produced water on benthic communities have been limited to lotic systems, however macroinvertebrates are still useful tools in biomonitoring for lentic systems. The difference between lotic and lentic systems necessitates different sampling devices and implies a change in overall community composition, but the ecological theory behind the application remains unchanged.

4.3.3.2 TASK 2.0 – SEDIMENT BIOASSAYS AND IN-SITU STUDIES

Toxicity testing of sediments, like water column tests, provides information on the potential biotic effects from sediment contaminants. Contaminant concentrations in sediments may exceed water column concentrations by several orders of magnitude due to preferential sorption and partitioning of contaminants to the sediment matrix (Burton 1991). This concentration difference can result in adverse effects to sediment dwelling organisms while water column-based tests indicate no observed effects (Chapman 1989). Sorption to the sediment depends on many factors including the individual chemical properties of the contaminant and sediment properties such as pH, redox, organic carbon, grain size, and inorganic particles (Di-Toro et al. 1991). Chemical analyses of sediment contaminants can only address potential bioavailability, whereas benthic community analyses can be confounded by numerous factors unrelated to sediment contamination. Sediment toxicity tests therefore serve as an intermediate method between chemical analyses and community sampling by using biota as indicators of toxicity and allowing direct links to be made between chemical composition, bioavailability, and benthic biotic effects (US EPA 2000).

As with other standardized laboratory test procedures, sediment toxicity tests are usually conducted under controlled conditions which may undermine the "real world" applicability of toxicity test results. *In situ* toxicity testing provides an intermediate level of experimental control between laboratory experiments and field sampling of resident communities, while incorporating environmental variability, thus increasing applicability (Burton et al. 2005). Due to the potential for *in situ* studies to link and validate laboratory experiments with field surveys, *in situ* studies should consistently mirror results of other assessment methods.

The midge, *Chironomus dilutus* (formerly *C. tentans*; Shobanov et al. 1999), and the amphipod, *Hyallela azteca,* can be used for both laboratory sediment toxicity tests and *in situ* assays. Simultaneous use of a test organism for laboratory and field experiments allows greater experimental control and reduces errors made when extrapolating the results of traditional toxicity tests to the environment (Burton et al. 2005). Both of these organisms have been successfully used to evaluate sediment toxicity, with those evaluations confirmed using natural benthic populations (US EPA 2000). For example, experiments comparing the response of chronic *H. azteca* bioassays with field-collected benthic community data demonstrated similar responses to increasing concentrations of polycyclic aromatic hydrocarbons (Ingersoll et al. 2005). Tucker and Burton (1999) found that comparisons of laboratory and *in situ* exposures of both test organisms to agricultural and urban runoff, especially when combined with benthic community assessment, provided a more comprehensive and environmentally relevant assessment than either method alone.

Initial experiments with *C. dilutus* in laboratory sediment toxicity tests resulted in no significant differences in growth, although there does appear to be reduced survival at both oil production site locations compared to the reference site (Figure 19). Sediment experiments with the amphipod, *Hyallela azteca,* revealed no significant differences in mortality between any treatments in both lab and *in situ* experiments. Few differences in amphipod growth were found during laboratory experiments (Figure 20). In the *in situ* study with *H. azteca*, a significant triple interaction between sediment treatment, location, and experiment run was apparent which makes direct site comparisons difficult. Still, amphipods at both locations at the oil production site exhibited lower growth than those placed at the reference site, regardless of experiment run (Figure 21).

The invasive Asian clam, *Corbicula fluminea,* has been effectively employed to monitor both sediment and water quality in previously infested environments and found to accurately reflect resident benthic community responses (Doherty 1990; Cataldo et al. 2001; Soucek et al. 2001). Clam tissues may or may not accumulate contaminants, such as metals, and accumulation rates can vary greatly between contaminants and field sites, but growth consistently reflects field conditions (Cataldo et al. 2001). Asian clams have successfully been used alone (Soucek et al. 2001) and as a part of an integrated multiple approach assessment (Soucek et al. 2000) to detect and monitor the effects of acid mine drainage in southwestern Virginia streams. In the present study, *Corbicula fluminea* growth rates (Table 10) were significantly less in 2006 than in 2005 ($P = 0.0302$) and 2007 (1.14 mm, $P < 0.0001$) regardless of placement location. Clams placed at the reference site grew significantly larger compared to clams at all oil production site locations for the first two years of the study (2005 $P < 0.0001$; 2006 $P < 0.0001$; 2007 $P = 0.5948$ and transect two $P = 0.4498$). When all data from all years were combined, clams grown at the reference site grew more than those placed at the oil production site (submerged brine pit $P < 0.0001$ and transect 2 $P = 0.0100$), but no differences were observed between locations at the oil production site ($P = 0.4109$).

4.3.3.3 TASK 3.0 – COMPARISON OF LABORATORY AND FIELD DATA

The laboratory bioassays of the produced water and water from some of the groundwater wells indicated the samples were acutely toxic to the standard laboratory test species used in the study. A pattern of groundwater contamination in association with the brine injection well and the predicted and laboratory-generated toxicity of that groundwater did indicate the potential for biotic effects if a sufficient volume of groundwater reached the sediment surface. Laboratory bioassays on sediment taken from the reference site and two locations at the oil production site did not indicate any negative effects associated with exposure to sediment from the oil production site, although the results of the *in situ* bioassays suggest a consistent negative effect on growth of both the amphipod, *H. azteca*, and the Asian clam, *C. fluminea* at the oil production site as compared to the reference site. Slight differences in the macrobenthic communities resident at the reference and oil production sites were also apparent, but these differences were not consistent with a negative impact at the production site, and in fact some attributes of the benthic community at the oil production site were indicative of a healthier benthic community than found at the reference site.

A direct link between produced water bioassays and field conditions would have been easier to make if a point source of produced water release existed. However, no clear indication of groundwater upwelling into the littoral zone of the lake could be established, and if this upwelling is occurring, it would most likely be a diffuse "nonpoint" input. It may also be that, if any upwelling of produced water did occur, it was restricted to times when water levels in the sub-littoral zone of the lake were particularly low and overall effects may be transient. In this regard, the use of longer-term *in-situ* and field evaluations may be important to integrate the transient stressors that could arise from occasional upwelling events.

Based on the existing literature, the results of experiments comparing laboratory bioassays, *in situ* experiments, and field studies vary greatly. Compared to laboratory bioassays, *in situ* exposures have been found to underestimate toxicity responses (Hose and Van den Brink 2004), overestimate toxicity responses (Sasson-Brickson and Burton 1991; Kater et al. 2001; Anderson et al. 2004), both over-and underestimate responses depending on the stressor (Tucker and Burton 1999), or agree with laboratory results (Schroer et al. 2004). Comparative studies have also found that effects measured in field experiments more closely resembled benthic community responses than laboratory bioassays, though laboratory studies are still relevant to field conditions (Hose and Van den Brink 2004; Ingersoll et al. 2005).

4.4 CONCLUSIONS

Site produced water was found to be acutely toxic to aquatic test organisms at concentrations ranging from 1% to 10% of the whole produced water sample. As determined by toxicity identification evaluation and associated ion toxicity modeling, major ion salts and hydrocarbons were the primary mixture toxicants, and the salts were found to contaminate groundwater across the entire site. The standardized test species used in the laboratory bioassays (fathead minnows and daphnids) exhibited differences in sensitivity to these two general classes of contaminants, which underscores the importance of using multiple species when evaluating produced water toxicity.

Groundwater toxicity varied greatly between sampling wells (LC50's ranged from 2.12% to over 100%), but no seasonal pattern was detectable. The wells closest to the produced water injection well exhibited the greatest measured toxicity. Wells not near the injection well but located on a salt scar derived from an on-site evaporation pond also had toxicity greater than that found in wells located in a vegetated area, suggesting that contamination from both the injection well and evaporation pond contributed to measured groundwater effects. Toxicity modeling of groundwater ion profiles indicated the produced water injection well was probably the most significant contributor to groundwater toxicity. The use of the GIS model surfaces may facilitate study of produced water plumes at other contaminated oil production sites in which elevated ions are the primary source of effects, although toxicity testing is still necessary to detect the presence of other toxic groundwater constituents and validate predictions. In addition, another statistical technique, principle component analyses (PCA), proved useful for indicating dilution by lake water and possible biogeochemical reactions as possible factors that ameliorated groundwater toxicity.

Elevated concentrations of major ions were found in pore water from lake sediments, but toxicity from these ions appears limited to sediment depths of 10 cm or greater, which is outside of the primary zone of biological activity. Further, site sediments did not have any effects on test organisms during laboratory toxicity testing and macroinvertebrate communities did not indicate impairment at the oil production site as compared to a reference site. In situ experiments with amphipods and Asian clams did indicate a sublethal site effect, but these could not be definitively linked with produced water infiltration.

Due to the lack of clear evidence of produced water infiltration into the sub-littoral zone of the lake, it is not possible to assess whether the laboratory bioassays of produced water effectively indicate risk in the receiving system. However, the acutely toxic nature of the produced water and general lack of biological effects in the lake at the oil production site do further support the idea that the degree of produced water infiltration into surficial lake sediments and the near-shore water column is not very extensive. This study was able to demonstrate the utility of ion toxicity modeling to support data from toxicity identification evaluations aimed at identifying key toxic constituents in produced water. This information could be used to prioritize options for treating produced water in order to reduce toxic constituents and enhance options for reuse. The study also demonstrated how GIS, toxicity modeling, and toxicity assessment could be used to facilitate future site assessments.

REFERENCES

1. ALL Consulting. 2003. Handbook on Coal Bed Methane Produced Water: Management and Beneficial Use Alternatives. ALL Consulting, Tulsa, OK. Available online at: http://www.ALL-LLC.com.
2. Atkinson, J. F., M. R. Matsumoto, M. D. Bunn, and D. S. Hodge. 1992. Use of solar ponds to reclaim salt products from brine waters from oil and gas well operations in New York. In Produced Water, J. P. Ray and F. R. Engelhart, eds. Plenum Press, New York, Pp. 535-547.
3. Boelter, A. M., F. N. Lamming, A. M. Farag, and H. L. Bergman. 1992. Environmental effects of saline oil-field discharges on surface waters. Environmental Toxicology and Chemistry 11: 1187-1195.
4. Burton Jr., G. A. 1991. Assessment of freshwater sediment toxicity. Environmental Toxicology and Chemistry 10: 1585-1627.
5. Burton Jr., G. A., M. S. Greenberg, C. D. Rowland, C. A. Irvine, D. R. Lavoie, J. A. Brooker, L. Moore, D. F. N. Raymer, and R. A. McWilliam. 2005. In situ exposures using caged organisms: a multi-compartment approach to detect aquatic toxicity and bioaccumulation. Environmental Pollution 134: 133-144.
6. Cairns, J. Jr., J.R. Bidwell, and M.E. Arnegard. 1996. Toxicity testing with communities: microcosms, mesocosms and whole system manipulations. Reviews of Environmental Contamination and Toxicology 147:45-69.
7. Carter, J. L., V. H. Resh, M. J. Hannaford, and M. J. Meyers. 2006. Macroinvertebrates as biotic indicators of environmental quality. Chapter 30 in F. R. Hauer and G. A. Lamberti, eds. Methods in stream ecology. Academic Press, San Diego, CA.
8. Cataldo, D. H., D. Boltovskoy, J. Stripeikis, and M. Pose. 2001. Condition index and growth rates of field caged Corbicula fluminea (Bivalvia) as biomarkers of pollution gradients in the Paraná river delta (Argentina). Aquatic Ecosystem Health and Management 4: 187-201.
9. Chapman, P.M. 1989. Current approaches to developing sediment quality criteria. Environmental Toxicology and Chemistry 8: 589-599.
10. Clemens, H.P. and W.H. Jones. 1954. Toxicity of brine water from oil wells. Transactions of the American Fisheries Society 84:97-109.
11. Collins, A.G., 1985, Geochemistry of Oilfield Waters: Developments in Petroleum Science 1: New York, Elsevier Scientific Publishing Company. Cover, E. C. and R. C. Harrel. 1978. Sequences of colonization, diversity, biomass, and productivity of macroinvertebrates on artificial substrates in a freshwater canal. Hydrobiologia 59: 81-95.

12. Cover, E.C. and R.C. Harrel. 1978. Sequences of colonization, diversity, biomass, and productivity of macroinvertebrates on artificial substrates in a freshwater canal. Hydrobiologia 59: 81-95.
13. Dalmacija, B., E. Kalrlovic, Z. Tamas, and D. Miskovic. 1996. Purification of high-salinity wastewater by activated sludge process Water Research 30: 295-298. Daly, D.J., and Mesing, G.E., 1995, Atlas of Gas Related Produced Water for 1990: Chicago, Gas Research Institute Topical Report 95/0016.
14. Di Toro, D.M., C.S. Zarba, D.J. Hansen, W.J. Berry, R.C. Swartz, C.E. Cowan, S.P. Pavlou, H.E. Allen, N.A. Thomas, and P.R. Paquin. 1991. Technical basis for establishing sediment quality criteria for nonionic chemicals using equilibrium partitioning. Environmental Toxicology and Chemistry 10: 1541-1583.
15. Doherty, F. G. 1990. The Asiatic clam, Corbicula spp., as a biological monitor in freshwater environments. Environmental Monitoring and Assessment 15: 143-181.
16. Douglas, W. S. and J. A. Veil. 1996. Do produced water toxicity tests accurately measure produced water toxicity in marine environments? 3rd International Petroleum Environmental Conference, ANL/EA/CP--89257, CONF-9609237--4.
17. Elias-Samlalsingh, N. and J. B. R. Agard. 2004. Application of toxicity identification evaluation procedures for characterizing produced water using the tropical mysid, Metamysidopsis insularis. Environmental Toxicology and Chemistry 23: 1194-1203.
18. Epler, J. H. 2001. Identification manual for the larval Chironomidae (Diptera) of North and South Carolina. St. Johns River Water Management District Special Publication SJ2001-SP13.
19. Fucik, K. W. 1992. Toxicity identification evaluation and characteristics of produced water discharges from Colorado and Wyoming. In Produced Water, J. P. Ray and F. R. Engelhart, eds. Plenum Press, New York, Pp. 187-198.
20. Ganjegunte G. K., G. F. Vance, and L. A. King. 2005. Soil chemical changes resulting from irrigation with water co-produced with coalbed natural gas. Journal of Environmental Quality 34: 2217-2227.
21. Gulley, D. D., D. R. Mount, J. R. Hockett, and H. L. Bergman. 1992. A statistical model to predict toxicity of saline produced waters to freshwater organisms. In Produced Water, J. P. Ray and F. R. Engelhart, eds. Plenum Press, New York, Pp. 89-96.
22. GWPC. 2002. Ground Water Protection Council Produced Water Conference, Oct. 1617, Colorado Springs, CO. Papers available online at: http://www.gwpc.org/Meetings/PW2002/Papers-Abstracts.htm.
23. Henderson, S.B., S.W.J. Grigson, P. Johnson,and B.D. Roddie. 1999. Potential impact of production chemicals on the toxicity of produced water discharges from North Sea oil platforms. Marine Pollution Bulletin 38:1141-1151.
24. Herkelrath, W.N. and Y.K. Kharaka. 2002. Hydrologic controls on the subsurface transport of oil-field brine at the Osage-Skiatook Petroleum Research "B" Site, Oklahoma. Proceedings of the Ninth Annual International Petroleum Environmental Conference, Integrated Petroleum Environmental Consortium, Albuquerque, NM.
25. Ho, K. and D. Caudle. 1997. Ion toxicity and produced water, Letter to the Editor. Environmental Toxicology and Chemistry 16:1993-1995.
26. Hose G.C. and P.J. Van den Brink. 2004. Confirming the species-sensitivity distribution concept for endosulfan using laboratory, mesocosm, and field data. Archives of Environmental Contamination and Toxicology 47: 511-520.

27. Hudak, P.F. and S. Blanchard. 1997. Land use and groundwater quality in the Trinity group outcrop of north-central Texas, USA. Environment International 23: 507-517.
28. Ingersoll, C.G., N. Wang, J.M.R. Hayward, J.R. Jones, S.B. Jones, and D.S. Ireland. 2005. A field assessment of long-term laboratory sediment toxicity tests with the amphipod Hyallela azteca. Environmental Toxicology and Chemistry 24: 2853-2870.
29. Jackson, L. and J. Meyers. 2002. Alternative use of produced water in aquaculture and hydroponic systems at Naval Petroleum Reserve No. 3. Paper presented at the 2002 Ground Water Protection Council Produced Water Conference, Oct. 16-17, Colorado Springs, CO. (Paper available online at: http://www.gwpc.org/Meetings/PW2002/Papers-Abstracts.htm).
30. Kater, B.J., J.F. Postma, M. Dubbledam, and J.T.H.J. Prins. 2001. Comparison of laboratory and in situ sediment bioassays using Corphium volutator. Environmental Toxicology and Chemistry 20: 1291-1295.
31. Kefford, B. J. 2000. The effect of saline water disposal: implications for monitoring programs and management. Environmental Monitoring and Assessment 63: 313-327.
32. Kharaka, Y.K., J.J. Thordsen, E. Kakouros, and M.M. Abbott. 2002. Environmental impacts of petroleum production: fate of inorganic and organic chemicals in produced water from the Osage-Skiatook Petroleum Environmental Research sites, Osage County, Oklahoma. Proceedings of the Ninth Annual International Petroleum Environmental Conference, Integrated Petroleum Environmental Consortium, Albuquerque, NM.
33. Kharaka, Y. K., J. J. Thordsen, E. Kakouros, and W. N. Herkelrath. 2005. Impacts of petroleum production on ground and surface waters: results from the Osage-Skiatook petroleum environmental research A site, Osage county, Oklahoma. Environmental Geosciences 12: 127-138.
34. Krause, P.R. 1995. Spatial and temporal variability in receiving water toxicity near an oil effluent discharge site. Archives of Environmental Contamination and Toxicology 29: 523-529.
35. Mancini, E. R. and C. T. Stillwell. 1992. Biotoxicity characterization of a produced-water discharge in Wyoming. Journal of Petroleum Technology 44: 744-748.
36. Marshall N. A. and P. C. E. Bailey. 2004. Impact of secondary salinisation on freshwater ecosystems: effects of contrasting, experimental, short-term releases of saline wastewater on macroinvertebrates in a lowland stream. Marine and Freshwater Research 55: 509-523.
37. Mendelssohn, I.A., K.M. Flynn,and B.J. Wilsey. 1990. The relationship between produced water discharges and plant biomass and species composition in 3 Louisiana marshes. Oil & Chemical Pollution 7: 317-335.
38. Merrit, R. W. and K. W. Cummins. 1996. An introduction to the aquatic insects of North America, 3rd ed. Kendall/Hunt Publishing Company, Dubuque, IA.
39. Moraes, J. E. F., D. N. Silva, F. H. Quina, O. Chiavone-Filho, and C. A. O. Nascimento. 2004. Utilization of solar energy in the photodegredation of gasoline water and oil-fieldproduced water. Environmental Science and Technology 38: 3746-3751.
40. Mount, D. R., K. R. Drottar, D. D. Gulley, J. P. Fillo, and P. E. O'Neil. 1992. Use of laboratory toxicity data for evaluating the environmental acceptability of produced water discharge to surface waters. In Produced Water, J. P. Ray and F. R. Engelhart, eds. Plenum Press, New York, Pp. 175-185.

41. Mount, D. R., D. D. Gulley, J. R. Hockett, T. D. Garrison, and J. M. Evans. 1997. Statistical models to predict the toxicity of major ions to Ceriodaphnia dubia, Daphnia magna and Pimephales promelas (fathead minnows). Environmental Toxicology and Chemistry 16: 2009-2019.
42. Okandan, E., F. Gumrah, and B. Demiral. 2001. Pollution of an aquifer by produced oil field water. Energy Sources 23: 327-336.
43. Olive J. H., J. L.Jackson, D. Keller, and P. Wetzel. 1992. Effects of oil-field brines on biological integrity of 2 tributaries of the Little Muskingum Ruver, Southeastern Ohio. Ohio Journal of Science 92: 139-146.
44. O'Neil, P. E. S. C. Harris, M. F. Mettee, H. R. Isaacson, and J. M. Evans. 1992a. Biological fate and effect of coalbed methane produced waters discharged into streams of the Warrior Basin, Alabama. In Produced Water, J. P. Ray and F. R. Engelhart, eds. Plenum Press, New York, Pp. 315-327.
45. O'Neil, P. E., H. R. Isaacson, and J. M. Evans. 1992b. Surface discharge of coalbed methane produced waters in the Warrior Basin of Alabama, the Cedar Cove model. In Produced Water, J. P. Ray and F. R. Engelhart, eds. Plenum Press, New York, Pp. 329-341.
46. Peeters E. T. H. M., J. J. P. Gardeniers, and A. A. Koelmans. 2000. Contribution of trace metals in structuring in situ macroinvertebrate community composition along a salinity gradient. Environmental Toxicology and Chemistry 19: 1002-1010.
47. Produced Waters Database. 2002. Accessed online at: http://energy.cr.usgs.gov/prov/prodwat/ 15 April 2004.
48. Ramirez, P., Jr. 1993. Contaminants in oil field produced waters discharged into the Loch Katrine wetland complex, Park County, Wyoming and their bioconcentration in the aquatic bird food chain: U.S. Fish and Wildlife Service Contaminant Report Number R6/706C/93, 37 p.
49. Reed, D.C., R.J. Lewis, and M. Anghera. 1994. Effects of an open coast oil production outfall on patterns of giant kelp (Macrocyctis pyrifera) recruitment. Marine Biology 120: 25-31.
50. Rosenberg, D. M. and V. H. Resh. 1982. The use of artificial substrates in the study of freshwater benthic macroinvertebrates. In Artificial Substrates, J. Cairns Jr. ed., Ann Arbor Science, Ann Arbor, Pp.175-235.
51. Rosenberg, D. M. and V. H. Resh. 1993. Introduction to freshwater biomonitoring and benthic macroinvertebrates. In Freshwater Biomonitoring and Benthic Macroinvertebrates, D. M. Rosenberg and V. H. Resh, eds. Chapman and Hall, New York, pp. 1-9.
52. Sadiq, M. and I. Alam. 1997. Metal concentrations in a shallow groundwater aquifer underneath petrochemical complex. Water Resources 31: 3089-3097.
53. Sasson-Brickson, G. and G.A. Burton, Jr. 1991. In situ and laboratory sediment toxicity testing with Ceriodaphnia dubia. Environmental Toxicology and Chemistry 10: 201-207.
54. Sauer, T. C., H. J. Costa, J. S. Brown, and T. J. Ward. 1997. Toxicity identification evaluations of produced-water effluents. Environmental Toxicology and Chemistry 16: 2020-2028.
55. Schiff, K. C., D. J. Reish, J. W. Anderson, and S. M. Bay. 1992. A comparative evaluation of produced water toxicity. In Produced Water, J. P. Ray and F. R. Engelhart, eds. Plenum Press, New York, Pp. 199-207.

56. Schroer, A.F.W., J.D.M. Belgers, T.C.M. Brock, A.M. Matser, S.J. Maund, and P.J. Van den Brink. 2004. Comparison of laboratory single species and field population-level effects of the pyrethroid insecticide λ-Cyhalothrin on freshwater invertebrates. Archives of Environmental Contamination and Toxicology 46: 324-335.
57. Shobanov, N.A., I.I. Kiknadze, and M.G. Butler. 1999. Palearctic and Nearctic Chironomus (Camptochironomus) tentans (Fabricus) are different species (Diptera: Chironomidae). Entomologica Scandinavica 30: 311-322.
58. Short, T.M., J.A. Black, and W.J. Birge. 1991. Ecology of a saline stream: community responses to spatial gradients of environmental conditions. Hydrobiologia 226: 167-178.
59. Smith, D. G. 2001. Pennak's freshwater invertebrates of the United States, 4th ed. John Wiley and Sons, New York.
60. Smith, J. P., A. O. Tyler, and Z. A. Sabeur. 1998. Ecological assessment of produced waters in Indonesia. Environmental Toxicology and Water Quality 13: 323-336.
61. Soucek, D. J., D. S. Cherry, R. J. Currie, H. A. Latimer, and G. C. Trent. 2000. Laboratory to field validation in an integrative assessment of an acid mine drainage-impacted watershed. Environmental Toxicology and Chemistry 19: 1036-1043.
62. Soucek, D. J., T. S. Schmidt, and D. S. Cherry. 2001. In situ studies with Asian clams (Corbicula fluminea) detect acid mine drainage and nutrient inputs in low-order streams. Canadian Journal of Fisheries and Aquatic Sciences 58: 602-608.
63. Stephenson, M. T. 1992. A survey of produced water studies. In Produced Water, J. P. Ray and F. R. Engelhart, eds. Plenum Press, New York, Pp. 1-11.
64. Stromgren, T., S. E. Sorstrom, L. Schou, I. Kaarstad, T. Aunaas, O. G. Brakstad, and O. Johansen. 1995. Acute toxic effects of produced water in relation to chemical composition and dispersion. Marine Environmental Research 40: 147-169.
65. Thurston, R. V., R.C. Russo, and G. A. Vinogradov. 1981. Ammonia toxicity to fishes. Effect of pH on the toxicity of the unionized ammonia species. Environmental Science and Technology 15: 837-840.
66. Tietge, J. E., J. R. Hockett, and J. M. Evans. 1997. Major ion toxicity of six produced waters to three freshwater species: application of ion toxicity models and TIE procedures. Environmental Toxicology and Chemistry 16: 2002-2008.
67. Tucker, K.A. and G.A. Burton, Jr. 1999. Assessment of nonpoint-source runoff in a stream using in situ and laboratory approaches. Environmental Toxicology and Chemistry 18: 2797-2803.
68. Environmental Protection Agency. 1991. Methods for aquatic toxicity identification evaluations: phase I toxicity identification procedures, 2nd ed. EPA/600/6-91/003. Washington, DC.
69. Environmental Protection Agency. 2000. Methods for measuring the toxicity and bioaccumulation of sediment-associated contaminants with freshwater invertebrates. EPA/600/R-99/064. Duluth, MN.
70. Environmental Protection Agency. 2002a. Methods for measuring the acute toxicity of effluents and receiving waters to freshwater and marine organisms. EPA/821/R-02/012. Washington, DC.
71. Environmental Protection Agency. 2002b. Short-term methods for estimating the chronic toxicity of effluents and receiving waters to freshwater organisms. EPA/821/R02/013. Washington, DC.

72. Veil, J. A., B. G. Langhus, and S. Belieu. 1999. DOWS reduce produced water disposal costs. Oil and Gas Journal 76-85.
73. Veil, J.A. 2003. Innovative technologies for managing oil field wastes. Journal of Energy Resources Technology, 125: 238-248.
74. Washburn, L., S. Stone, and S. MacIntyre. 1999. Dispersion of produced water in a coastal environment and its biological implications. Continental Shelf Research 19: 57-78.
75. Wiebe, A.H., J.G. Burr, and H.E. Faubion. 1924. The problem of stream pollution in Texas with special reference to salt water from the oil fields. Transactions of the American Fisheries Society 64: 81-85.
76. Woolard, C. R. and R. L. Irvine. 1995. Treatment of hypersaline wastewater in the sequencing batch reactor. Water Research 29: 1159-1168.

Bidwell, J. R., Fisher, J. C., and Cooper, N. L. Field Validation of Toxicity Tests to Evaluate the Potential for Beneficial Use of Produced Water. Final Scientific Report. Project # DE-FC26-04NT15544. U.S. Department of Energy National Energy Technology Laboratory. Submitting Organization: Department of Zoology, Oklahoma State University.

CHAPTER 5

FRACKING VS FAUCETS: BALANCING ENERGY NEEDS AND WATER SUSTAINABILITY AT URBAN FRONTIERS

MATTHEW FRY, DAVID J. HOEINGHAUS, ALEXANDRA G. PONETTE-GONZÁLEZ, RUTHANNE THOMPSON, and THOMAS W. LA POINT

Newly accessible shale deposits have dramatically increased global gas reserves and are touted as a bridge to a clean energy future. For example, in the U.S., where shale gas is projected to comprise 49% of national natural gas production by 2035, proponents argue that shale gas production can provide energy independence, create employment, and stimulate regional economies. Amidst this optimism, however, are growing concerns about the effects of shale gas extraction, and, in particular, hydraulic fracturing or "fracking", on water resources [2]—concerns that are magnified in urban areas where human populations and extractive operations overlap. We believe that water conflicts arising from expansion of the U.S. shale gas industry foreshadow developments in other countries with cities situated over large shale-gas deposits, including Diyarbakir, Turkey; Ahmedabad, India; and Chongqing, China.

While much recent controversy over fracking and water resources is directed toward potential contamination, here we use the Dallas-Fort Worth (DFW) Metroplex, Texas, to illustrate challenges associated with balancing energy needs and water sustainability in cities with semiarid to arid climates. Along with one of the highest urban population growth rates in the U.S. (23.4%), the Barnett Shale underlying DFW represents the largest staging ground for shale gas extraction in the world.(3) Ninety percent of DFW's drinking water is supplied by surface runoff captured in 34 reservoirs, with municipalities and gas extractors consuming 86% (1.87 billion

m^3) and <3% of the total water supply, respectively. DFW also experiences recurrent drought, heightening anxieties over water availability. In 2011, we randomly surveyed 1000 DFW residents to evaluate public perceptions and knowledge of watersheds. Six questions dealt specifically with regional shale-gas extraction. Our survey reveals considerable public uncertainty regarding the effects of shale-gas extraction on fresh water availability: after drought, nearly one-third (27.7%) of respondents ranked gas drilling as the greatest hazard to the water supply in 2011. Unfortunately, public uncertainty diverts attention from the primary factor affecting water supply in expanding urban areas: increasing municipal water use.

We believe that DFW's situation holds important lessons for other cities faced with the prospect of expansive shale gas extraction, growing populations, and/or increasing aridity. First, in urban areas where municipal water use dominates water consumption, reduction in residential and commercial use has the potential to play a major role in urban water sustainability. Notwithstanding, strategies that figure prominently into plans to meet future water needs are often not sustainable (either for water provisioning or ecological systems) and merely increase resource capture rather than improve efficiencies or decrease per capita consumption. In Texas, water management strategies aim to add 2.96 billion m^3 of water to the region's supply by 2060, at a price of $21.5 billion (for comparison, China's Three Gorges Dam cost $25 billion). These include the construction of four new reservoirs, importation of surface water from other basins and districts, and water reuse (Figure 1). Together, these strategies are anticipated to comprise 87.5% of future water gains. Conservation initiatives account for just 12.3%, or 364 123 844 m^3 of additional water in 2060. [4]

Second, although municipal conservation is key to water sustainability, most urban residents have a limited understanding of the urban water cycle. For example, when asked if they lived in a watershed, 9.7% of our survey respondents answered "yes", 63.3% said "no", and 27.7% said "do not know". Respondents also were "not sure" (11.5%) or "did not know" (40.9%) the source of drinking water supplied to their home. This is likely due to "quality" rather than "quantity" of education: 54% of survey participants had a bachelor's degree or higher and yet were unfamiliar with the watershed concept. Clearly, as long as residents do not understand how

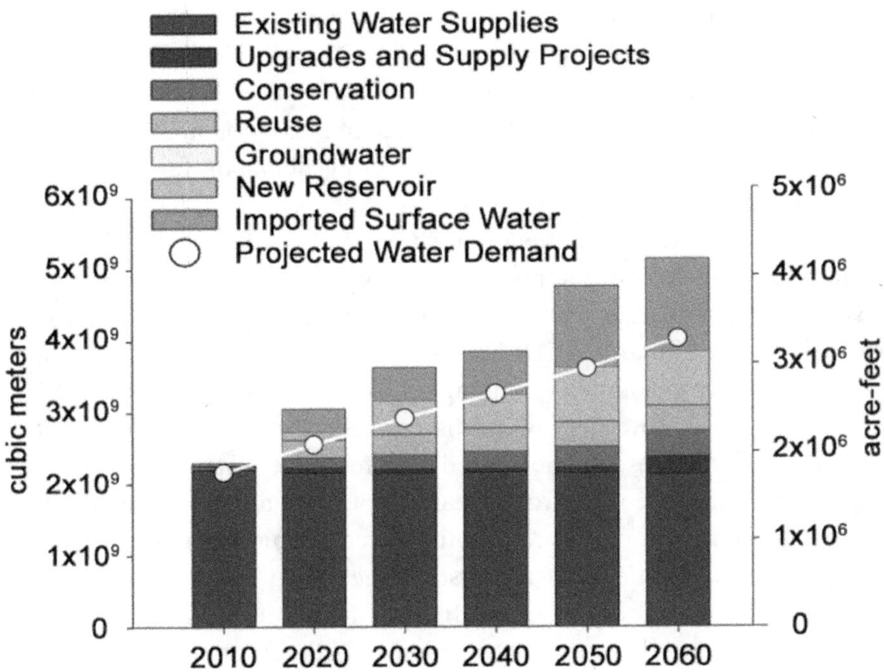

FIGURE 1. Existing water supplies, projected water demand, and proposed water augmentation strategies for the Dallas-Fort Worth (DFW) Metroplex. The Texas Water Development Board proposes $21.5b in water projects to add over 2.96 billion m3 of water to the region by 2060. (Modified from ref 4).

they fit into the urban water cycle and fully recognize their role in water conservation, it is unlikely that they will reduce their own water use.

Conservation education is thus essential to change household environmental behavior. In a recent study, 94% of school-aged children (5–10 years) signed commitment cards to turn water faucets off when brushing teeth after learning that this simple act can save 3.8 to 11.4 L of water per day per individual. Assuming all students in the study keep their commitment, between 44,399 m^3 and 134,366 m^3 of water could be conserved annually. Thus, small reductions in individual water consumption can lead to large water savings, particularly in arid urban areas with disproportionately high levels of residential water consumption. Scaling up to the DFW Metroplex, a one-quarter reduction in daily water consumption by all 6.5 million residents could save approximately 225,387,500 m^3, or decrease current municipal water demand by 12%. This is more than four times the estimated water use by shale-gas extractors in the region in 2010.

Conflicts arising as a result of competing demands for energy and water are of increasing global concern, especially in expanding urban areas. The international diffusion of shale gas extractive technologies is certain to heighten this potential in arid regions, where large cities must also contend with recurrent water shortage episodes. For this reason, we believe the situation in DFW, where uncertainty about the sustainability of water resources is entangled with residents' anxieties about gas extraction,(3) foreshadows probable scenarios in other global cities. Nevertheless, at present, the heated debate about water use for fracking is detracting attention from the primary issue affecting water availability in DFW and other metropolitan areas: rapid population growth and increasing municipal water use. Educational programs dedicated to improving residents' understanding of their role in the urban water cycle and enhanced water conservation efforts are urgently needed to refocus public attention toward actions and behaviors that may actually achieve water sustainability. A conservation-minded, water-consuming public intimately aware of its connection to the urban watershed is key to balancing energy needs and long-term water sustainability in urbanizing areas.

REFERENCES

1. The Railroad Commission of Texas. http://www.rrc.state.tx.us/ (accessed April 2012).
2. US Energy Information Administration. Technology Drives Natural Gas Production Growth from Shale Gas Formations, 2011. http://www.eia.gov/todayinenergy/detail.cfm?id=2170.
3. Nicot, J.-P.; Scanlon, B. R.Water use for shale-gas production in Texas Environ. Sci. Technol. 2012, 46, 3580– 3586.
4. Texas Water Development Board. Water for Texas 2012 State Water Plan, 2012. http://www. twdb.state.tx.us/publications/state_water_plan/2012/2012_SWP.pdf.

5. Thompson, R.; Coe, A.; Klaver, I.; Dickson, K.Design and implementation of a research-informed water conservation education program Appl. Environ. Educ. Comm. 2011, 10 (2) 91– 104.

Fry, M., Hoeinghaus, D. J., Ponette-González, A. G., Thompson, R., and La Point, T. W. Fracking vs Faucets: Balancing Energy Needs and Water Sustainability at Urban Frontiers. Reprinted with permission from Environmental Science & Technology 2012, Vol. 45, 7444-7445.

CHAPTER 6

RAPID EXPANSION OF NATURAL GAS DEVELOPMENT POSES A THREAT TO SURFACE WATERS

SALLY ENTREKIN, MICHELLE EVANS-WHITE, BRENT JOHNSON, and ELISABETH HAGENBUCH

Extraction of natural gas from hard-to-reach reservoirs has expanded around the world and poses multiple environmental threats to surface waters. Improved drilling and extraction technology used to access low permeability natural gas requires millions of liters of water and a suite of chemicals that may be toxic to aquatic biota. There is growing concern among the scientific community and the general public that rapid and extensive natural gas development in the US could lead to degradation of natural resources. Gas wells are often close to surface waters that could be impacted by elevated sediment runoff from pipelines and roads, alteration of streamflow as a result of water extraction, and contamination from introduced chemicals or the resulting wastewater. However, the data required to fully understand these potential threats are currently lacking. Scientists therefore need to study the changes in ecosystem structure and function caused by natural gas extraction and to use such data to inform sound environmental policy.

Natural gas drilling has dramatically expanded with advances in extraction technology and the need for cleaner burning fuels that will help meet global energy demands. Natural gas is considered a "bridge fuel" to renewable energy resources because its combustion releases fewer contaminants (eg carbon dioxide [CO_2], nitrogen oxide [NO_x], sulfur oxide [SO_x]) than compared with that of coal or petroleum. Horizontal drilling and hydraulic fracturing ("hydrofracking" or "fracking") now allow the extraction of vast shale gas reserves previously considered inaccessible or unprofitable. Shale gas production in the US is expected to increase threefold and will account for nearly half of all natural gas produced by 2035 (EIA 2011). This widespread proliferation of new gas wells and the use of modern drilling and extraction methods have now been identified as a global conservation issue (Sutherland et al. 2010). Here, we describe the threats to surface waters associated with increased natural gas development in shale basins and highlight opportunities for research to address these threats.

6.1 HORIZONTAL DRILLING AND HYDRAULIC FRACTURING

Gas-well drilling has historically used a single vertical well to access gas trapped in permeable rock formations (eg sandstone) where gas flows freely through pore spaces to the wellbore. Unlike these conventional sources, unconventional gas reservoirs are low permeability formations, such as coal beds, dense sands, and shale, that require fracturing and propping (addition of sand or other granular material suspended in the fracturing fluid to keep fractures open) before gas can travel freely to the wellbore. Hydrofracking uses high-pressure fracturing fluids, consisting of large volumes of water and numerous chemical additives, to create fractures, while added propping agents, such as sand, allow the gas to flow. Although hydrofracking was first used in the 1940s, the practice was not widely applied until the 1990s, when natural gas prices increased and advances in horizontal drilling made the technique more productive. Horizontal drilling increases the volume of rock a single well can access, thereby reducing the total number of wells required at the surface. The horizontal leg of a gas well is fractured in discrete lengths of 91–152 m, allowing up to 15 separate hydrofrack "events" along one horizontal well (Kargbo et al. 2010). Fracturing depth depends on target rock formations but varies from 150 m to more than 4000 m for the major shale formations in the US (US DOE 2009).

6.2 EXTENT OF RESOURCES

The US currently has 72 trillion cubic meters (tcm) of potentially accessible natural gas – enough to last 110 years, based on 2009 rates of

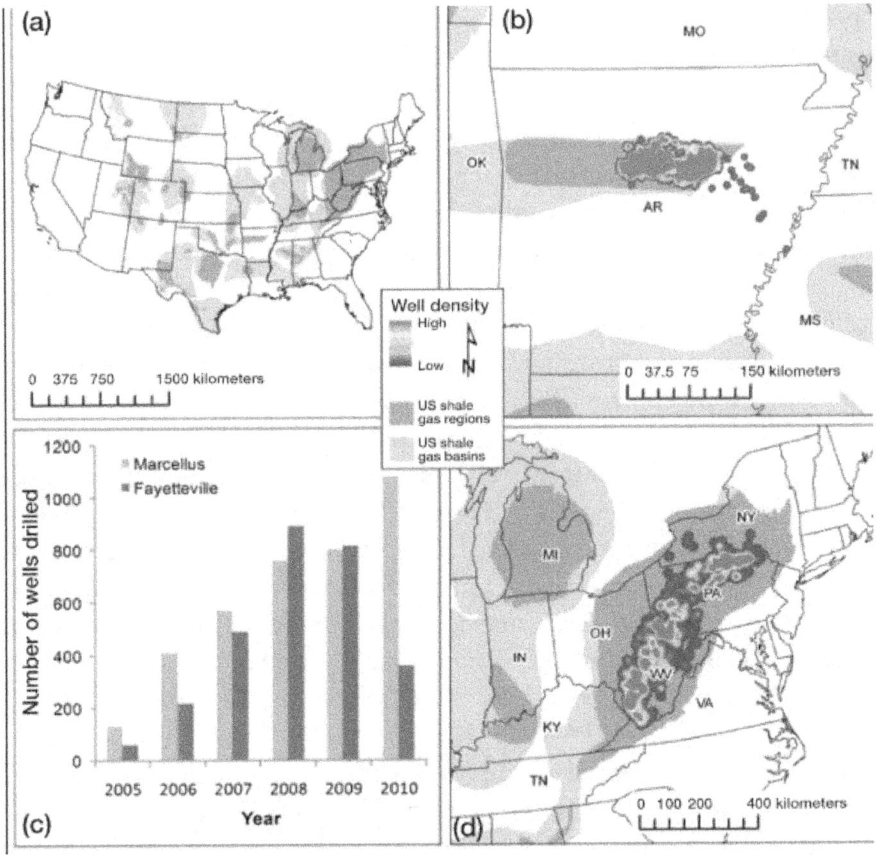

FIGURE 1. (a) National map of all recognized potential areas for unconventional natural gas exploration in the contiguous US; (b) density of wells in the Fayetteville unconventional natural gas basins; (c) number of gas wells installed in the Fayetteville and Marcellus basins from 2005 to 2010; and (d) density of wells in the Marcellus shale basins. We calculated densities using the kernel density tool in ArcMap 9.3.1.

consumption (EIA 2011). Approximately 23 tcm of that gas is found in unconventional (ie low permeability) gas reservoirs; development of such reservoirs has increased by 65% since 1998 (US DOE 2009). There are 29 known shale basins spanning 20 states, which are expected to contribute 45% of the total US gas produced by 2035 (EIA 2011; Figure 1a). Furthermore, the US gas supply represents only a fraction of the total global estimate of potentially accessible natural gas (~459 tcm) and, outside of North America, only 11% has so far been recovered (MIT 2010). Development of potentially accessible natural gas is expected to increase with rising global demand and the transfer of drilling technologies overseas.

Threats to surface waters

The rapid expansion in natural gas development threatens surface-water quality at multiple points, creating a need to assess and understand the overall costs and benefits of extracting this resource from shale reservoirs. Gas-well development of any type creates surface disturbances as a result of land clearing, infrastructure development, and release of contaminants produced from deep ground water (eg brines). However, the use of hydraulic fracturing poses additional environmental threats due to water withdrawals and contamination from fracking-fluid chemicals. Extraction of gas from shale formations may also produce considerably more methane (CH_4) than conventional wells and could have a larger greenhouse-gas footprint than other fossil-fuel development (Howarth et al. 2011). Furthermore, gas wells are often located adjacent to rivers and streams and may occur at high densities in productive shale basins, resulting in cumulative impacts within watersheds. Environmental and human health concerns associated with hydrofracking have stirred much debate, and the practice has received extensive attention from the media (Urbina 2011) and from researchers (US EPA 2004; Kargbo et al. 2010; Osborn et al. 2011; US EPA 2011; Colborn et al. in press). Research that addresses concerns regarding increased drilling and hydrofracking in shale basins has primarily focused on contaminants that threaten drinking water and groundwater, whereas data collection to address concerns associated with surface water and terrestrial ecosystems has largely been overlooked.

Our goal here is to provide background information on shale development in the US that may inform future ecological studies that assess the potential for environmental impacts. We use data from the Fayetteville and Marcellus shale formations to demonstrate the recent accelerated drilling activity, well proximity to streams, and well density relationships with stream turbidity. We also review other potential threats to aquatic freshwater ecosystems as a result of increased natural gas development.

Focus areas The Fayetteville and Marcellus shale basins are among the most productive in the US. The Fayetteville shale basin underlies more than 23 000 km^2 of Arkansas and eastern Oklahoma, at a depth of 300–2000 m (Figure 1a). The number of gas wells sited in this area has increased nearly 50-fold, from 60 to 2834 wells since 2005, in a concentrated area of north-central Arkansas (Figure 1, b and c). The Marcellus shale basin spans 240 000 km^2 at a depth of 1200–2500 m and underlies six states in the upper Mid-Atlantic, including much of the Appalachian region (Figure 1d). Estimates indicate natural gas reserves in the Marcellus to be 14 tcm, or 59% of the total estimated unconventional reserves in the US (US DOE 2009). As of summer 2010, the Marcellus had 3758 natural gas wells, with projections of up to 60 000 wells being constructed in the region over the next 30 years (Johnson 2011). The Marcellus formation also underlies sensitive watersheds, such as the threatened upper Delaware River, a designated wild and scenic river that supplies drinking water to > 15 million people (DRBC 2008). The rapid development of gas wells in relatively concentrated areas may increase the likelihood of ecological impacts on surrounding forests and streams.

6.3 PROXIMITY OF GAS-WELL DEVELOPMENT TO WATER RESOURCES

We initially assessed the proximity of active gas wells to water resources using state well-location data and the National Hydrography Dataset (NHD) flowlines (ie streams and rivers mapped from 1:24 000 Digital Line Graph hydrography data). Spatial analysis indicated that, for both the Fayetteville and Marcellus shale formations, gas wells were sited, on average, 300 m from streams, yet several hundred wells were located within 100 m of stream channels (Table 1). Gas wells were located, on average, 15 km from public surface-water drinking supplies, and 37 km and 123 km from public well water supplies in the Marcellus and Fayetteville shale reservoirs, respectively (Table 1). Although wells are generally constructed far from public drinking-water sources, there is potential for wastewater to travel long distances, given that many of the components of produced waters (ie a mixture of fracking fluids and natural geologic formation water flowing back out of the well), such as brines, will not settle out or be assimilated into biomass. Furthermore, the NHD underestimates the density of headwater stream channels (Heine et al. 2004), so our proximity measures probably underestimate the threat to streams. We therefore used geographic information system (GIS) tools to generate detailed drainage-area networks in portions

TABLE 1: Number of unconventional gas wells drilled each year since 2005 for Arkansas, New York, Ohio, Pennsylvania, and West Virginia

	State	Total wells	Total operators	Distance to NHD flowlines (mean ± SD, range ,m)	Total # (percent) of wells within			Distance to	
					100 m of NHD flowlines	200 m of NHD flowlines	300 m of NHD flowlines	public water wells (mean ± SD, range, m)	public drinking–water intakes (mean ± SD, range ,m)
Marcellus	PA	2091*	59	319 ± 171 (8–1172)	74 (4)	577 (28)	1141 (55)	25.83 ± 17.93 (0.32–79.60)	14.83 ± 10.06 (0.60–50.23)
	WV	1599⁺	86	214 ± 143 (1–850)	409 (26)	798 (50)	1198 (75)	52.32 ± 32.82 (0.55–125.42)	11.16 ± 5.36 (0.53–33.32)
	OH	42ᵗ	12	130 ± 153 (46–691)	8 (19)	23 (55)	33 (79)	71.85 ± 28.29 (26.46 – 138.17)	14.15 ± 8.38 (1.54 – 29.87)
	NY	26§	9	247 ± 182 (27–631)	9 (35)	12 (46)	14 (54)	10.47 ± 7.11 (2.58 – 34.19)	16.59 ± 9.06 (4.58 – 35.63)
	All four states combined	3758		273 ± 168 (1–1172)	500 (13)	1410 (38)	2386 (64)p	37.51 ± 28.88 (0.32 – 138.17)	13.27 ± 8.55 (0.53 – 50.23)
Fayetteville	AR	2854¶	21	353 ± 241 (7–1642)	269 (10)	900 (32)	1434 (51)p	123.67 ± 11.12 (78.94 – 156.12	15.15 ± 7.49 (0.66 – 133.43)

Notes: *PA: Pennsylvania Department of Environmental Protection Bureau of Oil and Gas (records available through 30 Sep 2010), www.dep.state.pa.us/dep/deputate/minres/ oilgas/reports.htm; †WV: West Virginia Geological and Economic Survey (records available through early Sep 2010), www.wvges.wvnet.edu/www/datastat/devshales.htm; ‡OH: Ohio Department of Natural Resources Division of Mineral Resources Management (records available through 30 Sep 2010), www.dnr.state.oh.us/ mineral/database/tabid/17730/Default.aspx; §NY: New York State of Environmental Conservation (records available through 30 Sep 2010), www.dec.ny.gov/energy/1603.html;¶AR: Arkansas Oil and Gas Commission (records available through 30 Sep 2010), www.aogc.state.ar.us/ (data downloaded from: ftp://www.aogc.state.ar.us/).

of the Fayetteville and Marcellus shale reservoirs where gas wells occur at high densities. The terrain processing tools in ArcHydro Tools 9 version 1.3 (an ArcGIS extension) were used to generate drainage area lines from 10-m digital elevation models (http:// seamless.usgs.gov/ned13.php) in a subset of drainage areas in each shale basin. A stream threshold of 500 (50 000 m^2) was used to define stream channels in the model. Gas-well proximity was analyzed again with a subset of modeled stream drainage areas and the same subset of NHD flowlines for comparison (Figure 2; Table 2). Active gas wells were an average of 130 m and 153 m from modeled drainage areas, as compared with 230 m and 252 m from NHD flowlines, in the Fayetteville and Marcellus shale reservoirs, respectively. Over 80% of the active gas wells were located within 300 m of modeled drainage areas (Table 2). Because the modeled drainage areas estimate some intermittent and ephemeral channels, the proximity of wells to stream channels (and the potential for downstream impacts) is greater than that reflected by NHD flowline data. This process may provide a more accurate assessment of potential stream impacts, particularly if shale gas development continues at its current rate. As gas-well densities continue to increase, the proximity of wells to stream channels may also increase, resulting in a greater risk of streamflow reductions from pumping, contamination from leaks and spills from produced waters or fracking fluids, and sedimentation from infrastructure development (eg pipelines and roads).

6.4 ENVIRONMENTAL REGULATION

Environmental regulation of oil and gas drilling is complex and varies greatly between states. The Safe Drinking Water Act (SDWA) provides federal laws for protecting surface and ground-waters and human health, but with the exception of diesel-fuel injection, hydraulic fracturing operations are exempt as a result of the 2005 Energy Policy Act. State agencies are therefore primarily responsible for regulation and enforcement of environmental issues associated with natural gas development. The rapid growth and expansion of US gas drilling has made regulation difficult, and violations are common; in Pennsylvania alone, there were more than 1400 drilling violations between January 2008 and October 2010 (PADEP 2010). Of these, nearly

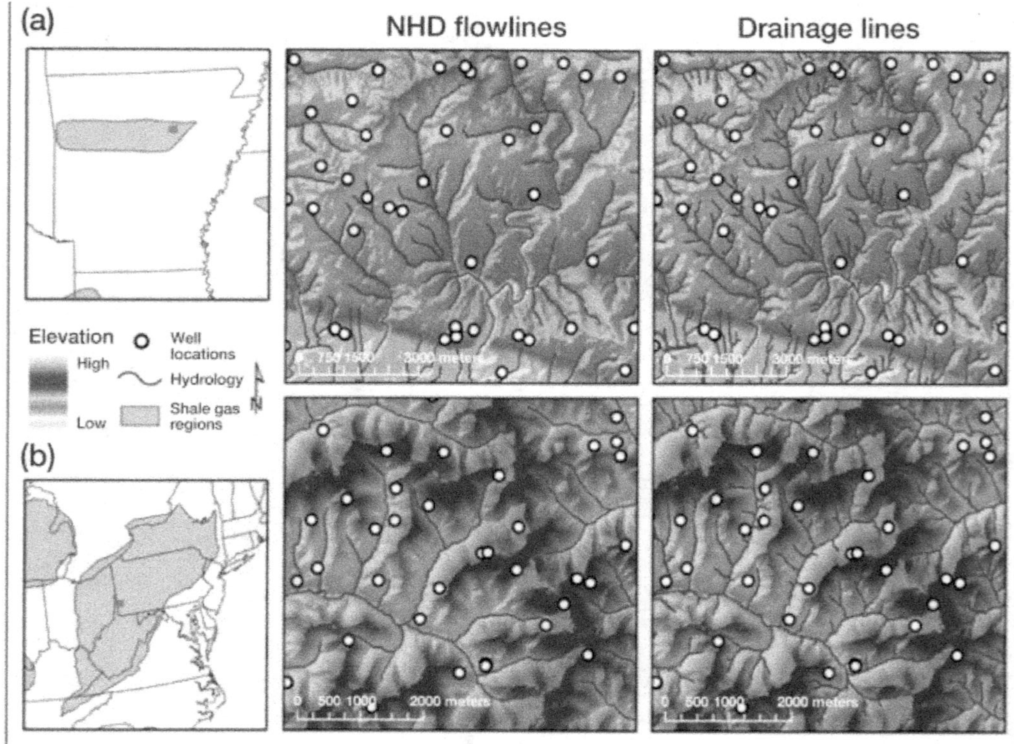

FIGURE 2: Proximity of gas wells to stream channels in a subset of the Fayetteville and Marcellus unconventional natural gas reservoirs. Blue squares represent the areas modeled by GIS in the Fayetteville shale ([a] drainage area modeled and represented by the blue square was 5809 km^2) and the Marcellus shale ([b] drainage area modeled and represented by the blue square was 4041 km^2). Topographic maps are example areas that demonstrate differences between the National Hydrography Dataset and modeled drainage area networks.

half dealt with surface-water contamination and included direct discharge of pollutants, improper erosion control, or failure to properly contain wastes. In contrast, the Arkansas Department of Environmental Quality cited only 15 surface-water violations in the Fayetteville shale in 2010; however, over half of these dealt with permitting and discharge violations associated with natural gas development (ADEQ 2010). The discrepancy in the numbers of violations between states demonstrates the variable degree of regulation at the state level and is probably based on differences in regulations as well as available regulatory resources. The number and proportion of violations associated with natural gas development indicates that sediments and contaminants associated with drilling are making their way into surface waters, and yet there are few studies examining their ecological effects. Primary threats to surface waters and potential exposure pathways (Figure 3) include sediments, water withdrawal, and release of wastewater.

6.4.1 SEDIMENTS

Excessive sediment levels are one of the primary threats to US surface waters (US EPA 2006) and have multiple negative effects in lotic (river, stream, or spring) food webs (Wood and Armitage 1999). Gas-well installation activities can negatively affect lotic ecosystems by increasing sediment inputs from well pads and supporting infrastructure (eg roads, pipelines, stream crossings), as well as loss of riparian area. Typically, at least 1.5–3.0 ha of land must be cleared for each well pad, depending on the number of wells per pad; where these occur in high densities, well pads can cumulatively alter the landscape. Land clearing and stream disturbance during well and infrastructure development can increase sediments in surface-water runoff (Williams et al. 2008), resulting in increased suspended and benthic sediments in surface waters. Nutrients, such as phosphorus, bound to these sediments may also have negative impacts on surface waters by contributing to eutrophication.

We identified seven streams in the Fayetteville shale with a variety of different well densities within their drainage areas, to test the prediction that stream turbidity would be positively related to the density of gas wells. The seven stream drainages were delineated through the use of the ArcHydro

TABLE 2: Proximity of natural gas wells to stream channels modeled by terrain processing tools in ArcHydro Tools 9 (version 1.3) to generate drainage area lines from a 10-m digital elevation model (http://seamless.usgs.gov ned13.php) as compared with well proximity to National Hydrography Dataset flowlines

		Subset		Previous distances (in Marcellus, PA only)		Subset			Previous distances (in Marcellus, PA only)		
		range (m)	mean ± SD (m)	range (m)	mean ± SD (m)	within 100m	within 200m	within 300m	within 100m	within 200m	within 300m
Marcellus	Drainage area lines	4–316	153 ± 56	–	–	17%	80%	100%	–	–	–
	NHD flowlines	48–681	252 ± 114	8–1172	319 ± 171	5%	39%	70%	4%	28%	55%
Fayetteville	Drainage area lines	0–420	130 ± 70	–	–	32%12%	71%	82%	–	–	–
	NHD flowlines	1–933	230 ± 136	7–1642	353 ± 241		43%	61%	10%	32%	51%

Notes: *Processed for 615 of 3758 wells (16%), processed 42 of 559 HUC-12 Units containing well point locations (8%). **Processed for 2372 of 2834 wells (84%), processed 55 of 84 HUC-12 Units containing well point locations (65%).

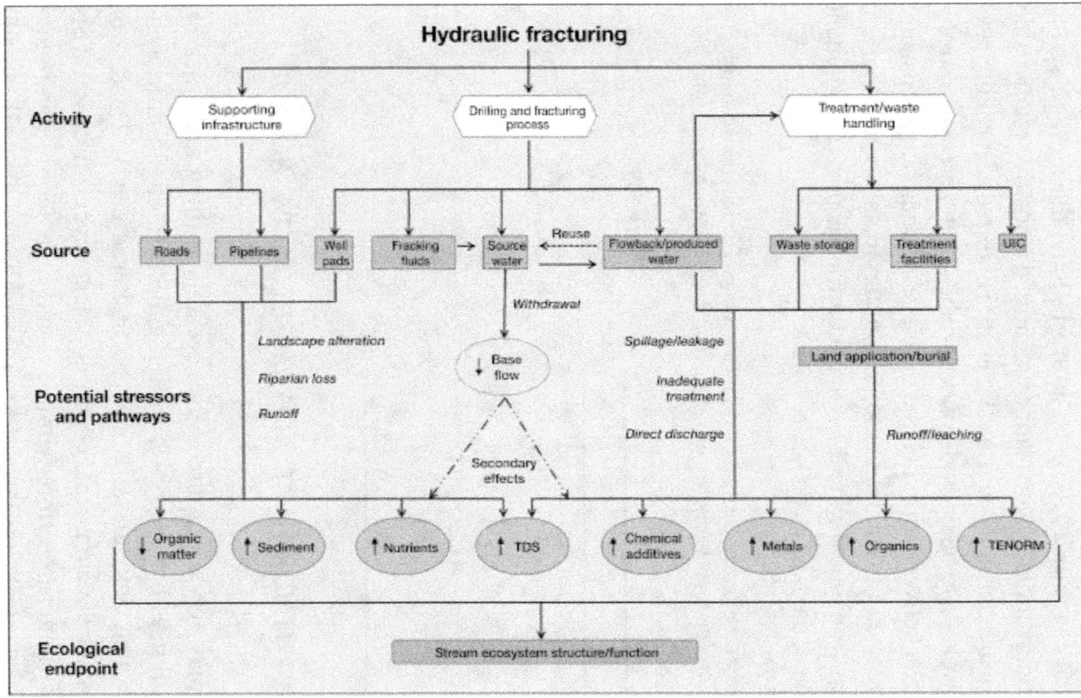

FIGURE 3: Simplified diagram of potential threats due to natural gas development through coupled horizontal drilling with hydraulic fluid fracturing in unconventional natural gas reservoirs. Exposure pathways that may result in structural and functional alterations to aquatic ecosystems will vary, depending on geographic location and rigor of best management practices applied. UIC = underground injection control; TDS = total dissolved solids; TENORM = technologically enhanced naturally occurring radioactive materials. Dotted lines indicate secondary effects from gas development. "Flowback" is underlined to indicate that it may be recycled and reused.

extension in ArcMap (version 9.3.1 ESRI). Using gas-well location data obtained from the Arkansas Oil and Gas Commission (ftp://www.aogc.state.ar.us/GIS_Files/), we quantified well density within each drainage area as the total number of wells divided by the drainage area. Turbidity was measured with a Hach Lamotte 2020 meter in April 2009, during high spring flow. Pearson product moment correlations identified a positive relationship between stream-water turbidity and well density (Figure 4). Turbidity was not positively correlated to other land-cover variables, but there was a strong negative correlation between turbidity and drainage area and percent pasture cover in the watershed (Table 3). These preliminary data suggest that the cumulative effects from gas well and associated infrastructure development may be detectable at the landscape scale.

TABLE 3: Pearson product moment correlations* and associated P values between turbidity (NTU) and other landscape-level variables, including land cover (Gorham and Tullis 2007) and drainage area.

Correlates	*r*	*P* value
Well density	0.91	0.003
Drainage area	-0.86	0.01
Low-impact urban	0.35	0.44
Wood/herbaceous	-0.63	0.12
Forest	-0.36	0.42
Pasture	-0.88	0.008

6.4.2 *WATER WITHDRAWAL MAY ALTER FLOW REGIME*

Surface waters may serve as sources for necessary drilling and fracking fluids – each well uses between 2–7 million gallons (~7.5–26 million liters) of source water. Several wells may be fractured per well pad over the life span of well development, which may last several decades. This concentration of fracturing effort within a small area should compound water use. Many gas wells are installed in regions where water is already being withdrawn for agriculture, and thus may further stress the resource. Streamflow may be negatively affected if streams are dammed to create holding ponds or if water is directly extracted for the fracturing process.

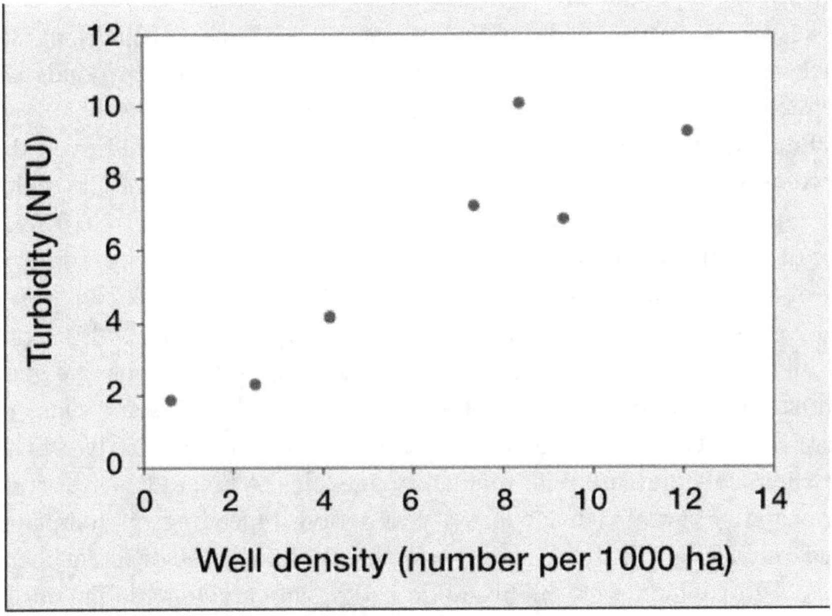

FIGURE 4: Well density and stream turbidity measured in April 2009 during high flows in seven stream drainages. NTU = nephelometric turbidity unit.

The rapid and concentrated extraction of water could create regional shortages during periods of drought, resulting in an altered flow regime and the further degradation of critical habitat for aquatic biota, particularly if low-order streams are primary sources. A reduction in streamflow may also result in secondary effects, such as increased contaminant concentrations and reduced downstream water quality, because less water is available for dilution.

6.4.3 RELEASE OF WASTEWATERS

Surface-water contamination from hydrofracking fluids and produced water is most likely to occur during hydrofracking or treatment and disposal processes, when the potential for accidental spills and leaking is greatest. Contamination from hydrofracking wastes can also occur through inadequate waste treatment practices, improper waste storage, inadequately constructed impoundments or well casings, and improper disposal of solid wastes (eg in poorly lined impoundments that are buried onsite) that may leach into nearby surface waters. Wastewater impoundment ponds can therefore also pose a threat to wildlife and livestock.

Fracturing fluids typically include a combination of additives that serve as friction reducers, cross-linkers, breakers, surfactants, biocides, pH adjusters, scale inhibitors, and gelling agents (NYSDEC 2010). The aim of additives is to achieve an ideal viscosity that encourages fracturing of the shale and improves gas flow, but discourages microbial growth and corrosion that can inhibit recovery efficiency (US DOE 2009). Composition of the fracturing fluids can vary greatly among wells and shale formations. Specific content is often proprietary, although some states require disclosure of constituents and companies may voluntarily register the chemicals they use with regulatory agencies. A recent Congressional investigation revealed that, over a 4-year period, 14 leading gas companies used over 2500 hydrofracking products that contained 750 different chemicals, 29 of which were highly toxic or known carcinogens. Fracturing fluids used over the period totaled 780 million gallons or ~2.9 billion liters (not including dilution water), and included lead, ethylene glycol, diesel, and formaldehyde, as well as benzene, toluene, ethylbenzene, and xylene compounds (US House of Representatives Committee on Energy and Commerce 2011). The volume of fracking fluids recovered is also highly variable, but unrecovered amounts can be substantial. Only 10–30% of fracture fluids are typically recovered from wells in portions of the Marcellus shale (NYSDEC 2010); there is currently no information on the fate and transport of the unrecovered chemicals. Produced waters pose a threat to surface waters because they typically contain not only fracking additives but also elevated levels of metals, dissolved solids (eg brine), organics, and radionuclides that occur naturally in deep groundwaters. Onsite waste impoundments or evaporation ponds could overflow, spill, or leach into groundwater and contaminate nearby streams. Even after treatment, total dissolved solids (TDS) in produced waters are very high and remaining salts are often disposed of through land application or used as road salts, which are known to enter surface waters and contribute to increased stream salinization (Kaushal et al. 2005). Recovered wastewaters are most often transported offsite for deep-well injection or to a domestic wastewater treatment plant (WWTP) and/or conventional waste treatment facility. After fracturing, initially recovered flowback water is sometimes reused as fracking fluid for other wells. Reuse of recovered fluids is becoming more common, but still requires a substantial amount of fresh water because of low recovery volumes and the need to dilute flowback water containing high concentrations of chlorides, sulfates, barium, and other potentially harmful substances. Domestic WWTPs are not capable of treating the high TDS (5000 to >100 000 mg L^{-1}) typical of recovered wastewater. Many WWTPs have therefore been forced to limit their intake of recovered hydrofracking waste to remain in compliance with effluent limitations (Veil 2010). Industrial WWTPs are better equipped to treat recovered wastes using reverse osmosis, filtration, or chemical precipitation, but such facilities are costly and not widely available. Therefore, although billions of liters of produced water are being generated annually on a national scale by hydrofracking (Clark and Veil 2009), water treatment options are limited, and the potential ecological impacts of wastes on terrestrial and aquatic ecosystems are not well studied.

6.5 CHALLENGES AND POTENTIAL FOR NEW RESEARCH

Quantifying the effects of natural gas development on surface waters in shale basins is difficult because multiple companies often work in the same geographical area and use different fracturing techniques (eg varied and often proprietary composition of fracturing fluids), resulting in uncoordinated timing of infrastructure development and well fracturing. In addition, the degree to which these companies adhere to best management practices, such as buffer strips and erosion control devices, varies among companies as a result of the differing regulations among states and agencies. Furthermore, wells occur across human-impacted watersheds with characteristics that may confound our ability to attribute effects from gas-well development.

Most studies that examine the effects of sediments on biological communities focus on shifts in abundance, biomass, diversity, or community composition (Wood and Armitage 1999); few studies have analyzed how sediments alter species' roles and their interactions (but see Hazelton and Grossman 2009). In addition, contaminant effects are often assessed through single-species laboratory acute and chronic toxicity tests with standardized test organisms (eg Daphnia, fathead minnows [Pimephales promelas]; Cairns 1983) and with single contaminants. Studies are therefore needed to assess the toxicity of contaminant mixtures (eg produced water and fracking fluids) and their effects on more complex communities and ecosystems, to predict effects in the real world (Clements and Newman 2002). Sediment and contaminants associated with recovered wastewater will likely affect organism behavior and alter ecological interactions at sublethal levels (Evans-White and Lamberti 2009). Reductions in feeding efficiencies (Sandheinrich and Atchison 1989) can lead to negative effects on reproduction (Burkhead and Jelks 2001) and growth (Peckarsky 1984), and may alter the magnitude or sign (+ or –) of species' effects, causing changes in community structure. Ecologists studying the environmental effects of natural gas extraction can therefore contribute to scientific understanding by examining the effects of sediment and contaminants from natural gas development on species and community interactions.

In addition to the need for traditional bioassessments, the inevitable alteration in land use that will occur as a result of rapid and expanded drilling offers a template for conducting novel experiments in an ecosystem context. Ecosystem functions, such as decomposition rates, are affected by multiple abiotic and biotic factors, making them well-suited for detecting large-scale alterations (Bunn et al. 1999). For example, reduced streamflows, contaminants from produced wastewater and fracking fluids, and elevated sediment inputs would alter ecosystem functions, such as whole-stream metabolism, decomposition of organic matter, and accrual of macroinvertebrate biomass over time. However, it is not known how natural gas development could influence biological processing rates. The potential effects may stimulate or inhibit specific ecosystem functions. For example, excessive sedimentation or chemical contamination associated with natural-gas-well development could stimulate macroinvertebrate production by expanding habitat for tolerant, multivoltine (species that produce several broods per season) taxa (Stone and Wallace 1998) or lead to a decline in production by eliminating sensitive taxa representing a majority of community growth and/or biomass (Woodcock and Huryn 2007).

A move to incorporate ecosystem functions into mainstream biological assessment and restoration protocols is currently underway (Fritz et al. 2010), yet few studies have been conducted to inform their implementation and interpretation in the context of concurrent structural changes (Young and Collier 2009). The rapid expansion of gas development across the US could provide a framework for the implementation of concurrent structural and ecosystem experiments to inform process-based ecological assessment. Furthermore, ecological studies relating to natural gas extraction could be combined with similar studies for surface mining (Fritz et al. 2010; Bernhardt and Palmer 2011), to gain a more holistic view of the environmental costs associated with fossil-fuel extraction.

The distinct elemental composition and isotopic signatures of produced water provide unique opportunities for tracer studies that could indicate aquatic system exposure. Stable isotopes of strontium and carbon have been used to trace water from coalbed natural gas production wells to surface waters and hyporheic zones (Brinck and Frost 2007). Osborn et al. (2011) used isotopes of water, carbon, boron, and radium to test for hydraulic fracturing contamination of shallow aquifers overlying the Marcellus and Utica shale formations in Pennsylvania and New York, respectively, and found significant changes in CH_4 concentrations in drinking-water wells near locations where gas wells have been drilled. Limited research has also suggested that CH_4-derived carbon is assimilated into stream food webs (Kohzu et al. 2004; Trimmer et al. 2010). Many gas-bearing geological formations also contain elevated levels of naturally occurring radioactive materials, such as radon (^{222}Rn) and radium (^{226}Ra, ^{228}Ra), that can be used as hydrological tracers (Genereux and Hemond 1990). The extent to which metals, organics, or other contaminants from the drilling and hydrofracking process may ultimately enter aquatic and terrestrial food webs remains unknown.

6.6 CONCLUSIONS

Natural gas exploration will continue to expand globally. In addition to the potential threats to groundwater and drinking-water sources, increasing environmental stress to surface-water ecosystems is of serious concern. Scientific data are needed that will inform ecologically sound development and decision making and ensure protection of water resources. Elevated sediment runoff into streams, reductions in streamflow, contamination of streams from accidental spills, and inadequate treatment practices for recovered wastewaters are realistic threats. Gas wells are often sited close to streams, increasing the probability of harm to surface waters, and preliminary data suggest the potential for detectable effects from sedimentation. Regulations that consider proximity of natural gas development to surface waters may therefore be needed. Further ecological research on impacts from developing natural-gas-well infrastructure are sorely needed, and will inform future regulatory strategies and improve our understanding of the factors affecting community structure and ecosystem function.

REFERENCES

1. ADEQ (Arkansas Department of Environmental Quality). 2010. Water violations database. www.adeq.state.ar.us/legal/cao_info.asp. Viewed 25 Aug 2010.
2. Bernhardt ES and Palmer MA. 2011. The environmental costs of mountaintop mining valley fill operations for aquatic ecosystems of the central Appalachians. Ann NY Acad Sci 1223: 39–57.
3. Brinck EL and Frost CD. 2007. Detecting infiltration and impacts of introduced water using strontium isotopes. Ground Water 45: 554–68.
4. Bunn SE, Davies PM, and Mosisch TD. 1999. Ecosystem measures of river health and their response to riparian and catchment degradation. Freshwater Biol 41: 333–45.
5. Burkhead NM and Jelks HL. 2001. Effects of suspended sediment on the reproductive success of the tricolor shiner, a crevice-spawning minnow. T Am Fish Soc 130: 959–68.
6. Cairns JJ. 1983. Are single-species toxicity tests alone adequate for estimating environmental hazard? Hydrobiologia 100: 47–57.
7. Clark CE and Veil JA. 2009. Produced water volumes and management practices in the United States. Argonne, IL: Argonne National Laboratory, Environmental Science Division; doi:10.2172/1007397.
8. Clements W and Newman M. 2002. Community ecotoxicology. West Sussex, UK: John Wiley & Sons Ltd.
9. Colborn TC, Kwiatkowski C, Shultz K, and Backstrom M. Natural gas operations from a public health perspective. Int J Hum Ecol Risk Assess. In press.
10. DRBC (Delaware River Basin Commission). 2008. 2008 Annual Report. Trenton, NJ: DRBC.
11. EIA (Energy Information Administration). 2011. Annual 382 energy outlook. Washington, DC: EIA, US Department of Energy. DOE/EIA-0383ER(2011).

12. Evans-White MA and Lamberti GA. 2009. Direct and indirect effects of a potential aquatic contaminant on grazer–algae interactions. Environ Toxicol Chem 28: 418–26.
13. Fritz KM, Fulton S, Johnson BR, et al. 2010. Structural and functional characteristics of natural and constructed channels draining a reclaimed mountaintop removal and valley fill coal mine. J N Am Benthol Soc 29: 673–89.
14. Genereux D and Hemond H. 1990. Naturally occurring radon 222 as a tracer for streamflow generation: steady state methodology and field sample. Water Resour Res 26: 3065–75.
15. Gorham BE and Tullis JA. 2007. Final report: 2006 Arkansas land use and land cover (LULC). Fayetteville, AR: Center for Adavanced Spatial Technologies (CAST), University of Arkansas.
16. Hazelton PD and Grossman GD. 2009. Turbidity, velocity and interspecific interactions affect foraging behaviour of rosyside dace (Clinostomus funduloides) and yellowfin shiners (Notropis lutippinis). Ecol Freshw Fish 18: 427–36.
17. Heine R, Lant C, and Sengupta R. 2004. Development and comparison of approaches for automated mapping of stream channel networks. Ann Assoc of Am Geogr 94: 477–90.
18. Howarth RW, Santoro R, and Ingraffea A. 2011. Methane and the greenhouse-gas footprint of natural gas from shale formations. Climatic Change 106: 679–90.
19. Johnson N. 2011. Pennsylvania energy impacts assessment: report
20. 1: Marcellus shale natural gas and wind. Philadelphia, PA: The Nature Conservancy.
21. Kargbo DM, Wilhelm RG, and Campbell DJ. 2010. Natural gas plays in the Marcellus shale: challenges and potential opportunities. Envir Sci Tech Lib 44: 5679–84.
22. Kaushal SS, Groffman PM, Likens GE, et al. 2005. Increased salinization of fresh water in the northeastern United States. P Natl Acad Sci USA 102: 13517–20.
23. Kohzu A, Kato C, Iwata T, et al. 2004. Stream food web fueled by methane-derived carbon. Aquat Microb Ecol 36: 189–94.
24. MIT (Massachusetts Institute of Technology). 2010. The future of natural gas: an interdisciplinary MIT study. Cambridge, MA: MIT. 978-0-9828008-0-5.
25. NYSDEC (New York State Department of Environmental Conservation). 2010. Well permit issuance for horizontal drilling and high-volume hydraulic fracturing to develop the Marcellus shale and other low-permeability gas reservoirs. www.dec.ny.gov/energy/58440.html. Viewed 5 Nov 2010.
26. Osborn SG, Vengosh A, Warner NR, and Jackson RB. 2011. Methane contamination of drinking water accompanying gas-well drilling and hydraulic fracturing. P Natl Acad Sci USA 108: 8172–76.
27. PADEP (Pennsylvania Department of Environmental Protection). 2010. Violations database. www.dep.state.pa.us/dep/deputate/ minres/OILGAS/OGInspectionsViolations/OGInspviol.htm. Viewed 8 Dec 2010.
28. Peckarsky B. 1984. Do predaceous stoneflies and siltation affect structure of stream insect communities colonizing enclosures? Can J Zoolog 63: 1519–30.
29. Sandheinrich M and Atchison G. 1989. Sublethal copper effects on bluegill, Lepomis macrochirus, foraging behavior. Can J Fish Aquat Sci 46: 1977–85.
30. Stone MK and Wallace JB. 1998. Long-term recovery of a mountain stream from clearcut logging: the effects of forest succession on benthic invertebrate community structure. Freshwater Biol 39: 151–69.

31. Sutherland WJ, Bardsley S, Bennun L, et al. 2010. Horizon scan of global conservation issues for 2011. Trends Ecol Evol 26: 10–16.
32. Trimmer M, Maanoja S, Hildrew AG, et al. 2010. Potential carbon fixation via methane oxidation in well-oxygenated riverbed gravels. Limnol Oceanogr 55: 560–68.
33. US DOE (US Department of Energy). 2009. Modern shale gas development in the United States: a primer. Washington, DC: US DOE. DoE-FG26-04NT15455.
34. US EPA (US Environmental Protection Agency). 2004. Evaluation of impacts to underground sources of drinking water by hydraulic fracturing of coalbed methane reservoirs study. Washington, DC: EPA. EPA 816-F-04-017.
35. US EPA (US Environmental Protection Agency). 2006. Wadeable stream assessment: a collaborative survey of the nation's streams. Washington, DC: EPA. EPA 841-B-06-002.
36. US EPA (US Environmental Protection Agency). 2011. Draft plan to study potential impacts of hydraulic fracturing on drinking water resources. Washington, DC: EPA. EPA/600/D-11/001.
37. US House of Representatives Committee on Energy and Commerce. 2011. Chemicals used in hydraulic fracturing. http:// democrats.energycommerce.house.gov/sites/default/files/ documents/Hydraulic%20Fracturing%20Report%204.18.11.pdf. Viewed 31 Aug 2011.
38. Urbina I. 2011. Drilling down (series). New York Times. http://topics. nytimes.com/top/news/us/series/drilling_down/index.html. Viewed 27 Apr 2011.
39. Veil JA. 2010. Final report: water management technologies used by Marcellus shale gas producers. Washington, DC: US DOE. National Energy Technology Laboratory Award No FWP 49462.
40. Williams HFL, Havens DL, Banks KE, and Wachal DJ. 2008. Field-based monitoring of sediment runoff from natural gas well sites in Denton County, Texas, USA. Environ Geol 55: 1463–71.
41. Wood PJ and Armitage PD. 1999. Sediment deposition in a small lowland stream – management implications. Regul River 15: 199–210.
42. Woodcock TS and Huryn AD. 2007. The response of macroinvertebrate production to a pollution gradient in a headwater stream. Freshwater Biol 52: 177–96.
43. Young RG and Collier KJ. 2009. Contrasting responses to catchment modification among a range of functional and structural indicators of river ecosystem health. Freshwater Biol 54: 2155–70.

CHAPTER 7

SILICA GEL BEHAVIOR UNDER DIFFERENT EGS CHEMICAL AND THERMAL CONDITIONS: AN EXPERIMENTAL STUDY

JONATHAN HUNT, SOUHEIL EZZEDINE, WILLIAM BOURCIER, and SARAH ROBERTS

7.1 INTRODUCTION

In enhanced geothermal systems (EGS) the reservoir permeability is often enhanced or created using hydrofracking. In hydrofracking, high fluid pressures are applied to confined zones in the subsurface usually using packers to fracture the host rock. This enhances rock permeability and therefore conductive heat transfer to the circulating geothermal fluid (e.g. water or supercritical carbon dioxide). The ultimate goal is to increase or improve the thermal energy production from the subsurface by either optimal designs of injection and production wells or by altering the fracture permeability to create different zones of circulation that can be exploited in geothermal heat extraction. Moreover, hydrofracking can lead to the creation of undesirable short-circuits or fast flow-paths between the injection and extraction wells leading to a short thermal residence time, low heat recovery, and thus a short-life of the EGS.

A potential remedy to these problems is to deploy a cementing (blocking, diverting) agent to minimize short-cuts and/or create new circulation cells for heat extraction. A potential diverting agent is the colloidal silica by-product that can be co-produced from geothermal fluids. Silica gels are abundant in various surface and subsurface applications, yet they have not been evaluated for EGS applications. In this study we are investigating the benefits of silica gel deployment on thermal response of an EGS, either by blocking short-circuiting undesirable pathways as a result of diverting the geofluid to other fractures; or creating, within fractures, new circulation cells for harvesting heat through newly active surface area contact. A significant advantage of colloidal silica is that it can be co-produced from geothermal fluids using an inexpensive membrane-based separation technology that was developed previously using DOE-GTP funding.

This co-produced silica has properties that potentially make it useful as a fluid diversion agent for subsurface applications. Colloidal silica solutions exist as low-viscosity fluids during their "induction period" but then undergo a rapid increase in viscosity (gelation) to form a solid gel. The length of the induction period can be manipulated by varying the properties of the solution, such as silica concentration and colloid size. We believe it is possible to produce colloidal silica solutions suitable for use as diverting agents for blocking undesirable fast-paths which result in short-circuiting the EGS once hydrofracking has been deployed. In addition, the gels could be used in conventional geothermal fields to increase overall energy recovery by modifying flow.

We believe there may be additional advantages for using colloidal silica as blocking agents. It can be inexpensively produced on site or at other geothermal sites (Figure 1); it is inorganic and environmentally friendly as opposed to organic gels often used in oil/gas industry, and unlike conventional blocking agents, the gel material might be hydraulically removed after emplacement if needed. Furthermore, colloidal silica gelation can be triggered externally, for example by mixing with salt solutions or changing the pH, there should be fewer environmental restrictions and permitting requirements for its use given that the material originates in the same place it is to be injected, and silica removal in itself benefits the power plant in terms of silica scale control.

To accurately predict when an injected geothermal fluid will set up and turn to gel, a quantitative understanding of the kinetics of silica gelation is required. Although silica gelation has been studied for decades, even the most comprehensive resources (Iler, 1979; Bergna and Roberts, 2006) provide only a qualitative understanding of the various factors (pH, salt concentration, temperature, colloid diameter, SiO_2 concentration, etc.) that influence gelation times. We did not find any quantitative method or model

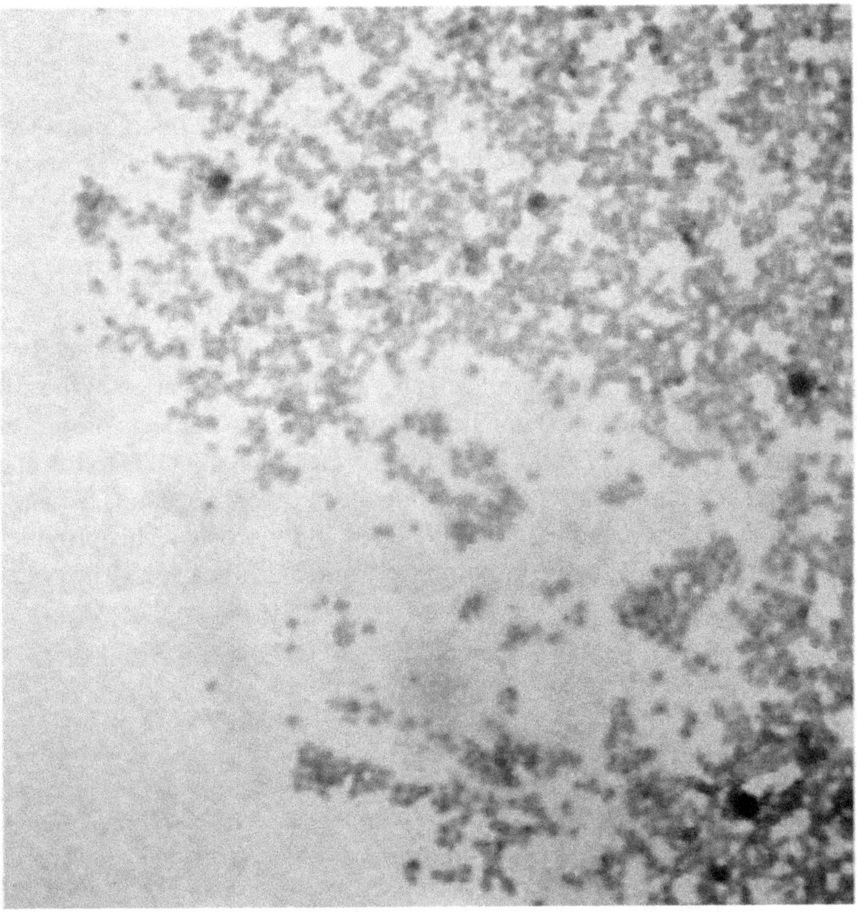

FIGURE 1: Silica gel produced from the Mammoth Lakes geothermal fluid. The gel is made up of a network of 10nm silica colloids.

for predicting gel times from colloidal silica compositions. Such a method is needed to develop formulations for colloidal silica solutions for given geothermal applications. In addition, there is very little information on gel times at elevated temperatures. Because of these needs, we began an effort to acquire additional data that could be combined with the existing data to develop a more comprehensive quantitative model for use in our geothermal application.

7.2 METHODS

For our initial experiments at 25 °C, we used a Sunshine Instruments Gel Time Meter (Figure 2). To use the Gel Time Meter, a sample is placed into a small polyethylene centrifuge tube which is in turn is set in a water bath. The temperature can be controlled within 1 °C. A stir rod is hung from a torsion wire and immersed into the sample, which will rotate when the power to the meter is turned on. Turning the power on also starts a time counter. The lower and upper electrical contacts are set 3/8" apart. The viscosity of the sample creates a drag on the rotation of the stir rod and the lower electrical contact, bringing the two contacts closer together. At a certain threshold viscosity, the electrical contacts touch and the meter and counter shut off, recording the number of seconds between turning the meter on and reaching the threshold viscosity. This provides an accurate report of the gelation time of aqueous silica sols. The water

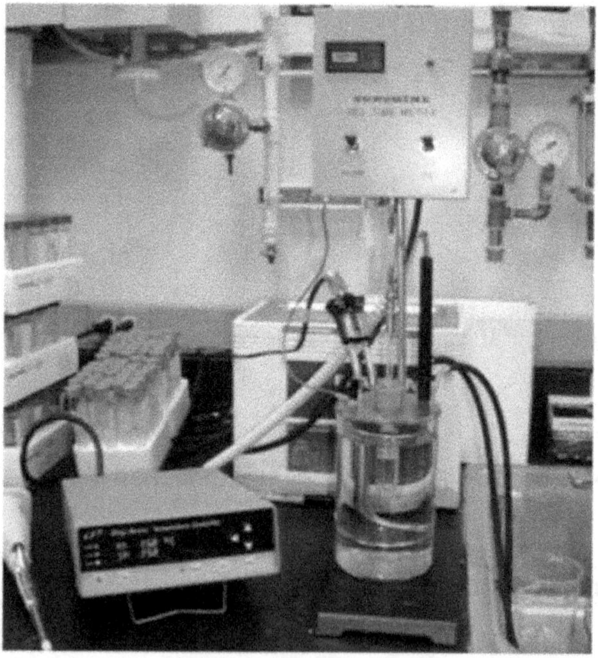

FIGURE 2: Sunshine Instruments Gel Time Meter (http://www.davis.com)

FIGURE 3: PVS Rheometer from Brookfield Engineering (http://www.brookfieldengineering.com)

bath can be used to bring the samples above ambient temperature, but because the stir rod must be lowered into the sample for the duration of the test, the water in the sample may escape due to evaporation or boiling.

The second instrument used in our experiments is a PVS rheometer from Brookfield Engineering, for experiments at elevated conditions (Figure 3). The PVS rheometer has an enclosed sample chamber, preventing sample boil-off. It can be operated up to ~200 °C and 1000 psi, and can provide an accurate, quantitative measure of sample viscosity as a function of time and/or shear rate, by measuring the torque exerted on an inner cylinder inside the sample chamber. The gel time meter provided only the time to gel formation and no viscosity vs. time data. The instrument enables measurement of both simple gelation times as well as non-Newtonian behavior, e.g., thixotropy, of colloidal silica sols and gels.

7.3 RESULTS AND DISCUSSION

7.3.1 GEL TIME METER, 25 °C

The gel time meter was used to quantitatively determine the effects of SiO_2 concentration, pH, and salt concentration on the gelation time of colloidal silica sols made with commercially available LUDOX SM-30 at 25 °C, to provide a foundation for experiments at higher temperature. The major finding of this investigation is that there is a simple and clear relationship between gelation time and SiO_2 concentration. Previous investigations varied SiO_2 concentration while keeping pH and NaCl concentration constant. However, if the dilution is done by adding different amounts of water to identical mixtures of colloidal silica sol, NaCl, and HCl, a linear

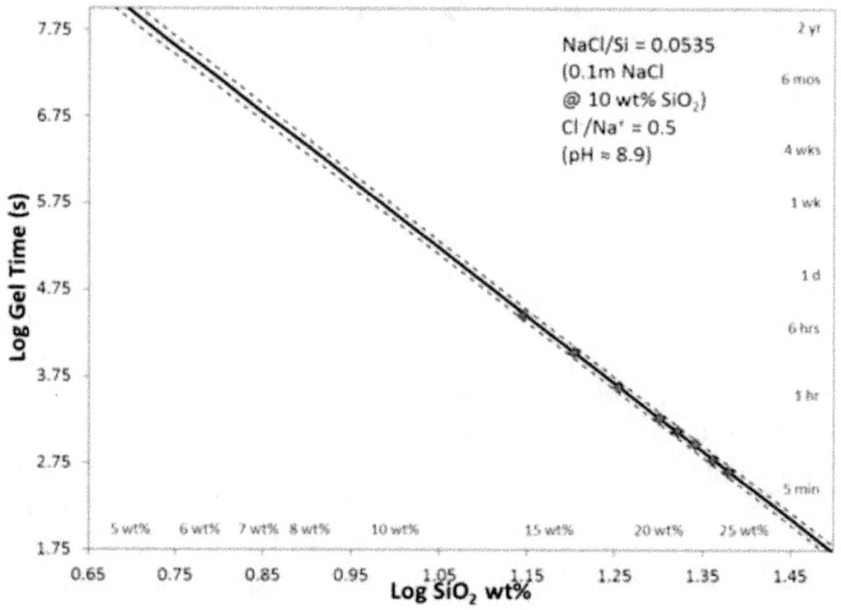

FIGURE 4: Log gel time vs. log silica concentration. Dashed lines are 95% confidence intervals over the entire model.

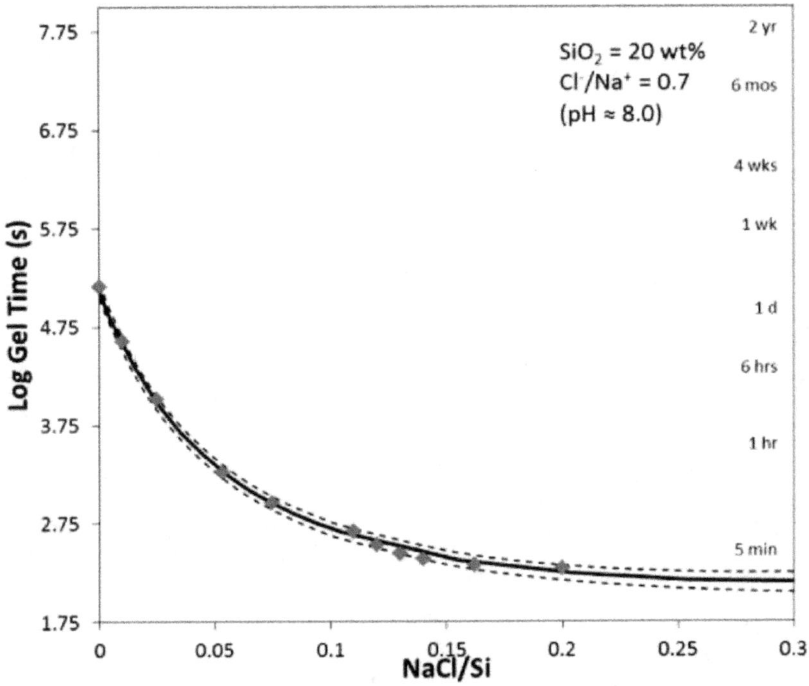

FIGURE 5: Log gel time vs. NaCl/silica ratio (NaCl created from the neutralization of Na^+ stabilizer with HCl is not included in this ratio). Dashed lines are 95% confidence intervals over the entire model.

relationship arises between the logarithm of gelation time and the logarithm of the silica concentration (Figure 4). While the NaCl concentration and pH are changing as the mixtures become more diluted, the molar ratio of NaCl to SiO_2 and the molar ratio of added HCl to the N^{a+} stabilizer present in the colloidal sol (hereafter referred to as universal neutralization ratio or UN ratio) remains constant.

The relationship between gelation time and the molar NaCl/SiO_2 ratio is more complex. Increasing the amount of salt while keeping all other variables constant initially has a large, decreasing effect on gelation time, but as more salt is added, the effect is diminished. However, adding salt will always decrease the gelation time. This has been modeled as a modified hyperbolic relationship (Figure 5).

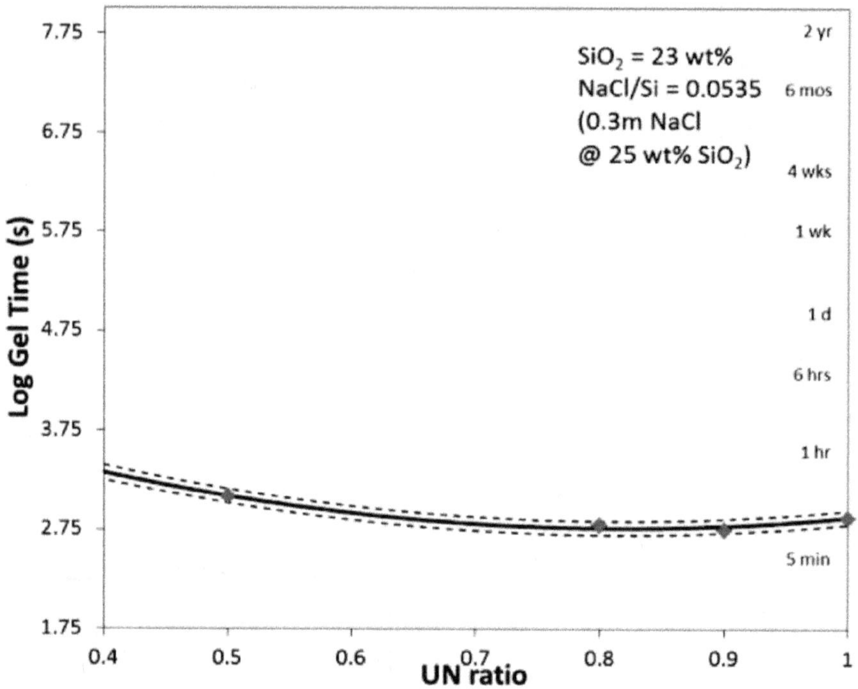

FIGURE 6: Log gel time vs. HCl/Na+ ratio (added NaCl is not included in this ratio). Dashed lines are 95% confidence intervals over the entire model.

The relationship between gelation time and UN ratio is more complex still, as there are competing ratelimiting steps at low pH and high pH. Nevertheless, at near-neutral pH (6-9; UN ratio = 0.5-1), the relationship can be modeled as a parabola (Figure 6).

The entire dataset (91 experiments) was fitted to a single equation with 11 regression parameters, as a function of SiO_2 wt%, molar NaCl/SiO2 ratio, and UN ratio. The dashed lines in each of Figures 4-6 represent the 2 standard deviation error for the entire regression. The regression equation ($R^2 = 0.9973$) is as follows:

$$\log(\textit{Gel Time (s)}) = A \times \log(SiO_2\ wt\%) + B \quad (1)$$

where A and B are defined as follows:

$$A = \frac{NaCl}{SiO_2}(A_1UN^2 + A_2UN + A_3) + A_4UN^2 + +A_5 \quad (2)$$

$$B = \left(\frac{NaCl}{SiO_2} + B_1\right)^{-1}[B_2UN^2 + B_3] + B_4\frac{NaCl}{SiO_2} + B_5UN + B_6 \quad (3)$$

The numerical values of A_1-A_5 and B_1-B_6 are given in Table 1, and the error between prediction and measurement for each experiment is shown in Figure 7. No systematic error was observed over SiO_2 concentration, UN ratio, NaCl/SiO_2 ratio, or observed gelation time. While most of the parameters were fitted by linear regression, it was not possible to fit parameter B_1 linearly. It was therefore fitted with non-linear regression (iterative use of Excel's solver and linear regression functions) to maximize the value of R^2, and therefore does not have a well-defined error.

7.3.2 PVS RHEOMETER

Preliminary results from the rheometer indicate that gelation times can be reliably obtained from plots of viscosity over time, and that shear rate does not affect gelation time. An example of a viscosity over time plot is shown on Figure 8.

The data in Figure 8 are for a colloidal silica solution with 17 wt% SiO_2, a NaCl/SiO_2 ratio of 0.11, and a universal neutralization (HCl/Na^+) ratio of 0.5, subjected to a continuous shear rate of $5s^{-1}$. The viscosity remained near that for water for approximately 1600 seconds, which agrees well with the gelation time of 1634 seconds recorded by the gel time meter for the same colloidal silica solution. During the gelation process, the colloidal solution starts thickening, the viscosity increases significantly, and reaches a maximum of 9000 cP at around 2400 seconds. After this point, the gel starts to slip past the inner cylinder rather than exerting a constant torque on it, causing the measured viscosity to drop and become irregular. However, the gel itself continues to stiffen and become brittle. Therefore, the information gathered after the maximum viscosity reached is irrelevant to this investigation.

TABLE 1: Fitting Parameters for Equation 1.

Parameter	Value	Error (1 σ)
A_1	-15.24	1.729
A_2	26.08	2.585
A_3	4.58	1.474
A_4	5.40	0.202
A_5	-11.25	0.184
B_1	0.066	N/A
B_2	-0.356	0.014
B_3	0.461	0.010
B_4	-18.03	1.415
B_5	-7.21	0.313
B_6	17.02	0.293

This experiment was repeated using several different shear rates. The measured gelation time for each shear rate was within 2-3 minutes of the measured gelation time in the gel time meter; however, the maximum viscosity was dependent on the shear rate. If the shear rate doubles, the maximum viscosity recorded will be halved. This is entirely due to the interfacial layer of gel/water at the inner cylinder, and not reflective of the actual viscosity of the bulk gel as a whole. Therefore, at least in this experiment geometry, shear rate is not a significant variable when determining the gelation time.

We have started investigating the effect of temperature on gelation time using the rheometer. It is well-known that gelation time follows an Arrhenius relationship, but the values for the activation energy associated with the relationship vary within the literature, between 5 and 20 kcal/mol. Preliminary results fall nearly in the middle of this range, at 12.8 kcal/mol.

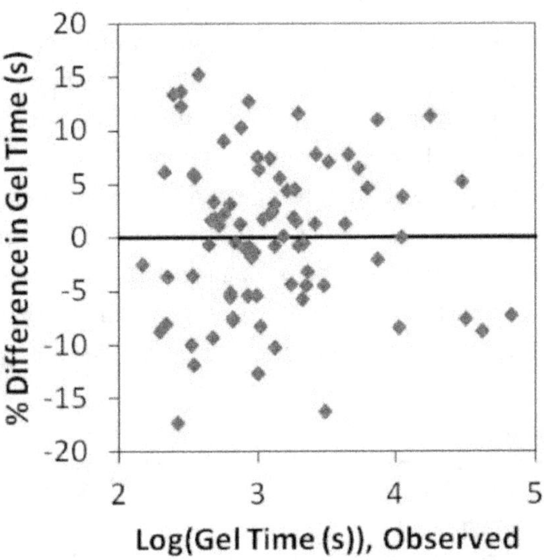

FIGURE 7: % Difference in Gel Time (s) between Predicted and Observed values vs. Observed Log(Gel Time).

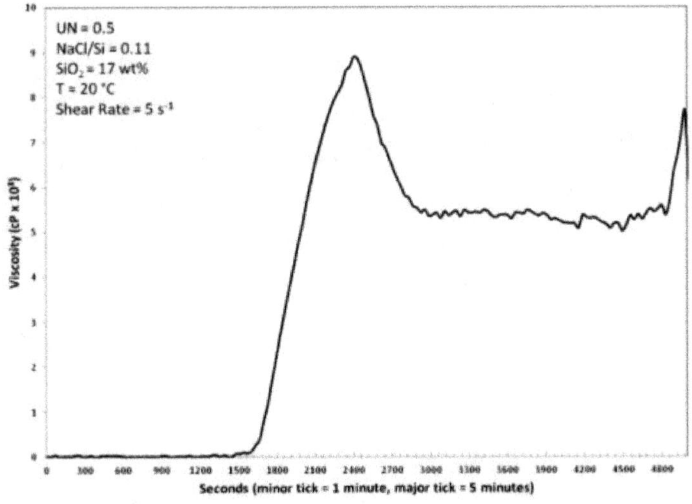

FIGURE 8: Viscosity (cP x 10^3) as a function of time (seconds), measured at room temperature at a constant shear rate of 5 s^{-1}.

Figure 9 shows gel times for a single solution composition measured at 25 and 75 °C (corrected to 63.5 °C due to temperature ramping time), with curves representing 5, 12.5 (bolded), and 20 kcal/mol activation energies.

It should be noted that the activation energy reported here is not the true activation energy. Iler (1979) reports that as colloidal solutions are heated, the colloids grow in size to an equilibrium value. The size of the colloids will affect the gelation time independent of the temperature. The true activation energy can only be measured for a colloidal solution with colloids larger than the ultimate equilibrium value in the experiment. Nevertheless, EGS applications would only require the total effect

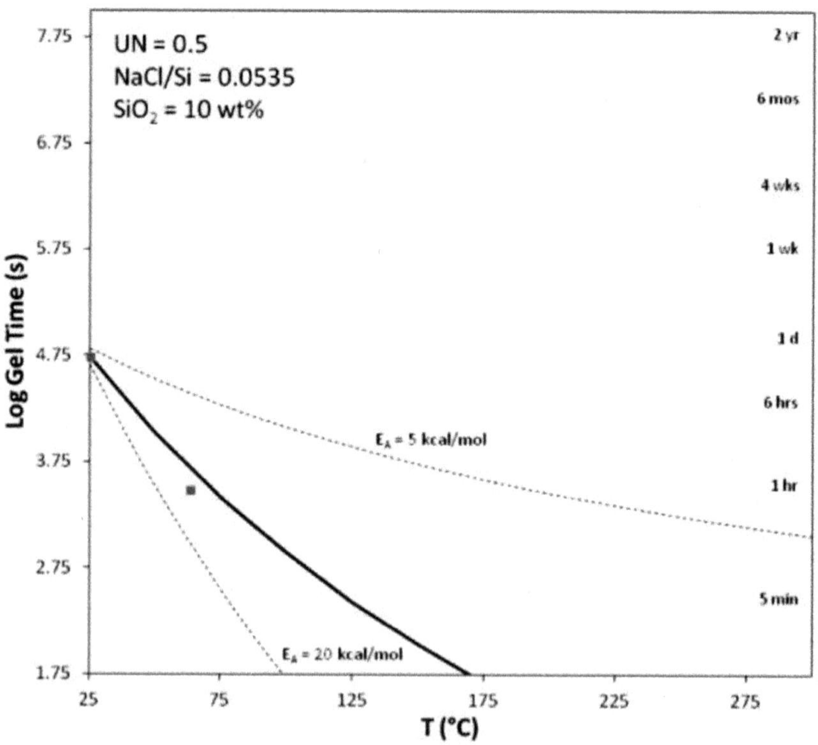

FIGURE 9: Arrhenius relationship of gel time vs. temperature (°C). Bold curve corresponds to 12.5 kcal/mol.

of temperature on a given initial colloidal solution composition, which is what is reported here.

One issue that has arisen in the course of our investigation is that in some of our two-phase high temperature tests, the silica precipitates on the walls of the sample chamber as scale, rather than forming a continuous gel. It may be that at these temperatures, the kinetics of amorphous silica deposition is faster than the kinetics of gelation. This may hinder usage of colloidal silica gels in some high-T environments, however, coatings of silica deposited from concentrated colloidal silica solutions onto fracture walls can also be used to modify subsurface flow fields, if not as completely as gelation. Colloidal solutions provide a method for transporting relatively large amounts of silica to the subsurface, much larger than is possible with saturated solutions of monomeric silica.

The model outlined by Eq. 1-3, especially with the observation that the dependence of the logarithm of the gelation time on the logarithm of silica concentration is linear, can be used to predict gelation times of compositions that would take a prohibitively long time to gel at ambient conditions. Because of the Arrhenius relationship between gelation time and temperature, such a composition is necessary to provide a reasonable gelation time at EGS conditions.

7.4 CONCLUSIONS

We have made significant progress in the determination of a new candidate agent for blocking, diverting, or modifying fracture flow networks in EGS systems. Although much work at high temperature still needs to be done, results at 25 °C indicate that it may be possible to choose formulations that will gel in a reasonable and predictable amount of time at the temperatures of EGS systems. Results from this study are informing concurrent work on modeling fluid flow in rough fractures. The modeling work, illustrated in Figure 10 and described in more detail by Ezzedine et al., is proof-of-concept work of the positive impact of silica gel deployment.

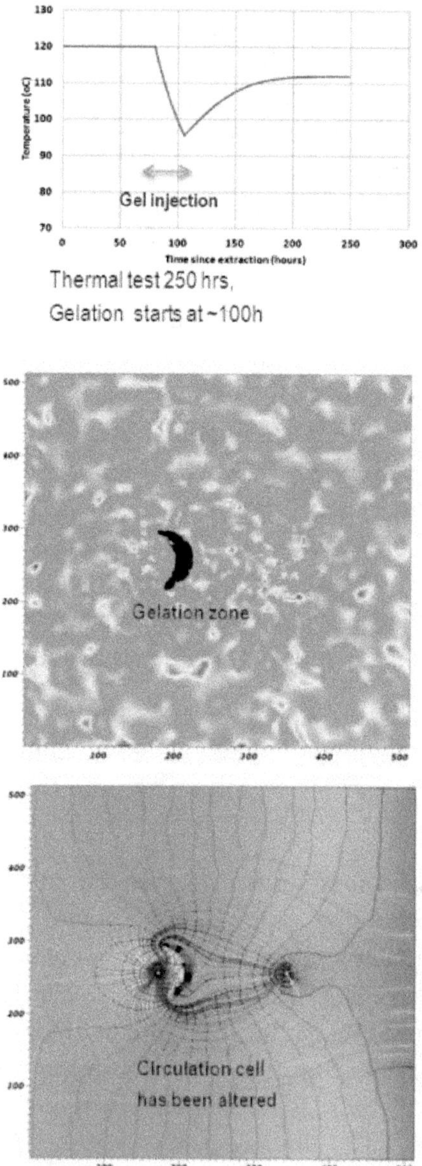

FIGURE 10: Two dimensional doublet flow and transport simulations of gel in a single rough fracture. Top: Temperature response at production well as a function of time. Center: Aperture field with gelled silica diverter. Bottom: Flow network showing the impact of silica gel on streamlines and flow cells.

REFERENCES

1. Iler R. K., 1979. The Chemistry of Silica. Wiley.
2. Bergna H. E., and Roberts W. O., 2006. Colloidal Silica Fundamentals and Applications. Taylor and Francis.
3. Ezzedine, S. M., Hunt, J. D., Bourcier, W., and Roberts, S. K. Impact of Silica Gel Deployment on Subsurface Flow and Heat Extraction from Enhanced Geothermal Systems, Annual Stanford Workshop on Geothermal Reservoir Engineering, Stanford, Feb 2012.

Hunt, J. D., Ezzedine, S., Bourgier, W., Roberts, S. and Roberts, J. Silica Gel Behavior Under Different EGS Chemical and Thermal Conditions: An Experimental Study. In: Proceedings from The Thirty-Seventh Workshop on Geothermal Reservoir Engineering, January 20- February 1, 2012; Stanford, CA: Stanford University; 2012: SGD-TR-194.

CHAPTER 8

STUDY OF THE POTENTIAL IMPACTS OF HYDRAULIC FRACTURING ON DRINKING WATER RESOURCES: PROGRESS REPORT

U. S. ENVIRONMENTAL PROTECTION AGENCY

8.1 SCENARIO EVALUATIONS

The objective of this approach is to use computer models to explore hypothetical scenarios across the hydraulic fracturing water cycle. The models include models of generic engineering andgeological scenarios and, where sufficient data are available, models of site-specific or region-specific characteristics. This chapter includes progress reports for the following projects:

8.1.1 SUBSURFACE MIGRATION MODELING

Lawrence Berkeley National Laboratory (LBNL), in consultation with the EPA, will simulate the hypothetical subsurface migration of fluids (including gases) resulting from six possible mechanisms using computer models. .

8.1.1.1 PROJECT INTRODUCTION

Stakeholders have expressed concerns about hydraulic fracturing endangering subsurface drinking water resources by creating high permeability transport pathways that allow hydrocarbons andother fluids to escape from hydrocarbon-bearing formations (US EPA, 2010b, d, e, f, g). Experts continue to debate the extent to which subsurface pathways could cause significant adverse consequences for ground water resources (Davies, 2011; Engelder, 2012; Harrison, 1983, 1985; Jackson et al., 2011; Myers, 2012a, b; Osborn et al., 2011; Warner et al., 2012). The segment of the population that receives drinking water from private wells may be especially vulnerable to health impacts from impaired drinking water. Unlike water distributed by public water systems, water from private drinking water wells is not subject to National Primary Drinking Water Regulations, and water quality testing is at the discretion of the well owner.

Lawrence Berkeley National Laboratory, in coordination with the EPA, is using numerical simulations to investigate six possible mechanisms that could lead to upward migration of fluids, including gases, from a shale gas reservoir and the conditions under which such hypothetical scenarios may be possible. The possible mechanisms include:

- Scenario A (Figure 1): Defective or insufficient well construction coupled with excessive pressure during hydraulic fracturing operations results in damage to well integrity duringthe stimulation process. A migration pathway is then established through which fluids could travel through the cement or area near the wellbore into overlying aquifers. In this scenario, the overburden is not necessarily fractured.
- Scenario B1 (Figure 2): Fracturing of the overburden because inadequate design of the hydraulic fracturing operation results in fractures allowing fluid communication, either directly or indirectly, between shale gas reservoirs and aquifers above them. Indirect communication would occur if fractures intercept a permeable formation between the shale gas formation and the aquifer. Generally, the aquifer would be located at a more shallow depth than the permeable formation.
- Scenario B2 (Figure 3): Similar to Scenario B1, fracturing of the overburden allowsindirect fluid communication between the shale gas reservoir and the aquifers after intercepting conventional hydrocarbon reservoirs, which may create a dual source of contamination for the aquifer.
- Scenario C (Figure 4): Sealed/dormant fractures and faults are activated by the hydraulicfracturing operation, creating pathways for upward migration of hydrocarbons and other contaminants.
- Scenario D1 (Figure 5): Fracturing of the overburden creates pathways for movement of hydrocarbons and other contaminants into offset wells (or

their vicinity) in conventionalreservoirs with deteriorating cement. The offset wells may intersect and communicate with aquifers, and inadequate or failing completions/cement can create pathways forcontaminants to reach the ground water aquifer.
- Scenario D2 (Figure 6): Similar to Scenario D1, fracturing of the overburden results in movement of hydrocarbons and other contaminants into improperly closed offset wells (or their vicinity) with compromised casing in conventional reservoirs. The offset well could provide a low-resistance pathway connecting the shale gas reservoir with the ground water aquifer.

The research focuses on hypothetical causes of failure related to fluid pressure/flow andgeomechanics (as related to operational and geological conditions and properties), and does not extend to investigations of strength of casing and tubing materials (an area that falls within the confines of mechanical engineering). Damage to the well casing due to corrosive reservoir fluids was one other scenario originally considered. Corrosion modeling requires a detailed chemical engineering analysis that is beyond the scope of this project, which focuses on geophysical and mechanical scenarios, so it is not a scenario pursued for this project. Additionally, hypothetical scenarios that would cause failure of well structural integrity (e.g., joint splits) are an issue beyond the scope of this study, as they involve material quality and integrity, issues not unique to hydraulic fracturing.

8.1.1.2 RESEARCH APPROACH

Objectives of the subsurface migration scenario evaluation research project include:

- Determining whether the hypothetical migration mechanisms shown in Figures 14 through 19 are physically and geomechanically possible during field operations of hydraulic fracturing and, if so, identifying the range of conditions under which fluid migration is possible.
- Exploring how contaminant type, fluid pressure, and local geologic properties control hypothetical migration mechanisms and affect the possible emergence of contaminants inan aquifer.
- Conducting a thorough analysis of sensitivity to the various factors affecting contaminant transport.

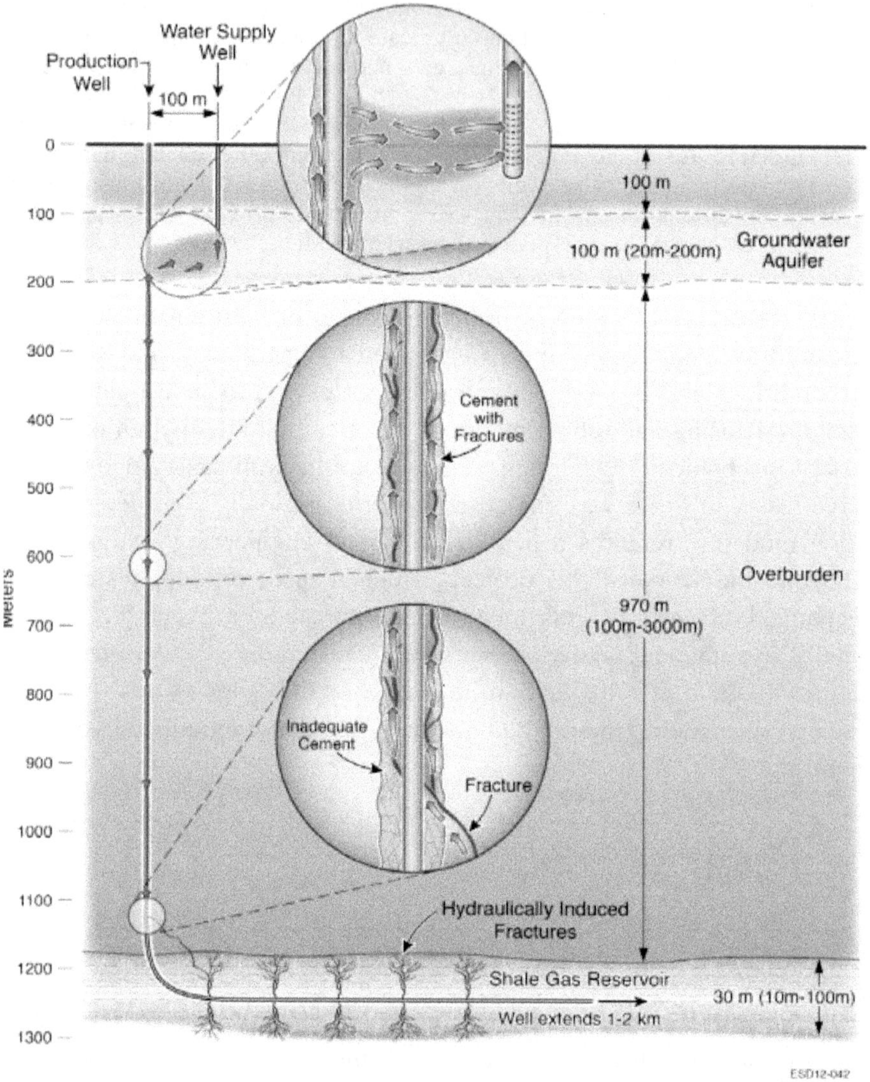

FIGURE 1: Scenario A of the subsurface migration modeling project. This scenario simulates a hypothetical migration pathway that occurs when a defective or insufficiently constructed well is damaged during excessive pressure from hydraulic fracturing operations. A migration pathway is established through which fluids could travel through the cement or area near the wellbore into overlying ground water aquifers.

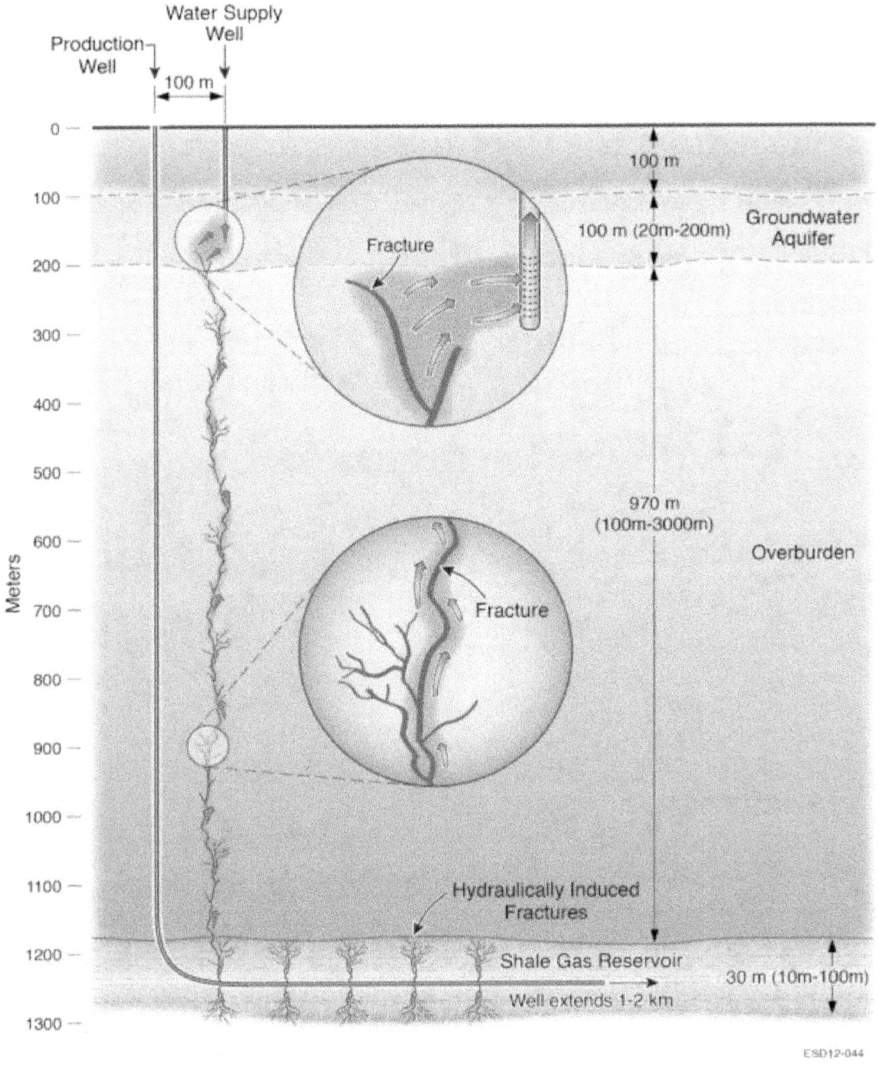

FIGURE 2: Scenario B1 of the subsurface migration modeling project. This hypothetical scenario simulates fluid communication, either directly or indirectly, between shale gas reservoirs and ground water aquifers as a result of the hydraulic fracturing design creating fractures in the overburden.

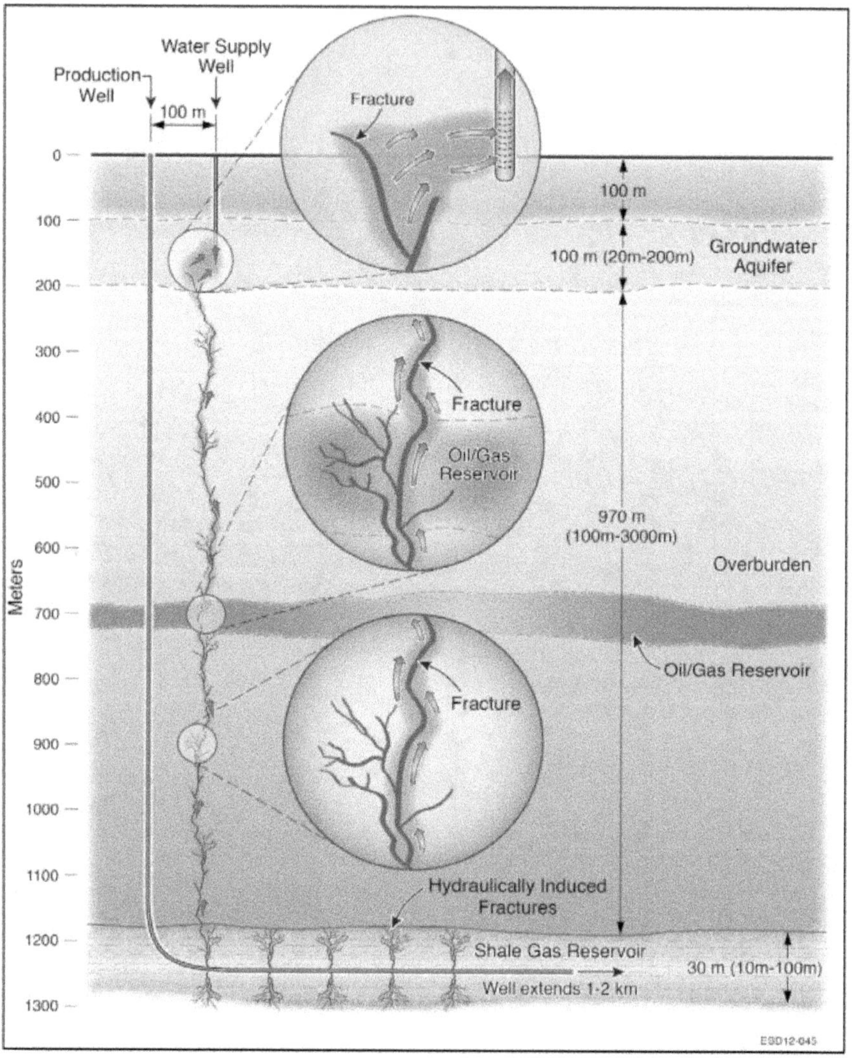

FIGURE 3: Scenario B2 of the subsurface migration modeling project. Similar to B1, this hypothetical scenario simulates fluid communication, either directly or indirectly, between shale gas reservoirs and ground water aquifers as a result of the hydraulic fracturing design creating fractures in the overburden. The fractures intercept a conventional oil/gas reservoir before communicating with the ground water aquifer, which may create a dual source of contamination in the aquifer.

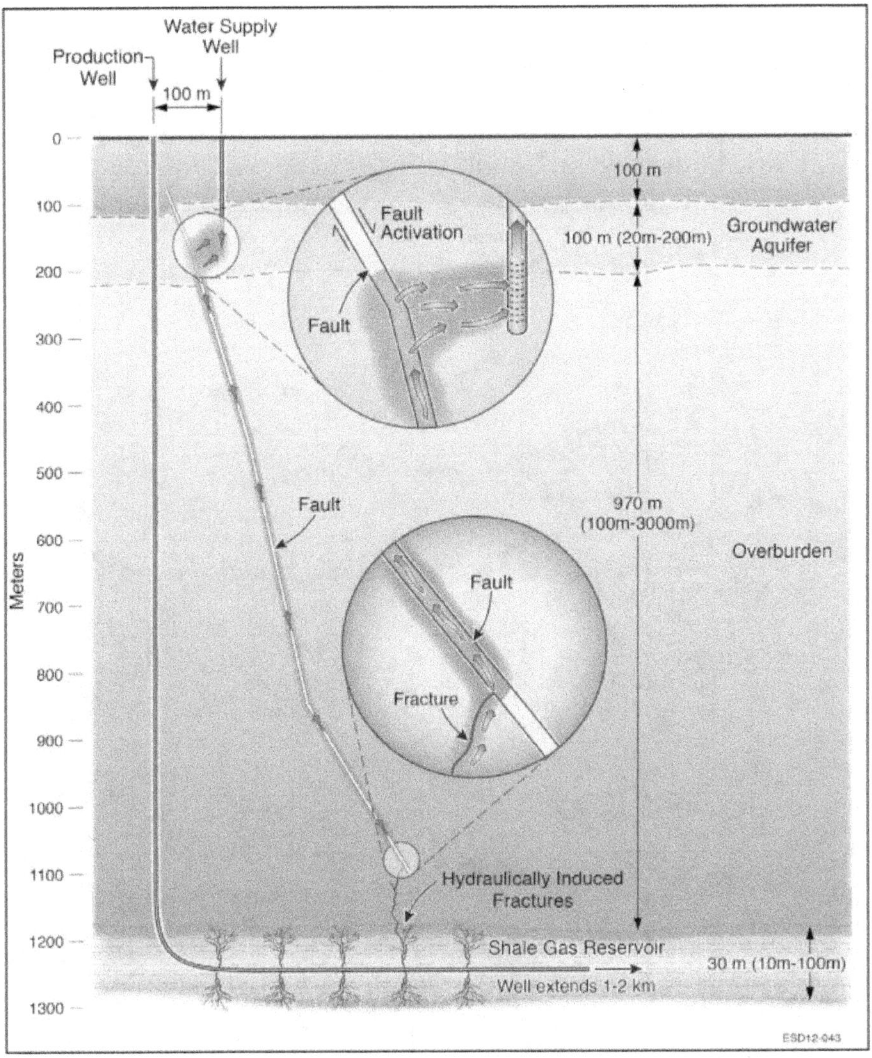

FIGURE 4: Scenario C of the subsurface migration modeling project. This hypothetical scenario simulates upward migration of hydrocarbons and other contaminants through sealed/dormant fractures and faults activated by the hydraulic fracturing operation.

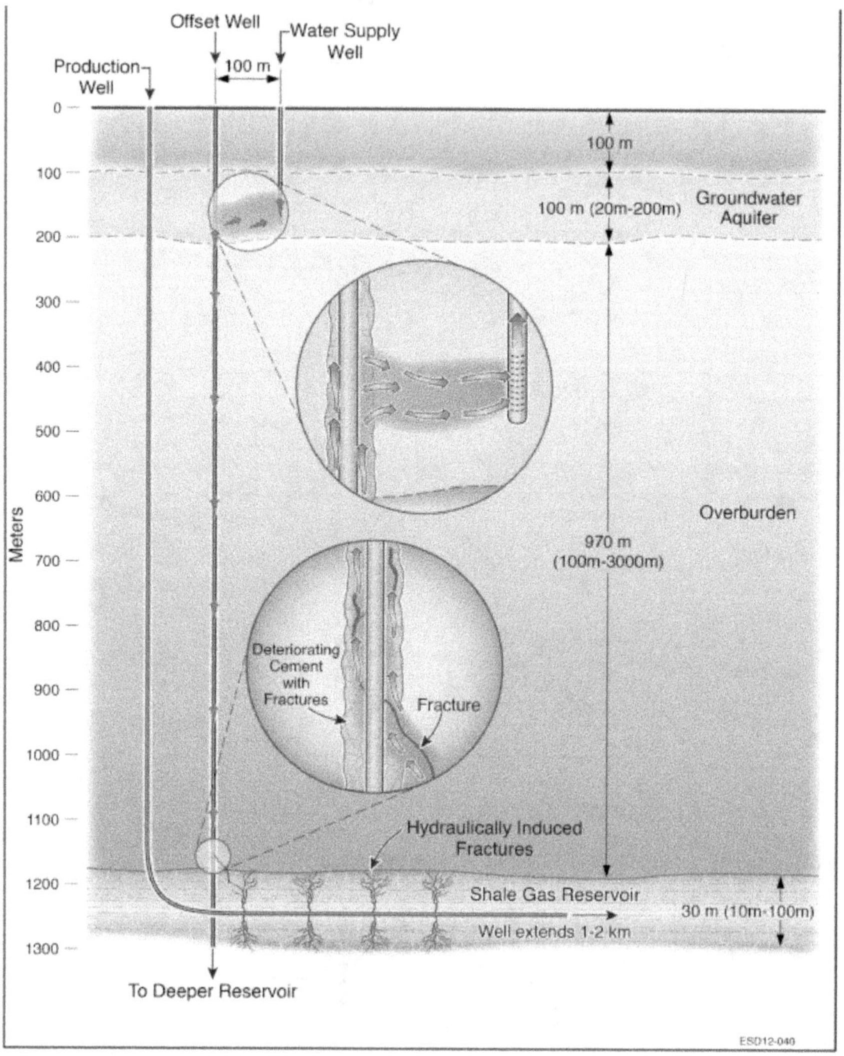

FIGURE 5: Scenario D1 of the subsurface migration modeling project. This hypothetical scenario simulates movement of hydrocarbons and other contaminants into offset wells in conventional oil/gas reservoirs with deteriorating cement due to fracturing of the overburden. The offset wells may intersect and communicate with aquifers, and inadequate or failing completions/cement can create pathways for contaminants to reach ground water aquifers.

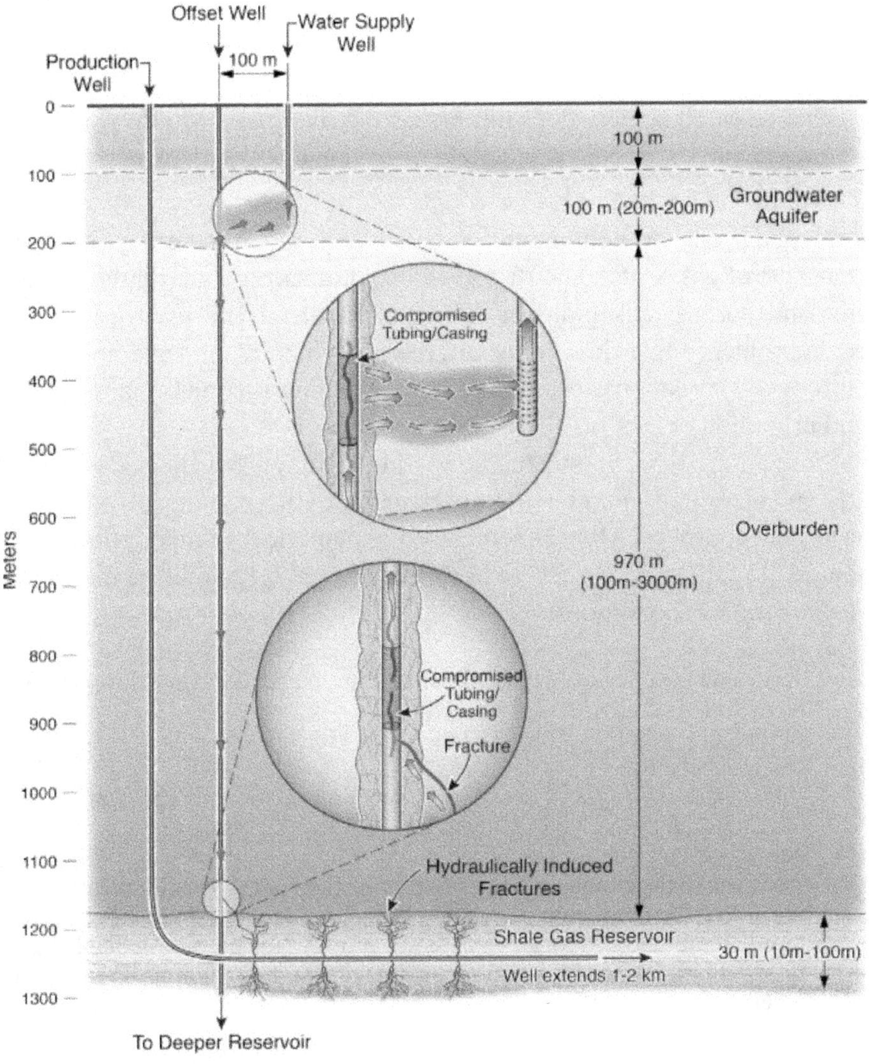

FIGURE 6: Scenario D2 of the subsurface migration modeling project. Similar to Scenario D1, this hypothetical scenario simulates movement of hydrocarbons and other contaminants into offset wells in conventional oil/gas reservoirs due to fracturing of the overburden. The offset wells in Scenario D2 are improperly closed with compromised casing, which provides a low-resistance pathway connecting the shale gas reservoir with the ground water aquifer

- Assessing the potential impacts on drinking water resources in cases of fluid migration.

This research project does not assess the likelihood of a hypothetical scenario occurring during actual field operations.

Computational Codes.

The LBNL selected computational codes able to simulate the flow andtransport of gas, water, and dissolved contaminants concurrently in fractures and porous rock matrices. The numerical models used in this research project couple flow, transport, thermodynamics, and geomechanics to produce simulations to promote understanding ofconditions in which fluid migration occurs.

Simulations of contaminant flow and migration began in December 2011 and identified a number of important issues that significantly affected the project approach. More specifically, the numerical simulator needed to include the following processes in order to accurately describe the hypothetical scenario conditions:

- Darcy and non-Darcy (Forchheimer or Barree and Conway) flow through the matrix andfractures of fractured media
- Inertial and turbulent effects (Klinkenberg effects)
- Real gas behavior
- Multi-phase flow (gas, aqueous, and potentially an organic phase of immiscible substancesinvolved in the hydraulic fracturing process)
- Density-driven flow
- Mechanical dispersion, in addition to advection and molecular diffusion
- Sorption (primary and secondary) of ions introduced in hydraulic fracturing-relatedprocesses and gases onto the grains of the porous media, involving one of three possible sorption models (linear, Langmuir, or Freundlich) under equilibrium or kinetic conditions

Thermal differentials between ground water and shale gas reservoirs are substantial and may significantly impact contaminant transport processes. Thus, the simulator needed to be able toaccount for the following processes in order to fully describe the physics of the problem:

Coupled flow and thermal effects, which affect fluid viscosity, density, and buoyancy and, consequently, the rate of migration.

Effect of temperature on solubility. Lower temperatures can lead to supersaturation ofdissolved gases or dissolved solids. The latter can result in halite formation stemming from salt precipitation, caused by lower temperatures and pressures as naturallyoccurring brines ascend toward the ground water. Halite precipitation can have a pronounced effect on both the specific fractures and the overall matrix permeability.

There is currently no single numerical model that includes all of these processes. Thus, the LBNL chose the Transport of Unsaturated Groundwater and Heat (TOUGH) family of codes (Moridis et al., 2008) in combination with existing modules to create a model that better simulates the subsurface flow and geomechanical conditions encountered in the migrationscenarios.

The TOUGH+ code includes equation-of-state modules that describe the non-isothermal flow of real gas mixtures, water, and solutes through fractured porous media and accounts for all processesinvolved in flow through tight and shale gas reservoirs (i.e., gas-specific Knudsen diffusion, gas and solute sorption onto the media, non-Darcy flow, salt precipitation as temperature and pressure drop in the ascending reservoir, etc.) (Freeman, 2010; Freeman et al., 2011; Freeman et al., 2009a,b; Freeman et al., 2012; Moridis et al., 2010; Olorode, 2011). The LBNL paired relevant modules with TOUGH+ code: one code, TOUGH+$R_{gas}H_2O_{Cont}$ (Moridis and Freeman, 2012), addresses the physics and chemistry of flow and transport of heat, water, gases, and dissolved contaminants in porous/fractured media; a second code, TOUGH+$R_{gas}H_2O$ (Moridis and Webb, 2012), describes the coupled flow of a gas mixture and water and the transport of heat; a third code, TOUGH+R_{gas} (Moridis and Webb, 2012), is limited to the coupled flow of a real gas mixture and heat in geologicmedia.

A geomechanical model, ROCMECH, was also coupled with the TOUGH+ code and modules and describes the interdependence of flow and geomechanics including fracture growth andpropagation (Kim and Moridis, 2012a, b, c, d, e). The ROCMECH code is designed for the rigorous analysis of either pure geomechanical problems or, when fully coupled with the TOUGH+ multi-phase, multi-component, non-isothermal code, for the simulation of the coupled flow andgeomechanical system behavior in porous and fractured media, including activation of faults and fractures. The coupled TOUGH+ ROCMECH codes allow the investigation of fracture growth during fluid injection of water (after their initial development during hydraulic fracturing) using fully dynamically coupled flow and geomechanics and were used in a series of fracture propagation studies (Kim and Moridis, 2012a, b, c, d, e). The ROCMECH code developed by the LBNL for thisstudy includes capabilities to describe both tensile and shear failure based on the Mohr-Coulomb model, multiple porosity concepts, non-isothermal behavior, and transverse leak-off (Kim andMoridis, 2012a).

Input Data.

Input data supporting the simulations are being estimated using information from the technical literature, data supplied by the EPA, and expert judgment. Input data include:

- Site stratigraphy
- Rock properties (grain density, intrinsic matrix permeability, permeability of natural fracture network, matrix and fracture porosity, fracture spacing and aperture)
- Initial formation conditions (fracture and matrix saturation, pressures)
- Gas composition
- Pore water composition
- Gas adsorption isotherm
- Thermal conductivity and specific heat of rocks
- Parameters for relative permeability
- Hydraulic fracturing pressure
- Number of hydraulic fracturing stages
- Injected volumes
- Pressure evolution during injection
- Volumes of fracturing fluid recovered

Uncertainty in the data will be addressed by first analyzing base cases that involve reasonable estimates of the various parameters and conditions and then conducting sensitivity analyses that cover (and extend beyond) the possible range of expected values of all relevant parameters.

8.1.1.3 STATUS AND PRELIMINARY DATA

The subsurface migration modeling project is proceeding along two main tracks. The first addresses the geomechanical reality of the mechanisms and seeks to determine whether it is physically possible (as determined and constrained by the laws of physics and the operational quantities and limitations involved in hydraulic fracturing operations) for the six migration mechanisms (Scenarios A to D2) to occur. The second axis focuses on contaminant transport, assuming that a subsurface migration has occurred as described in the six scenarios, and attempts to determine a timeframe for contaminants (liquid or gas phase) escaping from a shale gas reservoir to reach the ground water aquifer.

Analysis of Consequences of Geomechanical Wellbore Failure (Scenario A).
A large database of relevant publications has been assembled, and several important well design parameters and hydraulic fracturing operational conditions have been identified as a foundation for the simulation. Two pathways for migration have been considered using TOUGH+$R_{gas}H_2O_{Cont}$: cement separation from the outer casing or a fracture pattern affecting the entire cement, from the producing formation to the point where the well intercepts the ground water formation.

A separate geomechanical study using TOUGH+RealGasH_2O and ROCMECH will also assess the feasibility of either a fracture developing in weak cement around a wellbore or a cement-wellbore separation during the hydraulic fracturing process. The numerical simulation of the fracture propagation considered fracture development in the cement near the "heel" of a horizontal well during stimulation immediately after creation of the first fracture using varied geomechanicalproperties of gas-bearing shales. The work also involves sensitivity analyses of factors that are known to be important, as well as those that appear to have secondary effects (for completeness). Recent activities have focused mainly on such sensitivity analyses.

Analysis of the Consequences of Induced Fractures Reaching Ground Water Resources and after Intercepting Conventional Reservoirs (Scenarios B1 and B2).
A high-definition geomechanical study, involving a complex fracture propagation model that incorporates realistic data and parameters (as-gleaned from the literature and discussions with industry practitioners) was completed. A sensitivity analysis of the fracture propagation to the most important geomechanical properties and conditions is partially completed and will be included in the final publication.

Simulations of gas and contaminant migration from the shale gas reservoir through fractures into ground water are also in progress. The simulation domain is subdivided up to 300,000 elements and up to 1.2 million equations, which requires very long execution times that can range from several days to weeks. Work continues to streamline the processing of the simulation to significantly reduce the execution time requirements.

Scoping calculations are in development to provide time estimates for the migration of gas and dissolved contaminants from the shale gas reservoir to the drinking water resource through aconnecting fracture. As illustrated in Figure 15, the simulated system is composed of a 100-meterthick aquifer (from 100 to 200 meters below the surface), a fracture extending from the bottom of the gas reservoir at 1,200 meters below surface to the base of the aquifer, which is 1,000 metersabove the gas reservoir. These scoping studies indicated that the most important parameters and conditions were the permeability of the gas reservoir (matrix), the fracture permeability, the distance between the aquifer and the shale reservoir, and the pressure regimes in the aquifer and the shale. Results from this work are being analyzed and will be published when complete.

Analysis of Consequences of Activation of Native Faults and Fractures (Scenario C)

The simulation conditions for the analysis of contaminant transport through native fractures and faults in response to the stimulation process have been determined, and the variations used to conduct a sensitivity analysis are being developed.

A geomechanical study using the TOUGH-FLAC44 simulator began in March 2012 to investigate the possibility that hydraulic fracturing injections may create a pathway for transport through fault reactivation. The simulation input represents the conditions in the Marcellus Shale. Scopingcalculations were developed to study the potential for injection-induced fault reactivation associated with shale gas hydraulic fracturing operations. From these scoping calculations, theLBNL simulation results suggest that the hydraulic fracturing stimulation, under conditions reported in published literature, does not appear to activate fault rupture lengths greater than 40 to 50 meters and could only give rise to microseismicity (magnitude <1), which is consistent withwhat has been observed in the field (NAS, 2012). Therefore, preliminary simulations suggest that the possibility of fault reactivation creating a pathway to shallow ground water resources is remote. A more detailed analysis to better resolve local conditions and mechanical response at the injection point is underway and a manuscript is in development (Rutqvist et al., 2012).

Analysis of the Consequences of Induced Fractures Intercepting Offset Unplugged Wells (Scenarios D1 and D2).

A geomechanical study is in progress to assess the feasibility of a fracture extending through the shale gas reservoir into the weak/fractured cement around, or the unplugged wellbore of offset wells (Figures 5 and 6). The LBNL is investigating two mechanisms for fluidcommunication. In the first case, the fractures extend across the shale stratum into a nearby depleted conventional reservoir with abandoned defective wells in the overburden orunderburden. The energy for the lift of contaminants in this case is most likely provided by the higher pressure of the fluids in the shale (as the abandoned reservoir pressure is expected to be low) and by buoyancy; the main contaminant reaching the ground water is expected to be gas. Inthe second case, fractures extend from a deeper over-pressurized saline aquifer through the entire thickness of the shale to an overburden (a depleted conventional petroleum reservoir withabandoned unsealed wells). The energy for the lift of contaminants in this case is most likely provided by the higher pressure of the fluids in the shale and in the saline aquifer in addition tobuoyancy, and the contaminants reaching the ground water are expected to include gas and solutes encountered in the saline aquifer.

8.1.1.4 QUALITY ASSURANCE SUMMARY

The QAPP, "Analysis of Environmental Hazards Related to Hydrofracturing (Revision: 0)," was accepted by the EPA on December 7, 2011 (LBNL, 2011).

A TSA of the work being performed by the LBNL was conducted on February 29, 2012. Thedesignated EPA QA Manager found the methods in use satisfactory and further recommendations for improving the QA process were unnecessary. Work performed and scheduled to be performed was within the scope of the project. Work is proceeding on Scenarios A through D2 as described in Section 4.1.3. Reports, when presented, will be subjected to appropriate QA review.

8.1.2 SURFACE WATER MODELING

8.1.2.1 RELATIONSHIP TO THE STUDY

The EPA is using established surface water transport theory and models to identify concentrationsof selected hydraulic fracturing-relevant chemicals at public water supply intakes located downstream from wastewater treatment facilities that discharge treated hydraulic fracturing wastewater to rivers. .

8.1.2.2 PROJECT INTRODUCTION

When an operator reduces the injection pressure applied to a well, the direction of fluid flow reverses, leading to the recovery of flowback and produced water, collectively referred to as "hydraulic fracturing wastewater." The wastewater is generally stored onsite before being transported for treatment, recycling or disposal. Most hydraulic fracturing wastewater is disposedin UIC wells. In Pennsylvania, however, wastewater has been treated in wastewater treatment facilities (WWTFs), which subsequently discharge treated wastewater to surface water bodies.

The extent to which common treatment technologies used in WWTFs effectively remove chemicals found in hydraulic fracturing wastewater is currently unclear.46 Depending in part on the concentration of chemicals in the effluent, drinking water quality and the treatment processes atpublic water systems (PWSs) downstream from WWTFs might be negatively affected. For example, bromide in source waters can cause elevated concentrations of brominated disinfection byproducts (DBPs) in treated drinking water (Brown et al., 2011; Plewa et al., 2008), which are regulated by the National Primary Drinking Water Regulations.To learn more about impacts to downstream PWSs, the Pennsylvania Department of the Environment asked 25 WWTFs that accept Marcelluswastewater to monitor effluent for parameters such as radionuclides, total dissolved solids (TDS), alkalinity, chloride, sulfate, bromide, gross alpha, radium-226 and -228, and uranium in March 2011 (PADEP, 2011). The department also asked

14 PWSs with surface water intakes downstream fromWWTFs that accept Marcellus wastewater to test for radionuclides, TDS, pH, alkalinity, chloride, sulfate, and bromide (PADEP, 2011). Bromide and radionuclides are of particular concern indischarges because of their carcinogenicity and reproductive and developmental affects.

The EPA will use computer models—mass balance, empirical, and numerical—to estimate generic impacts of bromide and radium in wastewater discharges, based on the presence of these chemicalsin discharge data from WWTFs in Pennsylvania, impacts to downstream PWSs' ability to meet National Primary Drinking Water Regulations for DBPs and radionuclides, and the potential human health impacts from the chemicals.49 Uranium, also a radionuclide, was frequently not detected byanalytical methods for the discharges and therefore not considered for simulations. The generic model results are designed to illustrate the general conditions under which discharges might causeimpacts on downstream public water supplies. The analysis will include the effect of distance to the PWS, discharge concentration, and flow rate in the stream or river, among others. The uncertaintiesin these quantities will be addressed through Monte Carlo analysis, as described below.

A steady-state mass balance model provides an upper-bound impact assessment of the transport simulation and a partially transient approach simulates the temporal variation of effluent concentration and discharge. Key data collected to model the transport of potential contaminantsinclude actual effluent data from WWTF discharges and receiving water body flow rates. Effluent data can be obtained from National Pollutant Discharge Elimination System (NPDES) monitoring data reported to states by the dischargers. NPDES information also documents the design of the industrial treatment plants, which can give insights into the capabilities of these and similarlydesigned treatment plants. The US Geological Survey (USGS) provides limited water quality and flow rate data from monitoring stations within the watersheds of the receiving water bodies. The surface water modeling results will directly address the applicable secondary research question by evaluating the possible impacts from a permitted release of treated effluent on both a downstream drinking water intake and in a watershed where there may be multiple sources and receptors.

8.1.2.3 RESEARCH APPROACH

Multiple approaches generate results on impacts: steady-state mass balance; transient empirical modeling; and a transient, hybrid empirical-numerical model developed by the EPA. The results of the mass balance model simulate possible impacts during a large volume, high concentration discharge without natural attenuation of contaminants. The empirical model and a hybrid empirical-numerical model estimate impacts in a more realistic setting with variable chemicalconcentrations, discharge volumes, and flow rates of the receiving surface water. The numerical model confirms the results of the empirical and hybrid models. The numerical modeling is based on an approach developed for this study from existing methods (Hairer et al., 1991; Leonard, 2002; Schiesser, 1991; Wallis, 2007). Application of these three types of models provides a panoramic view of possible impacts and enhances confidence in the study results.

Mass Balance Approach Estimates Impacts from an Upper-Bound Discharge Scenario

A simple,steady-state mass balance model simulates drinking water impacts from upper-bound discharge cases. This model assumes that the total mass of the chemical of interest is conserved duringsurface water transport and that the chemical concentration does not decrease due to reaction, decay, or uptake. The model estimates potential impacts to downstream PWSs using the maximum effluent concentration, maximum WWTF discharge volume, minimum flow rate in the receivingstream, and the distance to the downstream PWS intake. The EPA constructed generic discharge scenarios for rivers with varying flow regimes to determine the potential for adverse impacts atdrinking water intakes. Because the parameters describing transport are uncertain, Monte Carlo techniques will be used to generate probabilistic outputs of the model.

Empirical Model Estimates Impacts with Varying Discharge Volumes over Time

The upper-boundcase simulated in the steady-state mass balance model may be too conservative (by providing larger concentration estimates) to accurately represent downstream concentrations of chemicals since effluent concentrations, treatment plant discharge volumes, and flow rates change over time. Therefore, the EPA will also use an empirical transport model originally developed by the USGS (Jobson, 1996) to simulate impacts from varying monthly discharge volumes over time. The empirical approach is based on tracer studies performed around the United States since the early 1970s (e.g., Nordin and Sabol (1974)). The empirical equations address two major difficulties inapplying models to chemical transport scenarios: the inability to estimate travel times from cross-sectional data and the reduction of concentration due to turbulent diffusion. The empirical equationapproach gives an estimate of travel time and peak concentration so that the model does not need to be calibrated to tracer data.

Hybrid Empirical-Numerical Model Estimates Impacts for River Networks. The original empiricalapproach was suited for a single river segment, or reach, of spatially uniform properties. The hybrid empirical-numerical model being developed by the EPA to expand the capabilities of the just-described Jobson technique will easily account for multiple reaches that can form branching rivernetworks. Similar to all statistical relationships, the empirical equations do not always match tracer data exactly; therefore, the EPA is including the ability to perform Monte Carlo techniques in thesoftware being developed. The EPA will confirm the accuracy of the hybrid model with tracer data that fall within the range of Jobson's original set of inputs (taken from Nordin and Sabol (1974)) as well as later data from the Yellowstone River that provide a real-world test of this approach(McCarthy, 2009).

The numerical portion of the hybrid model provides a direct and automatic comparison with the empirical equations. The method is based on a finite difference solution to the transport equationusing recent developments in modeling to improve accuracy (Hairer et al., 1991; Leonard, 2002; Schiesser, 1991; Wallis, 2007). By including this numerical method, a hybrid empirical-numericalapproach can be achieved. The empirical travel times from Jobson (1996) can be used to parameterize velocity in the numerical method. Dispersion coefficients can be derived from empirical data or a method developed by Deng et al. (2002). Using these approaches providesimproved accuracy in the simulation results. The EPA will prepare a user's guide to the model and make both the computer model and user's guide widely available for duplicating the resultsprepared for this project and for more general use.

For the generic simulations described above, effluent concentrations and discharge volumes will be modeled directly as variable inputs based on the effluent data evaluation (as discussed next), while flow conditions will be modeled as low, medium, and high flow. Because the parameters describing transport are uncertain, statistical measures and Monte Carlo techniques will be used to generate probabilistic outputs from the model. To provide further assurance of the accuracy of the EPA hybrid model results, the Water Quality Simulation Package has been used to simulate tracer data and confirm the results (Ambrose et al., 1983; Ambrose and Wool, 2009; DiToro et al., 1981).

8.1.2.4 STATUS AND PRELIMINARY DATA

The models described above are being used to determine potential impacts of treated wastewaterdischarges on downstream PWSs. Enough data have been identified to perform generic simulations for the steady-state mass balance simulations and hybrid empirical-numerical models with variableeffluent concentration and plant discharge. For two WWTFs in Pennsylvania, USGS flow data have been compiled for segments of the rivers that reach downstream to drinking water intakes (50 to 100 miles downstream) for the two locations. These data will be used to generate realistic model inputs to assess, in a generic sense, the potential impacts of discharges from realistic treatment plants.

The EPA-developed hybrid empirical-numerical model has been favorably compared against a tracer experiment used by Jobson (1996) in developing the original empirical formulas. Calibration or other parameter adjustment was unnecessary for the hybrid model to produce accurate results. The EPA plans to compare the hybrid model to five more of the tracer experiments to cover the range of flow conditions used by Jobson (1996). Additionally, data from the more recentYellowstone River experiment (McCarthy, 2009) are being prepared for testing the hybrid model. Similar comparisons of empirical to tracer experiments were performed by Reed and Stuckey (2002) for streams in the Susquehanna River Basin. The EPA Water Quality Simulation Packagenumerical model was set up to simulate the same tracer experiment performed for the hybrid model. Additional calibration is planned to refine the results from the Water Quality Simulation Package. After completing the evaluation of the hybrid model, the WWTF simulations will be completed.

8.1.2.5 NEXT STEPS

A description of the EPA-developed empirical-numerical model and application of the empirical-numerical and mass balance models to tracer experiments is being developed by EPA scientists andare expected to be submitted for publication in a peer-reviewed journal. The results from testing of the models and the analysis of the WWTF effluent data will be included in another peer-reviewed journal article.

8.1.2.6 QUALITY ASSURANCE SUMMARY

The initial QAPP for "Surface Water Transport of Hydraulic Fracturing-Derived Waste Water" was approved by the designated EPA QA Manager on September 8, 2011 (US EPA, 2012s). The QAPP was subsequently revised and approved on February 22, 2012.

A TSA was conducted on March 1, 2012. The designated EPA QA Manager found the methods in use satisfactory and further recommendations for improving the QA process were unnecessary. An audit of data quality (ADQ) will be performed to verify that the quality requirements specified in the approved QAPP were met.

8.1.3 WATER AVAILABILITY MODELING

The EPA selected humid and semi-arid river basins as study areas for identifying potential impacts to drinking water resources from large volume water withdrawals (1 to 9 million gallons per well for the selected river basins) associated with hydraulic fracturing operations.

8.1.3.1 PROJECT INTRODUCTION

The volume of water needed in the hydraulic fracturing process for stimulation of unconventionaloil and gas wells depends on the type of formation (e.g., coalbed, shale, or tight sands), the well construction (e.g., depth, length, vertical or directional drilling), and fracturing operations (e.g.,fracturing fluid properties and fracture job design). Water requirements for hydraulic fracturing of CBM range from 50,000 to 250,000 gallons per well (Holditch, 1993; Jeu et al., 1988; Palmer et al., 1991; Palmer et al., 1993), although much larger volumes of water are produced during the lifetimeof a well in order to lower the water table and expose the coal seam (ALL Consulting, 2003; S.S. Papadopulos & Associates Inc., 2007a, b). The water usage for hydraulic fracturing in shale gasplays is significantly larger than CBM reservoirs—2 to 4 million gallons of water are typically needed per well (API, 2010; GWPC and ALL Consulting, 2009; Satterfield et al., 2008). The volume of water needed for well drilling is understood to be much less, from 60,000 gallons in theFayetteville Shale to 1 million gallons in the Haynesville Shale (GWPC and ALL Consulting, 2009). Water-based mud systems used for drilling vertical or horizontal wells generally require that freshwater (non-potable, potable, or treated) be used as makeup fluid, although wells can also be drilled using compressed air and oil-based fluids.

Water needed for hydraulic fracturing may come from multiple sources with varying quality. Sources may include raw surface and ground water, treated water from public water supplies, and water recycled from other purposes such as flowback and produced water from previous oil and gas operations or even acid mine drainage. The quality of water needed is dependent on the otherchemicals in the fracturing fluid formulations, availability of water source, and the chemical and physical properties of the formation. The goal of this project is to investigate the water needs andsources to support hydraulic fracturing operations at the river basin and county spatial scales and to place this demand in the watershed context in terms of annual, seasonal, and monthly water availability.

The EPA recognizes the unique circumstances of the geography and geology of every unconventional oil and gas resource and has chosen two study sites to initially explore and identify the potential differences related to water acquisition. The study areas includes two river basins:

the Susquehanna River Basin (SRB), located in the eastern United States (humid climate) and overlyingthe Marcellus Shale gas reservoir (Figure 7), and the Upper Colorado River Basin (UCRB), located in the western United States (semi-arid climate) and overlying the Piceance structural basin andtight gas reservoir (Figure 8). The EPA is calibrating and testing watershed models for the study; the SRB and UCRB watershed models were previously calibrated and tested in the EPA investigation of future climate change impacts on watershed hydrology (the "20 watersheds study") (Johnson et al., 2011).

8.1.3.1.1 SUSQUEHANNA RIVER BASIN

In both study areas, the river watershed and its subsurface basin include the river flows andreservoir and aquifer storages based on the hydrologic cycle, geography, geology, and water uses. The EPA's goal is to explore future hypothetical scenarios of hydraulic fracturing use in the easternand western study areas based on current understanding of hydraulic fracturing water acquisition and watershed hydrology. The EPA intends to characterize the significance, or insignificance, of hydraulic fracturing water use on future drinking water resources for the two study areas. The research will involve detailed representation of water acquisition supporting hydraulic fracturing in the Bradford County and Susquehanna County area in Pennsylvania and in the Garfield County and Mesa County areas of Colorado. These areas have concentrated hydraulic fracturing activity, as discussed below.

Geography, Hydrology, and Climate

The SRB has over 32,000 miles of waterways, drains 27,510square miles, and covers half of Pennsylvania and portions of New York and Maryland (Figure 7) (SRBC, 2006). On average, the SRB contributes 18 million gallons of water every minute (25,920million gallons per day, or MGD) to the Chesapeake Bay (SRBC, 2006). The humid climate of the region experiences long-term average precipitation of 37 to 43 inches per year (McGonigal, 2005).

Oil and Gas Resources and Activity

Large portions of the SRB watershed are underlain by the Marcellus Shale formation, which is rich in natural gas. Estimates of recoverable and undiscovered natural gas from this formation range from 42 to 144 trillion cubic

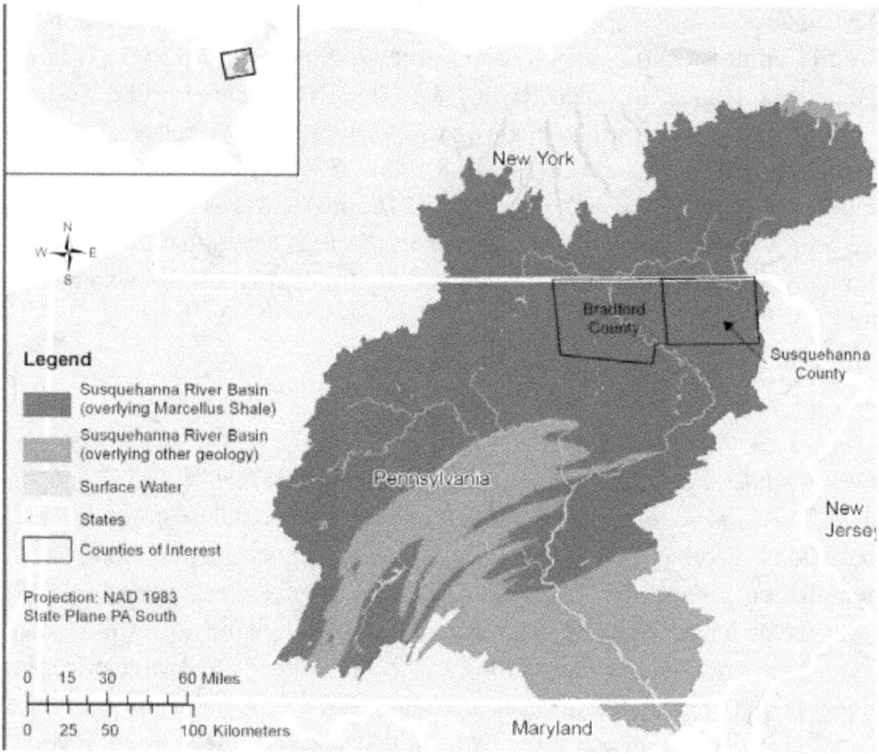

FIGURE 7: The Susquehanna River Basin, overlying a portion of the Marcellus Shale, is one of two study areas chosen for water availability modeling. Water acquisition for hydraulic fracturing will focus on Bradford and Susquehanna Counties in Pennsylvania. (GIS data obtained from ESRI, 2010a; US EIA, 2011e; US EPA, 2007.)

feet (Coleman et al., 2011) and production well development estimates for the next two decades range as high as 60,000 total wellsdrilled by 2030 (Johnson et al., 2010). The Pennsylvania Department of Environmental Protection reports that the number of drilled wells in the Marcellus Shale has been increasing rapidly. In 2007, only 27 Marcellus Shale wells were drilled in the state; in 2010 the number of wells drilled was 1,386. Data extracted from FracFocus indicate that the total vertical depth of wells in Bradford and Susquehanna Counties is between 5,000 and 8,500 feet

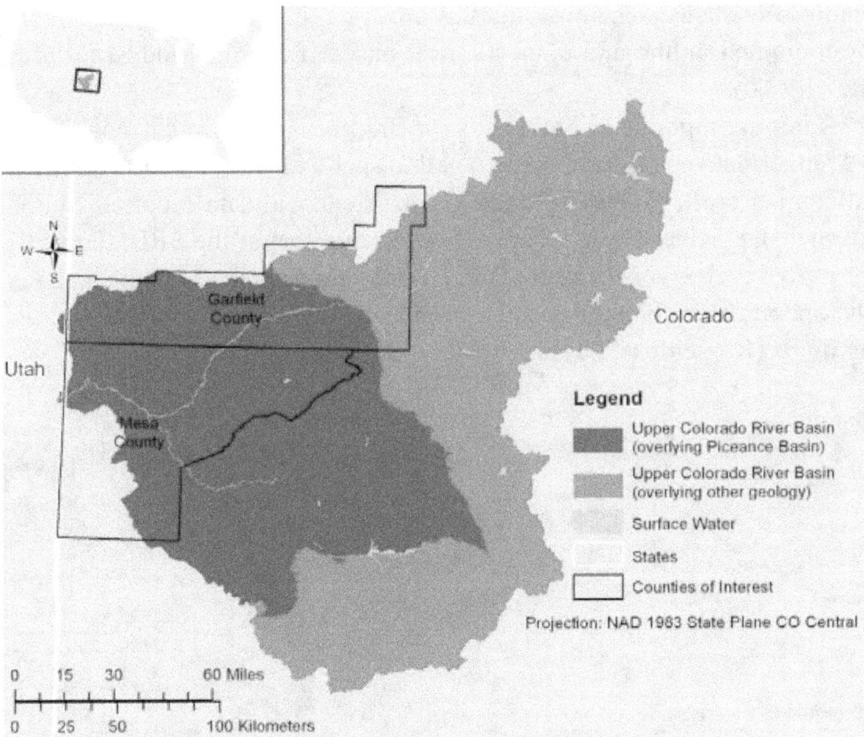

FIGURE 8: The Upper Colorado River Basin, overlying a portion of the Piceance Basin, is one of two river basins chosen for water availability modeling. Water acquisition for hydraulic fracturing will focus on Garfield and Mesa Counties in Colorado. (GIS data obtained from ESRI, 2010a; US EIA, 2011e; US EPA, 2007.)

(mean of 6,360 feet) below groundsurface, which implies that this depth range is the target production zone for the Marcellus Shale.

Water Use

The SRB supports a population of over 4.2 million people. The Susquehanna River Basin Commission estimates consumptive water use in five major categories, with PWSs consuming the greatest volume of water per day (325 MGD) followed by thermoelectric energy production (190 MGD) (Richenderfer, 2011). The greatest water withdrawals per day in

Bradford and Susquehanna Counties are for drinking water (8.25 MGD for combined public and domestic use) and self-supplied industrial uses (4.59 MGD).

Water use reported in FracFocus for Bradford and Susquehanna Counties ranges between 2 and 9 million gallons per well (median of 4.7 million gallons per well; (GWPC, 2012a)), consistent withdata reported by the Susquehanna River Basin Commission. In this part of the SRB, the wells are almost exclusively horizontal and producing from the Marcellus Shale. The operators are blendingtreated produced water into hydraulic fracturing fluids (Rossenfoss, 2011).

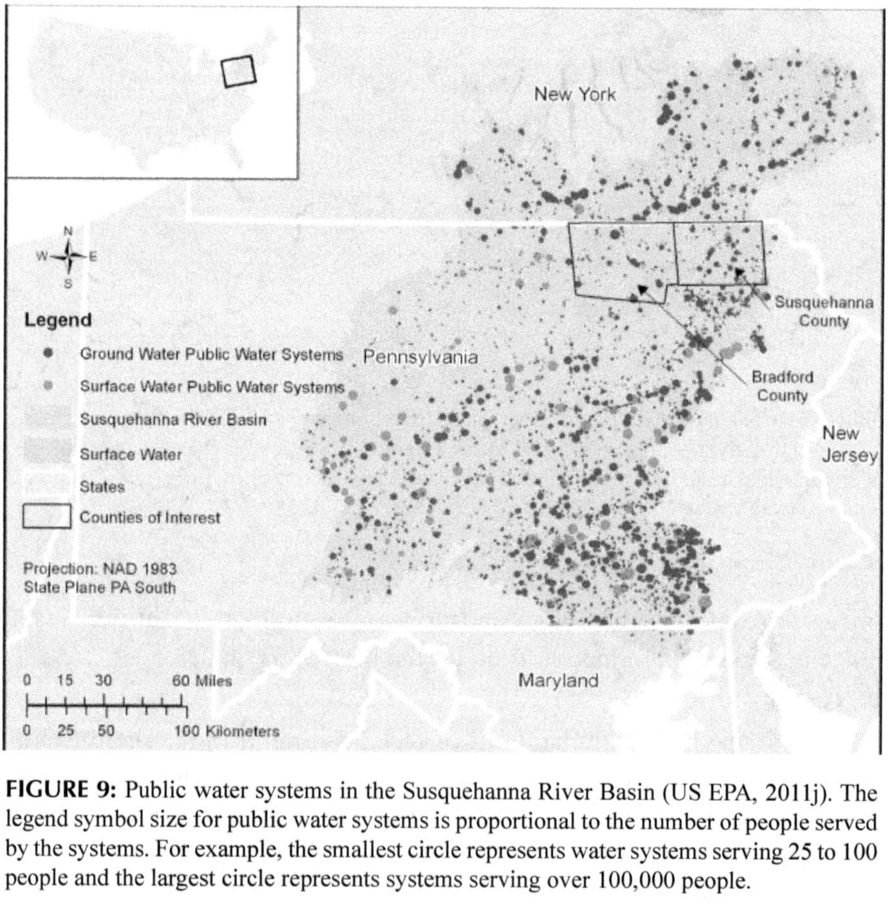

FIGURE 9: Public water systems in the Susquehanna River Basin (US EPA, 2011j). The legend symbol size for public water systems is proportional to the number of people served by the systems. For example, the smallest circle represents water systems serving 25 to 100 people and the largest circle represents systems serving over 100,000 people.

8.1.3.1.2 UPPER COLORADO RIVER BASIN

Geography, Hydrology, and Climate

The UCRB drains an area of 17,800 square miles and is characterized by high mountains in the east and plateaus and valleys in the west. The average discharge of the Colorado River near the Colorado-Utah state line is about 2.8 million gallons per minute (about 4,000 MGD) (Coleman et al., 2011). Precipitation ranges from 40 inches per year ormore in the eastern part of the basin to less than 10 inches per year in the western part of the basin (Spahr et al., 2000).

Oil and Gas Resources and Activity

The UCRB has a long history of oil, gas, and coal exploration. The Piceance Basin is a source of unconventional natural gas and oil shale. The basin was originallyexploited for its coal resources, and the associated CBM production peaked around 1992 (S.S. Papadopulos & Associates Inc., 2007a). The Upper Cretaceous Williams Fork Formation, a thicksection of shale, sandstone, and coal, has been recognized as a significant source of gas since 2004 (Kuuskraa and Ammer, 2004). The wells producing gas from the Williams Fork are either verticallyor directionally ("S"-shaped wells) drilled rather than horizontal. While the deeper Mancos Shale is considered a major resource for shale gas (Brathwaite, 2009), it must be exploited with horizontal drilling methods, and the economics are such that only prospecting wells are being drilled at thistime (personal communication, Jonathan Shireman, Shaw Environmental & Infrastructure, May 1, 2012). Estimated reserves in coalbeds and unconventional tight gas reservoirs are nearly 84 trillion cubic feet (Tyler and McMurry, 1995).

Gas production activities occur in the following counties within the UCRB: Delta, Eagle, Garfield, Grand, Gunnison, Hinsdale, Mesa, Montrose, Ouray, Pitkin, Routt, Saguache, and Summit (COGCC,2012b). Table 31 indicates that the greatest drilling activity has been in Garfield and Mesa Counties (Figure 8), where well completions increased steadily from 2000 (212 wells) to 2008 (2,725 wells), then dropped slightly to 1,160 wells in 2010 (COGCC, 2012b). The total vertical depth of wells in Garfield County and Mesa County as reported in FracFocus implies that the location of the target production zone(s) lies between 6,000 and 13,000 feet (mean of 8,000 feet) below ground surface.

Data extracted from FracFocus for Garfield and Mesa Counties shows water use per well between 1 and 9 million gallons (median 1.3 million gallons), which is consistent with the Colorado Oil and GasCompact Commission data (COGCC, 2012a; GWPC, 2012a). In this part of the Piceance Basin (Figure 21), the majority of wells are vertically drilled and producing gas from the Williams Fork tightsandstones. Based on conversations with Berry Petroleum, Williams Production, Encana Oil and Gas, and the Colorado Field Office of the US Bureau of Land Management, the water used to fracture wells in this area is entirely recycled formation water that is recovered during productionoperations. Fresh water is used only for drilling mud, cementing the well casing, hydrostatic testing, and dust abatement and is estimated to be about 251,000 gallons per well (US FWS, 2008).

8.1.3.2 RESEARCH APPROACH

Watershed Models

In order to assess the impact of hydraulic fracturing water withdrawals on drinking water availability at watershed and county spatial scales as well as annual, seasonal,monthly, and daily time scales, the EPA is developing separate hydrologic watershed models for the SRB and UCRB. The models are based in part on the calibrated and verified watershed models (hereafter called the "foundation" models) of the EPA Global Change Research Program (Johnson etal., 2011), namely the Hydrologic Simulation Program FORTRAN (HSPF) and the Soil and Water Assessment Tool (SWAT). Both HSPF and SWAT are physically based, semi-distributed watershedmodels that compute changes in water storage and fluxes within drainage areas and water bodies over time. Each model can simulate the effect of water withdrawals or flow regulation on modeledstream or river flows. Key inputs for the models include meteorological data, land use data, and time series data representing water withdrawals. The models give comparable performance at the scale of investigation (Johnson et al., 2011).

Modeling of the SRB will be completed using the calibrated and tested HSPF. Since its initialdevelopment nearly 20 years ago, HSPF has been

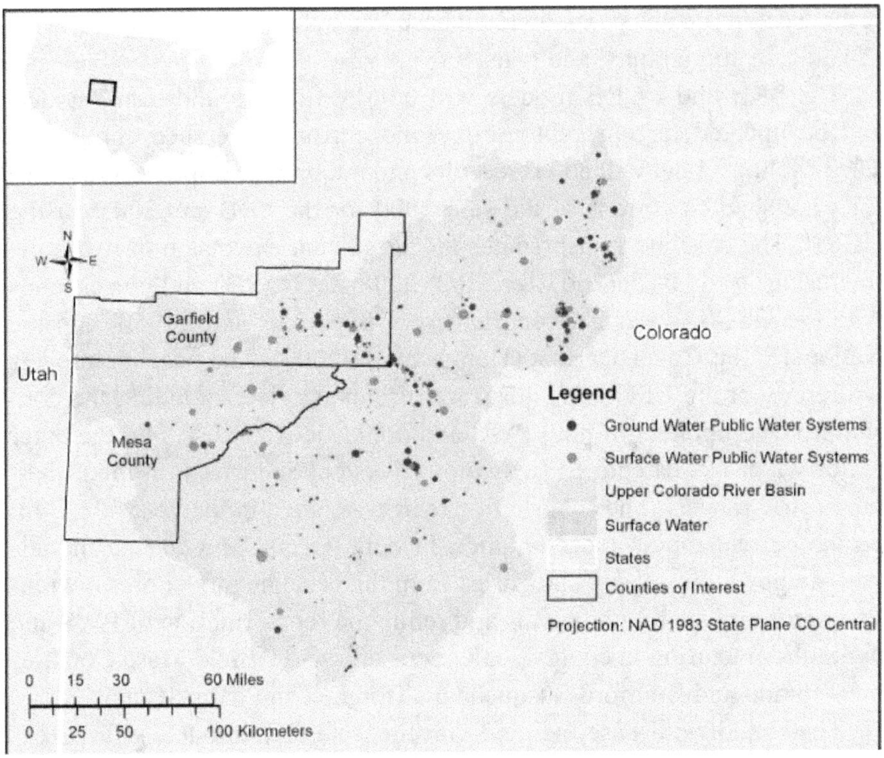

FIGURE 10: Public water systems in the Upper Colorado River Basin (US EPA, 2011j). The legend symbol size for public water systems is proportional to the number of people served by the systems. For example, the smallest circle represents water systems serving 25 to 100 people and the largest circle represents systems serving over 70,000 people.

applied around the world; it is jointly sponsored by the EPA and the USGS, and has extensive documentation and references (Donigian Jr., 2005; Donigian Jr. et al., 2011). The choice of HSPF in the SRB, a subwatershed within the larger Chesapeake Bay watershed, allows benchmarking to the peer-reviewed and community-accepted Chesapeake Bay Program watershed model.

Modeling of the UCRB will be completed using the calibrated and tested SWAT. The SWAT is acontinuation of over 30 years of modeling efforts conducted by the US Department of Agriculture's Agricultural Research Service and has extensive peer review (Gassman et al., 2007). SWAT is anappropriate choice in the less data-rich UCRB, where hydrological response units can be parameterized based on publicly available GIS maps of land use, topography, and soils.

The SRB and UCRB models will build on the "foundation" models and be updated to representbaseline and current watershed conditions. The baseline model will add reservoirs and major consumptive water uses for watershed conditions of the year 2000 for the SRB and 2005 for the UCRB. The baseline year predates the significant expansion of hydraulic fracturing in the basin(2007 for SRB, 2008 for UCRB) and corresponds with the USGS' water use reports (every five years since 1950) and the National Land Cover Dataset (Homer et al., 2007). The baseline models willrepresent the USGS's major water use categories, including the consumptive component of both PWS and domestic water use, and the other major water use categories (irrigation, livestock,industrial, mining, thermoelectric power). The snapshot of each watershed in the year 2010 will be the current model representation in both basins. The current models will include all water use categories from the baseline model plus hydraulic fracturing water withdrawals and refine the representation of PWS and hydraulic fracturing in county-scale focus areas—Garfield/Mesa Counties in Colorado and Bradford/Susquehanna Counties in Pennsylvania.

The foundation, baseline, and current watershed models will be exposed to the historical meteorology (precipitation, temperature) from National Weather Service gauges located within each watershed. The calibration and validation of the foundation, baseline, and current models will be checked by comparing goodness-of-fit statistics and through expert judgment of comparisons of observed and modeled stream discharges.

Key characteristics of model configuration include:

- Land use will be based on the 2001 National Land Cover Dataset (Homer et al., 2007). Land use data are used for segmenting the basin land area into multiple hydrologicresponse units, each with unique rainfall/runoff response properties. For the SWAT model, soil and slope data will also be used for defining unique hydrologic response units.
- Each basin will be segmented into multiple subwatersheds at the 10-digit hydrologic unitscale.
- Observed meteorological data for water years 1972 to 2004 for SRB and 1973 to 2003 for UCRB will be applied to assess water availability over a range of weather conditions.

- The effect of reservoirs on downstream flows will be simulated using reservoirdimensions/operation data from circa 2000 from the Chesapeake Bay Program watershed model (Phase 5.3; (US EPA, 2010a)).
- Point source dischargers with NPDES-permitted flow rates of at least 1 MGD will be represented as sources of water on the appropriate stream reaches.
- Surface water withdrawals will be simulated for three unique water use categories:hydraulic fracturing water use, PWSs, and other. For the "other" category, the magnitude of withdrawals from modeled stream reaches will be based on water use estimates developed by the USGS (year 2000 for SRB; year 2005 for UCRB).

Modeling Future Scenarios

The modeling effort will also simulate a snapshot of heightened annualhydraulic fracturing relative to the baseline and current condition models at levels that could feasibly occur over the next 30 years, based on recent drilling trends and future projections ofnatural gas production (US EIA, 2012; US EPA, 2012w). Because projections of future conditions are inherently uncertain, three separate scenarios will be simulated: business-as-usual, energy plus,and green technology. The scenarios assume distinct levels of natural gas drilling and hydraulic fracturing freshwater use and, therefore, apply distinct hydraulic fracturing water withdrawal time series to modeled stream reaches. Further, significant population growth is projected in Garfield/Mesa Counties, Colorado, over the next 30 years (US EPA, 2010c), where natural gas extraction in the UCRB has recently been concentrated. Therefore, the UCRB future scenarios alsoconsider a potential increase in PWS surface withdrawals in the basin. The balance between surface water availability and demand depicted in each scenario's annual snapshot of water use will be assessed across a range of weather conditions (i.e., drought, dry, wet, and very wet years based onthe historical record).

Future drilling patterns in the SRB and UCRB are assessed from National Energy Modeling System (NEMS) regional projections of the number of wells drilled annually from 2011to 2040 in shale gas (SRB) and tight gas (UCRB) plays (US EIA, 2012; US EPA, 2012w). Based on analysis of NEMS well projections and undiscovered resources in the Marcellus Shale (Coleman et al., 2011), peak annual drilling in the SRB could exceed the recent high in 2011 by as much as 50%. In the UCRB, analysis of NEMS well projections and undiscovered tight gas resources in the Piceance Basin (USGS, 2003) suggest that the 2008 peak level of drilling in the basin could be repeated in the late 2030s, when a growing population would exert a higher demand for freshwater. The future scenarios willincorporate these projections, with high-end estimates of the number of wells drilled/fractured applied in the energy plus scenario.

The volume of surface water required for drilling and hydraulic fracturing varies according to localgeology, well characteristics, and the amount of recycled water available for injection. In the SRB, 2008 to 2011 water use data (SRBC, 2012) show that, on average, 13% of total water injected for hydraulic fracturing is composed of recycled produced water or wastewater. Per well surface water use in the SRB business as usual and energy plus scenarios will therefore be established as 87% of the 4 million gallons of water used for hydraulic fracturing, or 3.5 million gallons. The SRB green technology scenario reflects a condition of increased water recycling, where the 90th percentile of current recycled water amount (29%) becomes the average. Per well surface water use in the SRB green technology scenario will therefore be established as 71% of the 4 million gallons of water used for hydraulic fracturing, or 2.8 million gallons.

In the UCRB, 100% recycled water use is typical for hydraulic fracturing of tight sandstones (personal communication, Jonathan Shireman, Shaw Environmental & Infrastructure, May 7, 2012). Surface water is acquired for well drilling and cementing (0.18 million gallons), dust abatement (0.03 million gallons), and hydrostatic testing (0.04 million gallons) only (US FWS, 2008). Per well surface water use in the UCRB business as usual and energy plus scenarios will therefore be 0.25million gallons. For the UCRB green technology scenario, surface water will be assumed to be acquired for well drilling and cementing only (0.18 million gallons per well).

Following the development of water withdrawal datasets for each scenario, model output will bereviewed to assess the impacts of water acquisition for hydraulic fracturing on drinking water supplies by evaluating annual and long-term streamflow and water demand, and identifying short-term periods (daily to monthly) in which water demand exceeds streamflow. Since many publicwater supplies originate from ground water sources, simulated ground water recharge will also be computed. Results will be compared among the three scenarios to identify noteworthy differences and their implications for future management of hydraulic fracturing-related water withdrawals.

8.1.3.3 STATUS AND PRELIMINARY DATA

Existing water use information for hydraulic fracturing has been collected from the Susquehanna River Basin Commission and the Colorado Oil and Gas Compact Commission by Shaw Environmental Technologies. The data underwent a QA review before submission to the modelingteams of The Cadmus Group, Inc. The models are being calibrated and validated. The future scenarios are being designed, with model simulations to follow. Work is underway and will be published in peer-reviewed journals when completed.

8.1.3.4 QUALITY ASSURANCE SUMMARY

The QAPP, "Modeling the Impact of Hydraulic Fracturing on Water Resources Based on Water Acquisition Scenarios (Version 1.0)," contracted through The Cadmus Group, Inc., was accepted onFebruary 8, 2012 (Cadmus Group Inc., 2012a). A technical directive/contract modification dated April 25, 2012, modifies the scope of the project but not the procedures. Additionally, there is a pending QAPP revision that adapts the scope to the contract modification.

A TSA of The Cadmus Group, Inc., contract was performed by the designated EPA QA Manager on June 14, 2012. The methods in use were found to be satisfactory and further recommendations forimproving the QA process were unnecessary. Work performed and scheduled to be performed was within the scope of the project.

The interim progress report "Development and Evaluation of Baseline and Current Conditions for the Susquehanna River Basin," received on June 19, 2012, was found to be concise but detailedenough to meet the QA requirements, as expressed in the QAPP, its revision, and the contract modification/technical directive. The same was true for the interim progress report "Impact ofWater Use and Hydro-Fracking on the Hydrology of the Upper Colorado River Basin," submitted on July 2, 2012.

8.2 LABORATORY STUDIES

The laboratory studies are targeted research projects designed to improve understanding of the ultimate fate and transport of selected chemicals, which may be components of hydraulic fracturing fluids or naturally occurring substances released from the subsurface during hydraulic fracturing. This chapter includes progress reports for the following projects:

The EPA is combining data collected from samples of wastewater treatment facility discharges and receiving waters with existing modeling programs to identify the proportion of hydraulic fracturingwastewater that may be contributing to contamination at downstream public water system intakes.

8.2.1 PROJECT INTRODUCTION

The large national increase in hydraulic fracturing activity has generated large volumes of hydraulic fracturing wastewater for treatment and disposal or recycling. In some cases, states have allowedhydraulic fracturing wastewater to be treated by WWTFs with subsequent discharge to rivers. Most WWTFs are designed to filter and flocculate solids, as well as consume biodegradable organic species associated with human and some commercial waste. Very few facilities are designed to manage the organic and inorganic chemical compounds contained in hydraulic fracturing wastewater.

Public water supply intakes may be located in river systems downstream from WWTFs and avariety of other industrial and urban discharges, and it is critical to evaluate sources of contamination at those drinking water intakes. Elevated bromide and chloride concentrations are of particular concern in drinking water sources due to the propensity of bromides to react with organic compounds to produce THMs and other DBPs during drinking water treatment processes(Plewa and Wagner, 2009). High

TDS levels—including bromide and chloride—have been detected in the Monongahela River in 2008 and the Youghiogheny River in 2010 (Lee, 2011; Ziemkiewicz, 2011). The source and effects of these elevated concentrations remains unclear.

This project's overall goal is to establish an approach whereby surface water samples may beevaluated to determine the extent to which hydraulic fracturing wastewaters (treated or untreated) may be present, and to distinguish whether any elevated bromide and chloride in those samplesmay be due to hydraulic fracturing or other activities. To accomplish this goal, the EPA is: (1) quantifying the inorganic chemical composition of discharges in two Pennsylvania river systems from WWTFs that accept and treat flowback and produced water, coal-fired utility boilers, acid mine drainage, stormwater runoff of roadway deicing material, and other industrial sources; (2) investigating the impacts of the discharges by simultaneously collecting multiple upstream anddownstream samples to evaluate transport and dispersion of inorganic species; and (3) estimating the impact of these discharges on downstream bromide and chloride levels at PWS intakes usingmathematical models.

8.2.2 RESEARCH APPROACH

The "Quality Assurance Project Plan for Hydraulic Fracturing Wastewater Source Apportionment"provides a detailed description of the research approach (US EPA, 2012q). Briefly, water samples are being collected at five locations on two river systems; each river has an existing WWTF that iscurrently accepting hydraulic fracturing wastewater for treatment. Source profiles for significant sources such as hydraulic fracturing wastewater, WWTF effluent, coal-fired utility boiler discharge, acid mine drainage, and stormwater runoff from roadway deicing will be developedfrom samples collected from these sources during the study. Computer models will then be used to compare data from these river systems to chemical and isotopic composition profiles obtainedfrom potential sources.

Three two-week intensive sampling events were conducted to assess river conditions under different flow regimes: spring, summer, and fall 2012. The amount of water inthe river has historically been highest in the spring, resulting in the dilution of pollutants, and the summer and fall seasons typically have decreased stream flow, which may result in elevated concentrations due to less dilution (USGS, 2011a, b). USGS gauging stations near the WWTFs will beused to measure the flow rate during the three sampling periods.

During each sampling event, automatic water samplers (Teledyne Isco, model 6712) at each site collect two samples daily—morning and afternoon—based on the PWS and WWTF operationsschedule. The samples are stored in the sampler for one to four days, depending on the site visit schedule. Each river is sampled in five locations. The first sampling device downstream of the WWTF is far enough downstream to allow for adequate mixing of the WWTF effluent and river water. The second downstream sampling device is between the first downstream sampling location and the closest PWS intake. The locations of the samplers downstream of the WWTF also take into account the presence of other significant sources, such as coal-fired utility boiler and acid mine drainage discharges, and allow for the evaluation of their impacts.

8.2.2.1 SAMPLE ANALYSES

The EPA will analyze the river samples and effluent samples according to existing EPA methods forthe suite of elements and ions listed in Table 39. Inorganic ions (anions and cations) are being determined by ion chromatography. Inorganic elements are being determined using a combination of inductively coupled plasma optical emission spectroscopy for high-concentration elements and high-resolution magnetic sector field inductively coupled plasma mass spectrometry for low concentration elements. Additionally, the characteristic strontium (Sr) ratios ($^{87}Sr/^{86}Sr$; 0.7101– 0.7121) in Marcellus Shale brines are extremely sensitive tracers, and elevated concentrations of readily water soluble strontium are present in the hydraulic fracturing wastewaters (Chapman etal., 2012). Isotope analyses for $^{87}Sr/^{86}Sr$ are being conducted on a subset (~20%) of samples by thermal ionization mass spectrometry to corroborate source apportionment modeling results.

Although the majority of the species that are being quantified in this study have been identified inflowback or produced water,60 the species relationships and relative quantities of the species in other sources (i.e., coal-fired utility boiler and acid mine drainage discharges) will differ (Chapman et al., 2012). This will allow the models described below to distinguish among the contributions from each source type.

8.2.2.2 SOURCE APPORTIONMENT MODELING

The EPA is using the data gathered through the analyses described above to support source apportionment modeling. This source apportionment effort will use peer-reviewed receptor modelsto identify and quantify the relative contribution of different contaminant source types to environmental samples.61 In this case, river samples collected near PWS intakes are being evaluatedto discern the contributing sources (e.g., hydraulic fracturing wastewater or acid mine drainage) of bromide and chloride to those stream waters. Receptor models require a comprehensive analysis of environmental samples to provide a sufficient number of constituents to identify and separate the impacts of different source types. Analysis of major ions and inorganic trace elements (Table 39) will accomplish the needs for robust receptor modeling. Contaminant sources may be distinguishedby unique ranges of chemical species and their concentrations, and the models provide quantitative estimates of the source type contributions along with robust uncertainty estimates.

EPA-implemented models and commercial off-the-shelf software are being used to analyze the data from this particular study (e.g., Unmix, Positive Matrix Factorization, chemical mass balance). These models have previously been used to evaluate a wide range of environmental data for air, soil, andsediments (Cao et al., 2011; Pancras et al., 2011; Soonthornnonda and Christensen, 2008), and are now being used for emerging issues, such as potential impacts to drinking water from hydraulic fracturing.

8.2.2.3 STATUS AND PRELIMINARY DATA

The EPA completed the two-week spring, summer, and fall intensive sampling periods beginning on May 16, July 20, and September 19, 2012, respectively. The EPA collected 206, 198, and 209 samples during the spring, summer, and fall intensives, consisting of WWTF-treated discharge, river samples, raw hydraulic fracturing wastewater, and acid mine drainage. The data qualityobjectives (US EPA, 2012q) of 80% valid sample collection were met for both the spring (>85%) and summer (>96%) measurement intensives. Preparation work for the extraction and filtrationof spring intensive samples for inductively coupled plasma optical emission spectroscopy and high-resolution magnetic sector field inductively coupled plasma mass spectrometry is ongoing.

During the audit, it was observed that the custody seals may not have offered a level of security necessary for the project. The field team had already identified this potential problem and hadordered different tamper-resistant seals before the field trip. The new seals (NIK Public Safety Tamperguard brand evidence tape) have been in use since they were received on May 10, 2012.The second observation during the audit was the need to document the reasoning of changes performed to standard operating procedures. The researchers have documented all the changes performed as well as the logic and reasoning of the changes in the field laboratory notebooks. Mostmodifications to the procedures were related to procedural adjustments made as a result of the field site characteristics, which were slightly different from the field site characteristics used tofield-test the procedures in North Carolina. The documents also included updates to points of contact, references, and added text for clarification (e.g., river velocity measurements). Revisions reflecting these changes have been made to the QAPP and four SOPs based on the spring intensive field experience and the TSA. The revised version of the QAPP and four SOPs were approved on June 29, 2012. These updates do not impact the original data quality objectives.

The researchers are following the QA procedures described in the QAPP and the standard operatingprocedures. In accordance to the QAPP, a TSA was performed on July 16 and 17, 2012, to evaluate laboratory operations. The designated EPA QA Manager reviewed the ion chromatography andhigh-resolution magnetic sector field inductively coupled plasma mass spectrometer analyses, data processing, storage, sample receiving and chain of custody procedures. The audit identified twoobservations and one best practice. One of the observations highlighted the need for a process that would ensure proper transcription of the data from the ion chromatography instrument to the report file. To reduce uncertainty and potential transcription errors, the analyst developed a process to export the data produced by the instrument in a text file instead of copying and pasting the data to a separate file. Another observation was the need to include performance evaluationsamples in the analytical set. The performance evaluation samples will be analyzed in addition to the other quality controls already in place, which include blanks, duplicates, standard reference materials, and continuing calibration verification. The performance evaluation audit is beingscheduled as specified in the QAPP. The blind performance evaluation samples will be analyzed with the regular samples and the data reported back to the QA Manager of the organizationproviding the blind performance evaluation samples. The best practice identified by the auditor was the tracking system, which uses a scanner and bar codes to track sampling bottles through thewhole process: preparation, deployment to/from the field, sample analysis, and data reporting. The quality control (QC) procedures described in the QAPP have been followed in all instances. Besides the two TSAs performed and the performance evaluation audit, an ADQ is being coordinated by thedesignated EPA QA Manager. The source apportionment modeling will be described in a separate modeling QAPP. A TSA will be scheduled in 2013 for the modeling component of the study.

8.2.2 WASTEWATER TREATABILITY STUDIES

8.2.2.1 RELATIONSHIP TO THE STUDY

The EPA is conducting laboratory experiments to assess the efficacy of conventional wastewatertreatment processes on selected chemicals found in hydraulic fracturing wastewater. The results of the water treatabilityexperiments also complement the surface water modeling research project.

8.2.2.2 INTRODUCTION

Hydraulic fracturing wastewater, including flowback and produced water, is generally disposed of through underground injection in Class II UIC wells or treatment by a WWTF followed by surface water discharge. A generalized diagram for the onsite flow of water is given in Figure 11. A US Department of Energy report provides a state-by-state description of costs, regulations, and treatment/disposal practices for hydraulic fracturing wastes, including wastewater (Puder and Veil,2006).

Wastewater may be treated at a WWTF, such as a POTW or centralized waste treatment facility (CWT). This project focuses on the efficacy of treatment processes at POTWs and CWTs, since discharge of treated wastewater to surface waters provides an opportunity for chemicals found inthe effluent to be transported to downstream PWS intakes. This project will also explore treatment processes used for reuse of hydraulic fracturing wastewater.

8.2.2.2.1 PUBLICLY OWNED TREATMENT WORKS TREATMENT PROCESSES

Conventional POTW treatment processes are categorized into four groups: primary, secondary, tertiary, and advanced treatment. Primary treatment processes remove larger solids and wastewater constituents that either settle or float. These processes include screens, weirs, grit removal, and/or sedimentation and flotation (e.g.,primary clarification). Secondary treatment processes typically remove biodegradable organics by using microbial processes (e.g., "bioreactor") in fixed media (e.g., trickling filters) or inthe water column (e.g., aeration basins). There is typically another settling stage in the secondary treatment process where suspended solids generated in the aeration basin are removed through settling ("secondary clarifier"). In some systems, tertiary or advanced treatment ("filter and UV disinfection") may be applied as a polishing step to achieve a particular end use water quality (e.g., for reuse in irrigation).The POTW then discharges the treated effluent to surface water, if recycling or reuse is not

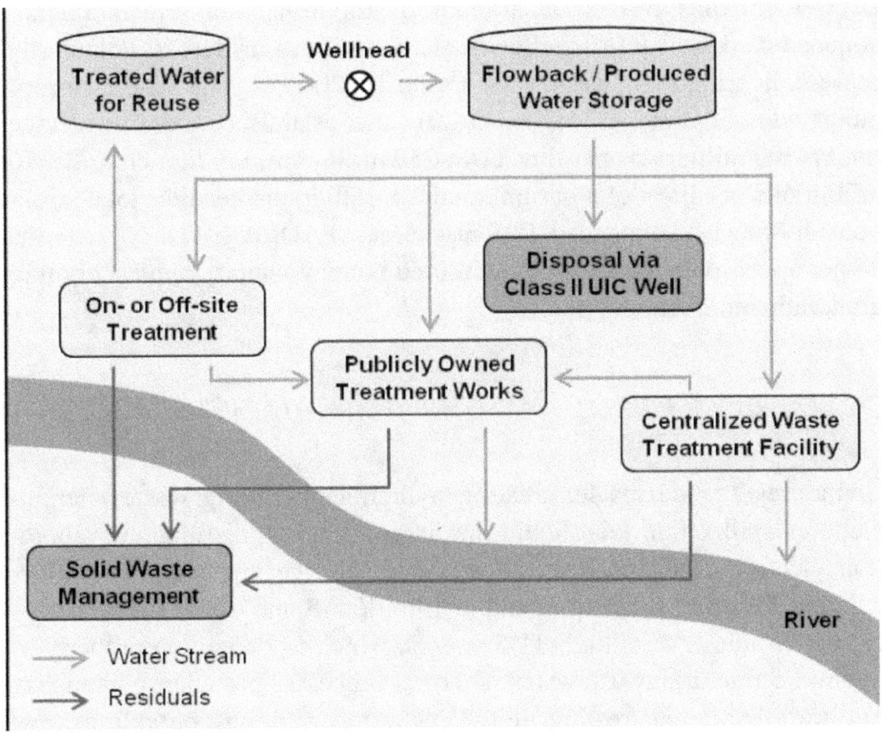

FIGURE 11: Hydraulic fracturing wastewater flow in unconventional oil and gas extraction. Flowback and produced water (collectively referred to as "hydraulic fracturing wastewater") is typically stored onsite prior to disposal or treatment. Hydraulic fracturing wastewater may be disposed of through Class II underground injection control (UIC) wells or through surface water discharge following treatment at wastewater treatment facilities, such as publicly owned treatment works or centralized waste treatment facilities. Wastewater may be treated on-or offsite prior to reuse in hydraulic fracturing fluids.

intended. Solid residuals formed as byproducts of the treatment processes may contain metals, organics, and radionuclides that were removed from the water. Residuals are typically de-watered and disposed of via landfill, land application, or incineration.

The exact number of POTWs currently accepting hydraulic fracturing wastewater is not known. InPennsylvania, where gas production from the Marcellus Shale is occurring, approximately 15 POTWs were accepting hydraulic fracturing wastewater until approximately May 2011. In April

2011, the Pennsylvania Department of Environmental Protection announced a request forMarcellus Shale natural gas drillers to voluntarily cease delivering their wastewater to the 15 POTWs. The state also promulgated regulations in November 2011 that established monthlyaverage limits (500 milligrams per liter TDS, 250 milligrams per liter chloride, 10 milligrams per liter total barium, and 10 milligrams per liter total strontium) for new and expanded TDS discharges (PADEP, 2011). These limits do not apply to the 15 facilities identified in the voluntary request or other grandfathered treatment plants.

5.2.2.2 COMMERCIAL WASTE TREATMENT FACILITY PROCESSES

Commercial processes for treating hydraulic fracturing wastewater include crystallization (zeroliquid discharge), thermal distillation/evaporation, electrodialysis, reverse osmosis, ion exchange, and coagulation/flocculation followed by settling and/or filtration. Some treatment processes are better able to treat high-TDS waters, which is a common property of hydraulic fracturingwastewater. Thermal processes are energy-intensive, but are effective at treating high-TDS waters and may be able to treat hydraulic fracturing wastewater with zero liquid discharge, leaving only a residual salt. Electrodialysis and reverse osmosis may be feasible for treating lower-TDSwastewaters. These technologies are not able to treat high-TDS waters (>45,000 milligrams per liter) and may require pre-treatment (e.g., coagulation and filtration) to minimize membrane fouling.

Centralized waste treatment facilities can be used for pre-treatment prior to a POTW or, under an approved NPDES permit, can discharge directly to surface water (Figure 11). Commercial waste treatment processes will also result in some residual material that will require management and disposal.

8.2.2.2.3 REUSE

Gas producers are accelerating efforts to reuse and recycle hydraulic fracturing wastewater in some regions in order to decrease costs associated with procuring fresh water supplies, wastewater transportation, and offsite treatment and disposal. The EPA requested information on current wastewater management practices in the Marcellus Shale region from six oil and gas operators in May 2011. Responses to the request for information indicated that reuse treatment technologies are similar, if not the same, to those used by WWTFs. Reuse technologies included direct reuse,onsite treatment (e.g., bag filtration, weir/settling tanks, third-party mobile treatment systems) and offsite treatment. Offsite treatment, in most instances, consisted of some form of stabilization, primary clarification, precipitation process, and secondary clarification and/or filtration. Specific details for offsite treatment methods were lacking as they are considered proprietary.

Innovation in coupling various treatment processes may help reduce wastewater volumes and fresh water consumed in hydraulic fracturing operations. A challenge facing reuse technologydevelopment is treating water onsite to an acceptable quality for reuse in subsequent hydraulic fracturing operations. Key water quality parameters to control include TDS, calcium, and hardness,all of which play a major role in scale formation in wells.

Recycling and reuse reduce the immediate need for treatment and disposal and water acquisition needs. There will likely be a need to treat and properly dispose of the final concentrated volumes of wastewater and residuals produced from treatment processes from a given area of operation, however.

8.2.2.3 RESEARCH APPROACH

The EPA is examining the fate and transport of chemicals through conventional POTW treatment processes and commercial chemical coagulation/settling processes. The objective of this work is toidentify the partitioning of selected chemicals between solid and aqueous phases and to assess the biodegradation of organic constituents. In addition, microbial community health will be monitored in the reactors to identify the point where biological processes begin to fail. Contaminants that can pass through treatment processes and impact downstream PWS intakes will be identified.

Fate and Transport of Selected Contaminants in Wastewater Treatment Processes

The EPA will initially analyze the fate and transport of selected hydraulic fracturing–related contaminants inwastewater treatment processes, including conventional processes (primary clarifier, aeration basin, secondary clarifier), commercial processes (chemical precipitation/filtration andevaporation/distillation), and water reuse processes (pretreatment and filtration). The initial phase of this work will involve bench-scale fate and transport studies in a primary clarifier followed by 10 liter chemostat reactors seeded with microbial organisms from POTW aeration basins. In bench-scale work relevant to CWTs, similar fate and transport studies will be performed in chemical coagulation, settling, and filtration processes.

A list of contaminants for initial treatability studies have been identified and are basedon the list of hydraulic fracturing-related chemicals identified for initial analytical method development. In addition to monitoring the fate of the contaminants listed in treatment settings, impacts on conventional wastewater treatment efficiency will be monitored by examining changes in chemical oxygen demand, biological oxygen demand, and levels of nitrate, ammonia, phosphorus, oxygen, TDS, and total organic carbon in the aeration basin.

Characterization of Contaminants in Hydraulic Fracturing Wastewater Treatment Residuals

The EPA will examine the concentrations and chemical speciation of inorganic contaminants in treatmentresiduals. Residuals generated from the research described above will be analyzed for inorganic contaminant concentrations via EPA Method 3051A (Microwave Assisted Digestion) andinductively coupled argon plasma-optical emission spectrometry. Samples will also undergo analysis via X-ray absorption spectroscopy in order to assess oxidation state and chemical speciation of target contaminants. Organic contaminants will be analyzed via liquid or gaschromatography-mass spectrometry after accelerated solvent extraction of the solids.

8.2.2.4 STATUS AND PRELIMINARY DATA

This research is currently in the planning stage.

8.2.2.5 NEXT STEPS

Initial studies will focus on establishing thresholds of TDS tolerance in chemostat bioreactors. Once the basic salt thresholds have been established, selected chemicals from the 26R forms will be added to the salt stock solutions. Salt concentrations will be kept below the thresholds where effects on the biological processes were observed. Potentially biodegradable pollutants (e.g., organics) will be measured, and the EPA will attempt to identify breakdown products.

Constituents that are not biodegradable (e.g., elements and anions) will be tracked through the treatment process by analyzing system effluent using the appropriate EPA Methods and by analyzing residuals from the primary clarifier and the bioreactors. The results of these bench-scale studies will be applied to a pilot-scale system that would target compounds identified in bench-scale studies as being the most problematic due to their lack of degradation or removal in the treatment process.

For studies on commercial treatment systems using chemical addition/settling, the EPA plans toconduct jar tests that employ coagulants/flocculants at appropriate contact and settling times. The jar tests will be conducted at the bench-scale using actual hydraulic fracturing wastewater samples. The EPA will also attempt to mimic evaporative/distillation processes by using thermal treatment on actual hydraulic fracturing wastewater samples. Both the jar test samples and residuals from thermal treatment will be analyzed for chemicals. Elements in the residuals will also be characterized via X-ray diffraction and X-ray absorption microscopy.

8.2.2.6 QUALITY ASSURANCE SUMMARY

The initial QAPP, "Fate, Transport and Characterization of Contaminants in Hydraulic FracturingWater in Wastewater Treatment Processes," was submitted on December 20, 2011, and approved in August 2012 (US EPA, 2012q).

Because project activities are still in an early stage, no TSA has been performed. A TSA will be performed once the project advances to the data collection stage.

As results are reported and raw data are provided from the laboratories, ADQs will be performed to verify that the quality requirements specified in the approved QAPP were met. Data will be qualified if necessary, based on these ADQs. The results of the ADQs will be reported with the summary of results in the final report.

8.2.3 BROMINATED DISINFECTION BYPRODUCT PRECURSOR STUDIES

The EPA is assessing the ability of hydraulic fracturing wastewater to contribute to DBP formationin drinking water treatment facilities, with a particular focus on the formation of brominated DBPs. This work is complemented by the analytical method development for DBPs

8.2.3.1 INTRODUCTION

Wastewaters from hydraulic fracturing processes typically contain high concentrations of TDS, including significant concentrations of chloride and bromide. These halogens are difficult to remove from wastewater; if discharged from treatment works, they can elevate chloride and bromide concentrations in drinking water sources. Upon chlorination at a drinking water treatment facility, chloride and bromide can react with naturally occurring organic matter (NOM) in the water and lead to the formation of DBPs. Because of their carcinogenicity and reproductive and developmental affects, the maximum contaminant levels (MCLs) of the DBPs bromate, chlorite, haloacetic acids, and total THMs in finished drinking water are regulated by the National Primary Drinking Water Regulations.

Increased bromide concentrations in drinking water resources can lead to greater total THM concentrations on a mass basis and may make it difficult for some PWSs to meet the regulatory limits of total THM in finished drinking water. As a first step, this project is examining the formation of brominated THMs, including bromoform ($CHBr_3$), dibromochloromethane ($CHClBr_2$), and bromodichloromethane ($CHCl_2Br$), during drinking watertreatment processes. The formation of haloacetic acids (HAAs) and nitrosamines during drinking water treatment processes is also being investigated.

Reactions of brominated biocides used in hydraulic fracturing operations with typical drinkingwater disinfectants associated with chlorination or chloramination are also being explored. Brominated biocides are often used in fracturing fluids to minimize biofilm growth. The objective ofthis work is to assess the contribution, if any, to brominated DBP formation and identify degradation pathways for brominated biocides.

It is important to note that hydraulic fracturing wastewater can potentially contain other contaminants in significant concentrations that could affect human health. The EPA identified the impacts of elevated bromide and chloride levels in surface water from hydraulic fracturingwastewater discharge as a priority for protection of public water supplies. This project will ultimately provide PWSs with information on the potential for brominated DBP formation in surface waters receiving discharges from WWTFs.

8.2.3.2 RESEARCH APPROACH

This research will (1) analyze and characterize hydraulic fracturing wastewater for presence of halides, (2) evaluate the effects of high TDS upon chlorination of surface water receiving discharges of treated hydraulic fracturing wastewater, and (3) examine the reactions of brominated biocides subjected to chlorination during drinking water treatment. Selected analytes for characterizinghydraulic fracturing wastewater include nitrosamines and the halide anions chloride, bromide, and iodide—ions that are the likeliest to form DBPs (Richardson, 2003), including THMs and HAAs.

Hydraulic fracturing wastewater samples have been obtained from several sources in Pennsylvania. The quantification of background concentrations of halides in the samples follows EPA Method 300.1 (rev. 1) and the modified version of the method using mass spectrometry detection forbromide and bromate. The samples are also being analyzed for the presence of DBPs, including THMs (EPA Method 551.1), HAAs (EPA Method 552.1), and Nnitrosamines (EPA Method 521), as well as elemental composition, anion concentration, TDS, and total organic carbon.

Three treatments are being applied to high-TDS wastewater samples: (1) samples will be blendedwith deionized water at rates that mimic discharge into varying flow rates of receiving water in order to account for dilution effects; (2) samples will be blended with deionized water with

NOM additions at concentration ranges typically found in surface waters; and (3) samples will be blended with actual surface water samples from rivers that receive treated hydraulic fracturingwastewater discharges. All samples will be subjected to formation potential experiments in the presence of typical drinking water disinfectants associated with chlorination or chloramination. Formation potential measures will be obtained separately for THMs, HAAs, and nitrosamines. Disinfection byproduct formation in surface water samples will be compared with DBP formation in deionized water as well as deionized water fortified with several NOM isolates from different water sources in order to examine the effects of different NOM on DBP formation.66

The brominated biocides 2,2-dibromo-3-nitropropionamide and 2-bromo-2-nitrol-1,3-propanediol, employed in hydraulic fracturing processes, are being subjected to chlorination conditions encountered during drinking water treatment. These experiments should provide insight on the potential formation of brominated THMs from brominated biocides. Effects of chlorination on the brominated biocides are also being monitored.

8.2.3.3 STATUS AND PRELIMINARY DATA

Work has begun on total THM formation studies to identify potential problems with analysis (EPA Method 551.1) due to the high TDS levels typical in hydraulic fracturing wastewater. Wastewaterinfluent and effluent samples were obtained from researchers involved in the source apportionment studies at two CWTs in Pennsylvania that are currently acceptinghydraulic fracturing wastewater for treatment via chemical addition and settling. For this preliminary research, samples were diluted 1:100 with deionized water and equilibrated withsodium hypochlorite until a 2 milligrams per liter concentration of sodium hypochlorite was achieved (a typical disinfectant concentration for finished water from a PWS). The samples are being analyzed for pH, metals, TDS, total suspended solids, total organic content, and selectedanions.

Efforts to identify and quantify the parent brominated biocides using liquid chromatography/mass spectrometry methods have been unsuccessful to date, possibly due to poor ionization of thebrominated molecules. The biocide samples subject to chlorination have been prepared for analysis of THMs.

8.2.3.4 NEXT STEPS

When the preliminary work on potential analytical effects from high TDS on total THM recovery is complete, a series of experiments to assess the potential formation of DBPs during chlorination will be run on the following samples:

- Deionized water
- Deionized water, varying concentrations of NOM
- Deionized water plus TDS
- Deionized water plus TDS and NOM
- Hydraulic fracturing wastewater

This series of samples will allow THM formation comparisons between hydraulic fracturing wastewater samples and less complex matrices. Dilutions will be made on the samples based on effluent discharge rates for existing WWTFs and receiving water flow rates. The samples will undergo chlorination and be sub-sampled over time (e.g., 0 to 120 minutes). Chloride to bromide ratios will be set at 50:1, 100:1, and 150:1 to encompass the range of conditions that may be found in surface waters impacted by varying concentrations of chloride and bromide. The sub-samples will be analyzed for individual THMs and formation kinetics will be determined. The EPAanticipates obtaining data for the formation of HAAs and nitrosamines, though THMs are the priority at this time.

8.2.3.5 QUALITY ASSURANCE SUMMARY

The initial QAPP, "Formation of Disinfection By-Products from Hydraulic Fracturing Fluids," was submitted on June 28, 2011, and approved on October 5, 2011 (US EPA, 2011h). On June 7, 2012, anaddendum was submitted and approved on June 28, 2012; this provided more details on modifications to EPA Method 300.1 for optimizing bromide/bromate recoveries in high-salt matrices. There are no deviations from existing QAPPs to report at this time.

A TSA was performed on March 15, 2012, for this research project. Five findings were observed, related to improved communication, project documentation, sample storage, and QA/QC checks. Recommended corrective actions were accepted to address the findings. Since the TSA wasperformed before data generation activities, no impact on future reported results is expected. It is anticipated that a second TSA will be performed as the project progresses.

As raw data are provided from the laboratories and results are reported, ADQs will be performed to verify that the quality requirements specified in the approved QAPP have been met. Data will be qualified if necessary based on these ADQs. Audits of data quality are scheduled for the first quarterof 2013 (none have been performed yet). The results of these ADQs will be reported with the summary of results in the final report.

8.2.4 ANALYTICAL METHOD DEVELOPMENT

8.2.4.1 RELATIONSHIP TO THE STUDY

Sample analysis is an integral part of the EPA's Plan to Study the Potential Impacts of Hydraulic Fracturing on Drinking Water Resources (US EPA, 2011e) and is clearly specified in research plans being carried out for the study's retrospective case studies, prospective case studies, and laboratorystudies. The EPA requires robust analytical methods to accurately and precisely determine the composition of hydraulic fracturing-related chemicals in ground and surface water, flowback and produced water, and treated wastewater.

8.2.4.2 PROJECT INTRODUCTION

Analytical methods enable accurate and precise measurement of the presence and quantities of different chemicals in various matrices. Since the quantification of the presence or absence of hydraulic fracturing-related chemicals will likely have substantial implications for the conclusions of the study, it is important that robust analytical methods exist for chemicals of interest.

In many cases, standard EPA methods that have been designed for a specific matrix or set ofmatrices can be used for this study. Standard EPA methods are peer-reviewed and officially promulgated methods that are used under different EPA regulatory programs. For example, EPAMethod 551.1 is being used to detect THMs as part of the Br-DBP research project (see Section 5.3) and EPA Method 8015D is being used to detect diesel range organics in ground and surface water samples collected as part of the retrospective case studies.

In other cases, standard EPA methods are nonexistent for a chemical of interest. In these situations, methods published in the peer-reviewed literature or developed by consensus standard organizations (e.g., the American Society for Testing and Materials, or ASTM) are used. However, these methods are rarely developed for or tested within matrices associated with the hydraulic fracturing process. In rare, but existing cases, where no documented methods exist, researchersgenerally develop their own methods for determining the concentrations of certain chemicals of interest. For these latter two situations, the analytical methods chosen must undergo rigorous testing, verification, and potential validation to ensure that the data generated they generate are of known and high quality. The EPA has identified selected chemicals found in hydraulic fracturing fluids and wastewater for the development and verification of analytical methods.

8.2.4.3 RESEARCH APPROACH

8.2.4.3.1 CHEMICAL SELECTION

Hydraulic fracturing-related chemicals include chemicals used in the injected fracturing fluid, chemicals found in flowback and produced water, and chemicals resulting from the treatment ofhydraulic fracturing wastewater (e.g., chlorination or bromination at wastewater treatment facilities).

Some of these chemicals are present due to the mobilization of naturally occurringchemicals within the geologic formations or through the degradation or reaction of the injected chemicals in the different environments (i.e., subsurface, surface and wastewater). The EPA hasidentified over 1,000 chemicals that are reported to be used in fracturing fluids or found in hydraulic fracturing wastewaters; these range from the inert and innocuous, such as sand and water, to reactive and toxic chemicals, like alkylphenols and radionuclides.

To help choose chemicals for analytical method testing, a group of EPA researchers and analytical laboratory chemists discussed the factors most important to their research needs and to the overallstudy. The following criteria were developed to identify a subset of the chemicals listed in Appendix A for initial analytical method testing activities:

- Frequency of occurrence in hydraulic fracturing fluids and wastewater
- Toxicity
- Mobility in the environment (expected fate and transport)
- Availability of instrumentation/detection systems for the chemical

8.2.3.2 ANALYTICAL METHOD TESTING AND DEVELOPMENT

Method Development

The EPA's process for analytical method development is shown in Figure 12. In the first step, and existing base method is identified for the specific chemical(s) of interest in a given matrix. Base methods may include prmulgated, standard methods or, if no standard methods are available, methods existing in peer-reviewed literature or developed through a consensus standard organization. Analytical methods may exist for specific chemicals or for a general class of chemicals (e.g., alcohols).

Once a candidate base method is selected, an initial QA/QC round of testing is conducted. Testing occurs first with spiked laboratory water samples to familiarize the analyst with the method procedure, eliminate any potential matrix interferences, and determine various QA/QC controlparameters, such as sensitivity, bias, precision, spike recovery, and analytical carry-over potential (sample cross-contamination). The results from the initial QA/QC testing are examined to determine if they meet the

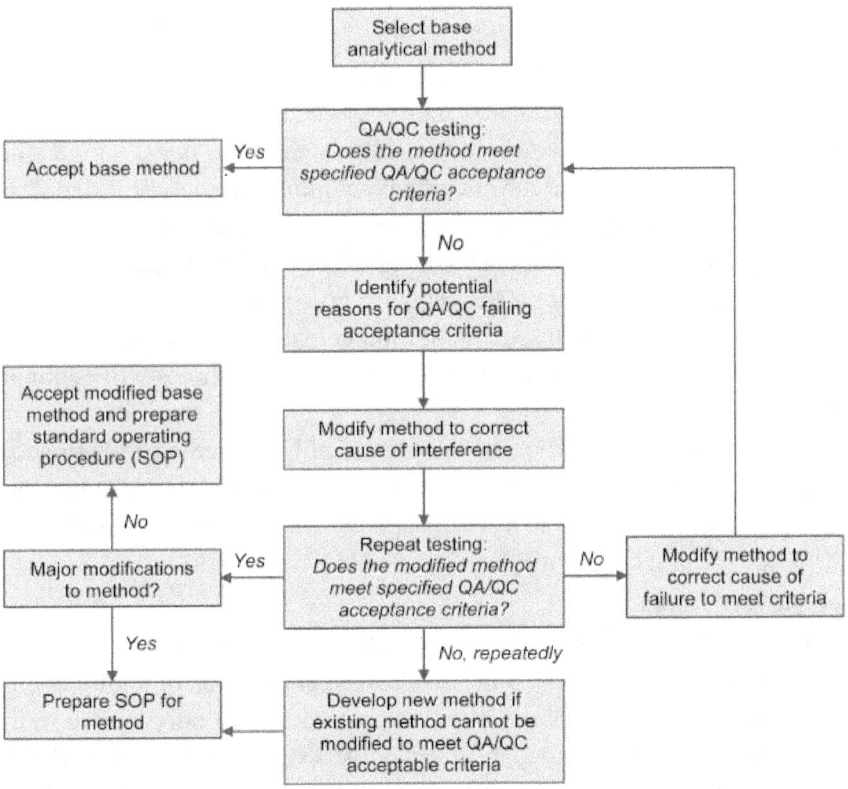

FIGURE 12: Flow diagram of hte EPA's process leading to the development of modified or new analytical methods.

acceptance criteria specified in the QAPP (US EPA, 2011g) and thus are sufficient to meet the needs of the research study. Some of the key QA/QC samples examined include:

- Standard and certified reference materials (where available) for bias
- Matrix and surrogate spikes for bias (when reference materials are not available) and matrix interferences
- Replicates for precision
- Blanks for analytical carry-over

If an acceptance criterion for any of the QA/QC samples is not met, the sample is typically re-run to ensure that the result is not a random event. If an acceptance criterion is repeatedly not met, a systematic problem is indicated, and method modification is undertaken to help reduce or eliminate the problem.

The method modification process can take many forms, depending on the specific circumstances,and may include changing sample preparation and cleanup techniques, solvents, filters, gas flow rates, temperature regimes, injector volumes, chromatographic columns, analytical detectors, etc. Once the method modification process is complete, the analysis is repeated as described above using spiked laboratory water samples. If the new QA/QC sample results meet the acceptance criterion, the method modification is deemed to have been successful for that matrix and an updated SOP is prepared. Additional testing in more complex water matrices will continue, if appropriate.

If testing and modification of the identified base method fails to accurately and precisely quantifythe chemical of interest and/or fails to have the sensitivity required by the research program, the EPA may undertake new method development activities.

Method Verification

Method verification determines the robustness of successfully tested andmodified analytical methods. This involves the preparation of multiple blind spiked samples (i.e., samples whose concentrations are only known to the sample preparer) by an independent chemist(i.e., one not associated with developing the method under testing and verification) and the submission of the samples to at least three other analytical laboratories participating in the verification process. Results from the method verification process can lead to either the acceptance of the method or re-evaluation and further testing of the method (US EPA, 1995).

Method Validation

The final possible step in analytical method testing and development is method validation. Method validation involves large, multi-laboratory, round robin studies and is generallyconducted by the EPA program offices responsible for the publication and promulgation of standard EPA methods.

8.2.4.4 STATUS, PRELIMINARY DATA, AND NEXT STEPS

Method development, testing, and verification are being conducted according to the procedures outlined in two QAPPs: "Quality Assurance Project

Plan for the Chemical Characterization of SelectConstituents Relevant to Hydraulic Fracturing" (US EPA, 2011g) and "Quality Assurance Project Plan for the Inter-Laboratory Verification and Validation of Diethylene Glycol, Triethylene Glycol, Tetraethylene Glycol, 2-Butoxyethanol and 2-Methoxyethanol in Ground and Surface Waters by Liquid Chromatography/ Tandem Mass Spectrometry" (US EPA, 2012r).

8.2.4.4.1 GLYCOLS AND RELATED COMPOUNDS

Glycols (diethylene glycol, triethylene glycol, and tetraethylene glycol) and the chemically relatedcompounds 2-butoxyethanol and 2-methoxyethanol are frequently used in hydraulic fracturing fluids and not naturally found in ground water. Thus, they may serve as reliable indicators of contamination of ground water from hydraulic fracturing activities. EPA Method 8015b is the gaschromatography-flame ionization detector method typically used to analyze for glycols; however, the sensitivity is not sufficient for the low-level analysis required for this project. Therefore, the EPA's Region 3 Environmental Science Center developed a method for the determination and quantification of these compounds using liquid chromatography-tandem mass spectrometry. The method is based on ASTM D7731-11e1 and EPA SW-846 Method 8321. The EPA is currentlyverifying this method to determine its efficacy in identifying and quantifying these compounds in drinking water and other water matrices associated with the hydraulic fracturing process.

8.2.4.4.2 ACRYLAMIDE

Acrylamide is often used as a friction reducer in injected hydraulic fracturing fluids (GWPC,2012b). EPA SW-846 Methods 8316 and 8032A are both suitable methods for the analysis of acrylamide. Method 8316 involves analysis by high-performance liquid chromatography withultraviolet detector at 195 nanometers, with a detection level of 10 micrograms per liter. This short wavelength, however, is not very selective for acrylamide (i.e., interferences are likely), andthe sensitivity is not adequate for measurements in water. Method 8032A involves the bromination of acrylamide, followed by gas chromatography-mass spectrometry analysis. This method is much more selective for acrylamide, and detection limits are much lower (0.03micrograms per liter). However, in complex matrices (e.g., hydraulic fracturing wastewater), the accuracy and precision of acrylamide analysis may be limited by poor extraction efficiency andmatrix interference.

To avoid reactions with other compounds present in environmental matrices and to lower the detection limit, the EPA is developing a new analytical method for the determination of acrylamideat very low levels in water containing a variety of additives. The method currently under development involves solid phase extraction with activated carbon followed by quantitation by liquid chromatography-tandem mass spectrometry using an ion exclusion column. The EPA has begun the multi-laboratory verification of the method.

8.2.4.4.3 ETHOXYLATED ALCOHOLS

Surfactants are often added to hydraulic fracturing fluids to decrease liquid surface tension and improve fluid passage through pipes. Most of the surfactants used are alcohols or some derivative of an ethoxylated compound, typically ethoxylated alcohols. Many ethoxylated alcohols and ethoxylated alkylphenols biodegrade in the environment, but often the degradation byproductsare toxic (e.g., nonylphenol, a degradation product of nonylphenol ethoxylate, is an endocrine disrupting compound) (Talmage, 1994). No standard method currently exists for the determination of ethoxylated alcohols; therefore, the EPA is developing a quantitative method for ethoxylated alcohols. ASTM Method D 7458-09 and USGS Method Number O1433-01 were used as starting points for this method development effort; both of these methods involve solid-phaseextraction followed by liquid chromatography-tandem mass spectrometry quantitation. These methods both allow the analysis of nonylphenol diethoxylate and alkylphenols, but there are currently no standard methods for the analysis of the full range of nonylphenol ethoxylate oligomers (EO3–EO20) or alcohol ethoxylate oligomers (C12–15EOx, where x = 2–20). This method SOP is being prepared and will be followed by method verification.

8.2.4.4.4 DISINFECTION BYPRODUCTS

Flowback and produced water can contain high levels of TDS, which may include bromide and chloride (US EPA, 2012d). In some cases, treatment of flowback and produced water occurs atWWTFs, which may be unable to effectively remove bromide and chloride from hydraulic fracturing wastewater before discharge. The presence of bromide ions in source waters undergoing chlorination disinfection may lead to the formation of brominated DBPs—including bromate, THMs, and HAAs—upon reaction with natural organic material (Richardson, 2003). Brominated DBPs are considerably more toxic than corresponding chlorinated DBPs (Plewa et al., 2004; Richardson et al., 2007) and have higher molecular weight. Therefore, on an equal molar basis, brominated DBPs will have a greater concentration by weight than chlorinated DBPs, hence leading to a greater likelihood of exceeding the total THM and HAA MCLs that are stipulated in weight concentrations (0.080 and 0.060 milligrams per liter, respectively). Accordingly, it is important to assess and quantify the effects of flowback and produced water on DBP generation.

Analytical methods for the measurement of bromide and bromate in elevated TDS matrices are currently being developed. EPA Method 300.1 is being modified to use a mass spectrometer rather that an electroconductivity detector, which is unable to detect bromide and bromate in the presence high anion concentrations (SO_4^{2-}, NO_2^-, NO_3^-, F^-, Cl^-). The mass spectrometer allows selected ion monitoring specifically for the two natural stable isotopes of bromine (^{79}Br and ^{81}Br), with minimal interference from other anions in the high-salt matrix. Interference of the bromideand bromate response in the mass spectrometer are being assessed by comparing instrument responses to solutions of bromide and bromate in deionized water with selected anions over a range of ratios typically encountered in hydraulic fracturing wastewater samples (US EPA, 2012d). Interference concentration thresholds are being established, and a suitable sample dilution methodis being developed for the quantification of bromide and bromate in actual hydraulic fracturing wastewater samples. Method detection limits and lowest concentration minimum reporting levels are being calculated for bromide and bromate in high-salt matrices according to EPA protocols (USEPA, 2010h).

8.2.4.4.5 RADIONUCLIDES

Gross α and β analyses measure the radioactivity associated with gross α and gross β particles that are released during the natural decay of radioactive elements, such as uranium, thorium, andradium. Gross α and β analyses are typically used to screen hydraulic fracturing wastewater in order to assess gross levels of radioactivity. This information can be used to identify waters needing radionuclide-specific characterization. The TDS and organic content characteristic of hydraulic fracturing wastewater, however, interferes with currently accepted methods for gross α and β analyses. The QAPP for testing and developing gross α and β analytical methods is in development, and, after it is approved, work will begin.

8.2.4.4.6 INORGANIC CHEMICALS

In addition to the potential mobilization of naturally occurring radioactive elements, hydraulic fracturing may also release other elements from the fractured shales, tight sands, and coalbeds, notably heavy metals such as barium and strontium. Inorganic compounds may also be added to hydraulic fracturing fluids to perform various functions (e.g., cross-linkers using borate salts, brinecarrier fluids using potassium chloride, and pH-adjusting agents using sodium carbonates) (US EPA, 2011e). Due to the injection or release of naturally occurring metals in unknown quantities, it isessential that analytical methods for the determination of inorganic elements in waters associated with hydraulic fracturing be robust and free from interferences that may mask true concentrations.

The EPA SW-846 Method 6010, employing inductively coupled plasma-optical emission spectrometry, will be used as a base method for major elements while SW-846 Method 6020 basedon inductively coupled plasma-mass spectrometry will be used as a base method for trace elements.70 These methods will be tested and potentially modified for detection of major and traceelements in hydraulic fracturing wastewater.

8.2.4.5 QUALITY ASSURANCE SUMMARY

Three QAPPs have been prepared for the analytical method testing research program. The firstQAPP, "Quality Assurance Project Plan for the Chemical Characterization of Select Constituents Relevant to Hydraulic Fracturing" (US EPA, 2011g), is the broad general QAPP for the methodsdevelopment research project. The QAPP was approved on September 1, 2011. In order to maintain high QA standards and practices throughout the project, a surveillance audit was performed on November 15, 2011. The purpose of the surveillance audit was to examine the processes associated with the in-house extraction of ethoxylated alcohols. Three recommendations were identified and have been accepted.

The second QAPP, "Formation of Disinfection By-Products from Hydraulic Fracturing FluidConstituents Quality Assurance Project Plan," (US EPA, 2011h), provides details on modifications to EPA Method 300.1 for optimizing bromide/bromate recoveries in high-salt matrices. The QAPP was approved on October 5, 2011, and the addendum for bromide/bromate analytic method development was approved on June 28, 2012. There are no deviations from existing QAPPs to report at this time. A surveillance audit was performed in March 2011 before the analytical methodaddendum (June 28, 2012); therefore, the analytical method development for bromide/bromate has not yet been audited.

The third QAPP, "Quality Assurance Project Plan for the Inter-Laboratory Verification andValidation of Diethylene Glycol, Triethylene Glycol, Tetraethylene Glycol, 2-Butoxyethanol and 2Methoxyethanol in Ground and Surface Waters by Liquid Chromatography/Tandem MassSpectrometry" (US EPA, 2012r), was prepared specifically for the verification of the EPA Region 3 SOP. The QAPP was approved on April 4, 2012. Since then, two surveillance audits and two internal TSAs have been performed, specifically looking at procedures related to glycol standardpreparation and analysis. The two surveillance audits resulted in one case of potentially mislabeled samples during stock solution preparation. The potential mislabeling was already identified anddocumented by the researchers involved and corrective action taken.

The designated EPA QA Manager found the methods in use satisfactory and further recommendations for improving the QA process were unnecessary. The internal TSAs also yielded no acts, errors, or omissions that wouldhave a significant adverse impact on the quality of the final product.

8.3 RESEARCH PROGRESS SUMMARY AND NEXT STEPS

This report describes the progress made for each of the research projects conducted as part of the EPA's Study of the Potential Impacts of Hydraulic Fracturing on Drinking Water Resources. This chapter provides an overview of the progress made for each research activity as well as the progress made for each stage of the water cycle. It also describes, in more detail, the report of results.

8.3.1 SUMMARY OF PROGRESS BY RESEARCH ACTIVITY

The EPA is using a transdisciplinary research approach to investigate the potential relationship between hydraulic fracturing and drinking water resources. This approach includes compiling andanalyzing data from existing sources, evaluating scenarios using computer models, carrying out laboratory studies, assessing the toxicity associated with hydraulic fracturing-related chemicals, and conducting case studies.

Analysis of Existing Data

To date, data from seven sources have been obtained for review and ongoing analysis, including:

- Information provided by nine hydraulic fracturing service companies.
- 333 well files supplied by nine oil and gas operators.
- Over 12,000 chemical disclosure records from FracFocus, the national hydraulic fracturing chemical registry managed by the Ground Water Protection Council and the Interstate Oil and Gas Compact Commission.
- Spill reports from four different sources, including databases from the National Response Center, Colorado, New Mexico, and Pennsylvania.

As part of its literature review, the EPA has compiled, and continues to search for, relevant literature. This includes documents provided by stakeholders and recommended by the Science Advisory Board during its review of the draft study plan. A Federal Register notice requesting peer-reviewed data and publications relevant to the study, including information on advances in industry practices and technologies, has recently been published (US EPA, 2012u).

Scenario Evaluations

Potential impacts to drinking water sources from withdrawing large volumes of water in both a semi-arid and a humid river basin—the Upper Colorado River Basin in the westand the Susquehanna River Basin in the east—are being assessed. Additionally, complex computer models are being used to explore the possibility of subsurface gas and fluid migration from deepshale formations to overlying aquifers in six different scenarios. These scenarios include poor well construction and hydraulic communication via fractures (natural and created) and nearby existing wells. As a first step, the subsurface migration simulations will examine realistic scenarios to assess the conditions necessary for hydraulic communication rather than the probability of migration occurring. In a separate research project, the EPA is using surface water transport models to estimate concentrations of bromide and radium at public water supply intakes downstream from wastewater treatment facilities that discharge treated hydraulic fracturing wastewater.

Laboratory Studies

The ability to analyze and determine the presence and concentration of chemicals in environmental samples is critical to the EPA's study. In most cases, standard EPA methods are being used for laboratory analyses. In other cases, however, standard methods do not exist for the low-level detection of chemicals of interest or for use in the complex matricesassociated with hydraulic fracturing wastewater. Where necessary, existing analytical methods are being tested, modified, and verified for use in this study and by others. Analytical methods are currently being tested and modified for several classes of chemicals, including glycols, acrylamides, ethoxylated alcohols, DBPs, radionuclides, and inorganic chemicals.

Laboratory studies focusing on the potential impacts of inadequate treatment of hydraulicfracturing wastewater on drinking water resources are being planned and conducted. The studies include assessing the ability of hydraulic fracturing wastewater to create brominated DBPs and testing the efficacy of common wastewater treatment processes on removing selectedcontaminants from hydraulic fracturing wastewater. Samples of surface water, raw hydraulic fracturing wastewater, and treated effluent have been collected for the source apportionmentstudies, which aim to identify the source of high chloride and bromide levels in rivers accepting treated hydraulic fracturing wastewater.

Toxicity Assessment

The EPA has evaluated data to identify chemicals reportedly used in hydraulicfracturing fluids from 2005 to 2011 and chemicals found in flowback and produced water. Appendix A contains tables of these chemicals, with over 1,000 chemicals identified. Chemical, physical, and toxicological properties have been compiled for chemicals with known chemical-structures. Existing models are being used to estimate properties in cases where information is lacking. At this time, the EPA has not made any judgment about the extent of exposure to these chemicals when used in hydraulic fracturing fluids or found in hydraulic fracturing wastewater, or their potential impacts on drinking water resources.

Case Studies

Two rounds of sampling at all five retrospective case study locations have beencompleted. In total, water samples have been collected from over 70 domestic water wells, 15 monitoring wells, and 13 surface water sources, among others. A third round of sampling is expected to occur this fall in Las Animas and Huerfano Counties, Colorado; Dunn County, North Dakota; and Wise County, Texas. Additional sampling in Bradford and Washington Counties, Pennsylvania, is projected to take place in spring 2013.

The EPA continues to work with industry partners to plan and begin research activities forprospective case studies.

8.3.2 SUMMARY OF PROGRESS BY WATER CYCLE STAGE

Figures 13 and 14 illustrate the research underway for each stage of the hydraulic fracturing water cycle. The fundamental research questions and research focus areas are briefly described below for each water cycle stage.

Water Acquisition

What are the possible impacts of large volume water withdrawals from ground and surface waters on drinking water resources? Work in this area focuses on understanding the volumes and sources of water needed for hydraulic fracturing operations, and the potential impactsof water withdrawals on drinking water quantity and quality. Effects of recently emerging trends in water recycling will be considered in the report of results.

Chemical Mixing

What are the possible impacts of surface spills on or near well pads of hydraulic fracturing fluids on drinking water resources? Spill reports from several databases are beingreviewed to identify volumes and causes of spills of hydraulic fracturing fluids and wastewater. Information on the chemicals used in hydraulic fracturing fluids and their known chemical,physical, and toxicological properties has been compiled.

Well Injection

What are the possible impacts of the injection and fracturing process on drinking water resources? Work currently underway is focused on identifying conditions that may be associatedwith the subsurface migration of gases and fluids to drinking water resources. The EPA is exploring gas and fluid migration due to inadequate well construction as well as the presence of nearby natural faults and fractures or man-made wells.

Flowback and Produced Water

What are the possible impacts of surface spills on or near well pads of flowback and produced water on drinking water resources? As with chemical mixing, research in this area focuses on reviewing spill reports of flowback and produced water as well as collectinginformation on the composition of hydraulic fracturing wastewater. Known chemical, physical, and toxicological properties of the components of flowback and produced water are being compiled.

Wastewater Treatment and Waste Disposal

What are the possible impacts of inadequate treatment of hydraulic fracturing wastewater on drinking water resources? Work in this area focuses on evaluating treatment and disposal practices for hydraulic fracturing wastewater. Since somewastewater is known to be discharged to surface water after treatment in POTWs or commercial treatment systems, the EPA is investigating the efficacy of common treatment processes at removing selected components in flowback and produced water. Potential impacts to downstream public water supplies from discharge of treated hydraulic fracturing wastewater are also being investigated, including the potential for the formation of Br-DBPs.

8.3.3 REPORT OF RESULTS

This is a status report, describing the current progress made on the research projects that make up the agency's Study of the Potential Impacts of Hydraulic Fracturing on Drinking Water Resources. Results from individual research projects will undergo peer review prior to publication either as articles in scientific journals or EPA reports. The EPA plans to synthesize results from the publishedreports with a critical literature review in a report of results that will answer as completely as possible the research questions identified in the Study Plan. The report of results has been determined to be a Highly Influential Scientific Assessment and will undergo peer review by the Science Advisory Board. Ultimately, the results of this study are expected to inform the public and provide policymakers at all levels with high-quality scientific knowledge that can be used indecision-making processes.

The report of results will also be informed by information provided through the ongoing stakeholder engagement process. This process is anticipated to provideagency scientists with updates on changes in industry practices and technologies relevant to the study. While the EPA expects hydraulic fracturing technology to develop between now and the publication of the report of results, the agency believes that the research described here willprovide timely information that will contribute to the state of knowledge on the relationship between hydraulic fracturing and drinking water resources. For example, some companies mayadopt new injection or wastewater treatment technologies and practices, while others may continue to use current technologies and practices. Many of the practices, including wastewatertreatment and disposal technologies used

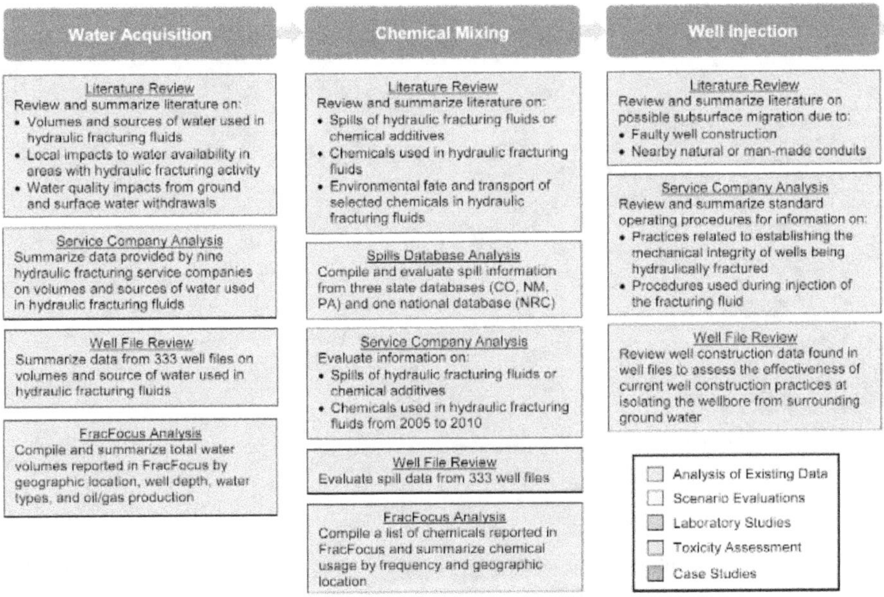

FIGURE 13A: Summary of research projects underway for the first three stages of the hydraulic fracturing water cycle.

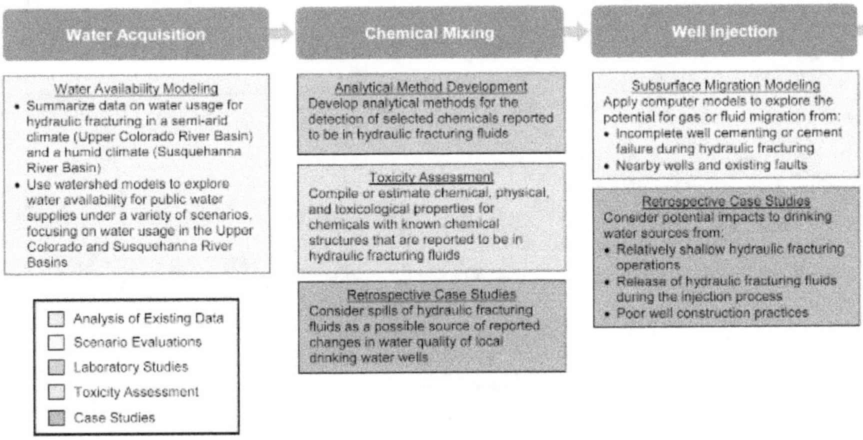

FIGURE 13B: Summary of research projects underway for the first three stages of the hydraulic fracturing water cycle.

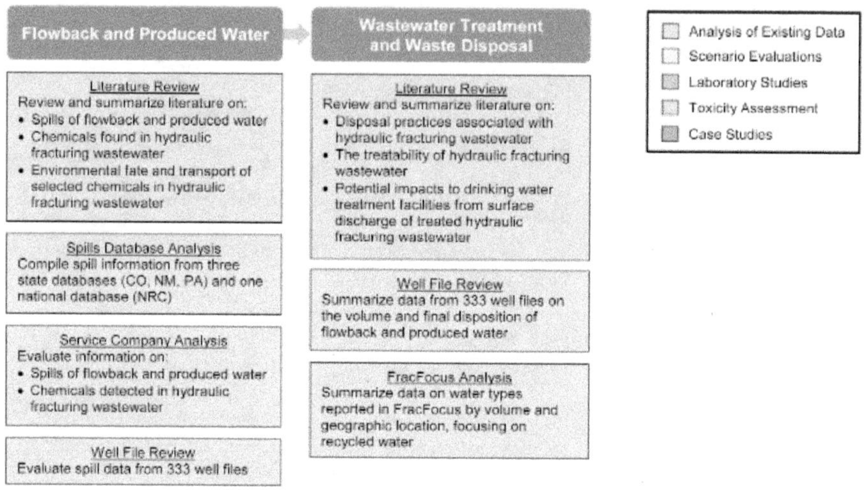

FIGURE 14A: Summary of research projects underway for the last two stages of the hydraulic fracturing water cycle.

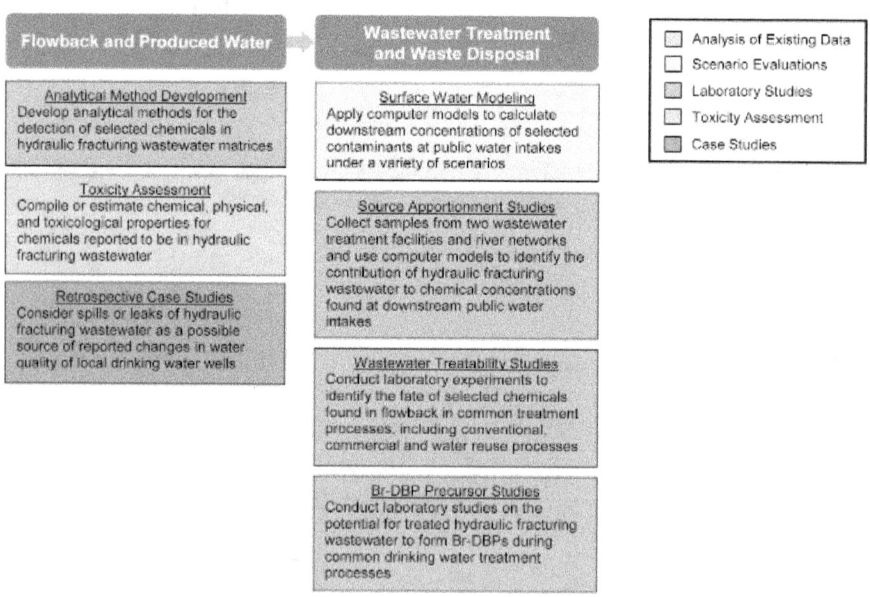

FIGURE 14B: Summary of research projects underway for the last two stages of the hydraulic fracturing water cycle.

by POTWs, are not expected to change significantly between now and the report of results.

Results from the study are expected to identify potential impacts to drinking water resources, if any, from water withdrawals, the fate and transport of chemicals associated with hydraulicfracturing, and wastewater treatment and waste disposal. Information on the toxicity of hydraulic fracturing-related chemicals is also being gathered. Although these data may be used to assess the potential risks to drinking water resources from hydraulic fracturing activities, the report of results is not intended to quantify risks. Results presented in the report of results will be appropriately-discussed and all uncertainties will be described.

The EPA will strive to make the report of results as clear and definitive as possible in answering all of the primary and secondary research questions, at that time. Science and technology evolve, however: the agency does not believe that the report of results will provide definitive answers onall research questions for all time and fully expects that additional research needs will be identified.

8.3.4 CONCLUSIONS

This report presents the EPA's progress in conducting its Study of the Potential Impacts of Hydraulic Fracturing on Drinking Water Resources. Each project progress report describes the project's relationship to the study, research methods, and status and summarizes QA activities. Information presented as part of this report cannot be used to draw conclusions about potential impacts to drinking water resources from hydraulic fracturing.

The EPA is committed to conducting a study that uses the best available science, independent sources of information, and a transparent, peer-reviewed process that ensures the validity andaccuracy of the results. The EPA will seek input from individual members of an ad hoc expert panel convened under the auspices of the EPA's Science Advisory Board. Information about this process is available at http://yosemite.epa.gov/sab/sabproduct.nsf/02ad90b136fc21ef85256eba00436459/ b436304ba804e3f885257a5b00521b3b!OpenDocument. The individual members of the ad hoc panel will consider public comment. The EPA will consider feedback from the individual experts, asinformed by the public's comments, in the development of the report of results.

REFERENCES

1. Abbott, P. O. 1985. Description of Water-Systems Operations in the Arkansas River Basin, Colorado. Water-Resources Investigations Report 85-4092. US Geological Survey. 79 p. Available athttp://pubs.er.usgs.gov/publication/wri854092. Accessed November 14, 2012.
2. Abbott, P. O., Geldon, A. L., Cain, D., Hall, A. P. and Edelmann, P. 1983. Hydrology of Area 61, Northern Great Plains and Rocky Mountain Coal Provinces, Colorado and New Mexico.Open-File Report 83-132. US Geological Survey. 105 p. Available at http://pubs.er.usgs.gov/ publication/ofr83132. Accessed November 14, 2012.
3. Toxicity Prediction by Komputer Assisted Technology (TOPKAT) version 3.5. 2012. AccelrysSoftware, Inc., San Diego, California.
4. ChemFolder version 12.00. 2008. ACD Labs, Toronto, Ontario.
5. ALL Consulting. 2003. Handbook on Coalbed Methane Produced Water: Management and Beneficial Use Alternatives. ALL Consulting for Ground Water Protection Research Foundation, USDepartment of Energy, and Bureau of Land Management. Available at http://www.allllc.com/publicdownloads/CBM_BU_Screen.pdf. Accessed November 30, 2012.
6. Ambrose, R. B., Hill, S. I. and Mulkey, L. A. 1983. User's Manual for the Chemical Transport and Fate Model (TOXIWASP), Version 1. EPA 600/3-83-005. US Environmental Protection Agency. 188 p. Available at http://www.epa.gov/nscep/index.html. Accessed November 30, 2012.
7. Ambrose, R. B. and Wool, T. A. 2009. WASP7 Stream Transport Model Theory and User's Guide. EPA/600/R-09/100. US Environmental Protection Agency. 43 p. Available athttp://www.epa.gov/nscep/index.html. Accessed November 30, 2012.
8. American Petroleum Institute. 2010. Water Management Associated with Hydraulic Fracturing. Available at http://www.api.org/Standards/new/api-hf2.cfm. Accessed January 20, 2011.
9. Ashley, G. H. and Robinson, J. F. 1922. The Oil and Gas Fields of Pennsylvania. Mineral Resource Report M1. Pennsylvania Bureau of Topographic and Geological Survey, Harrisburg, Pennsylvania.
10. Agency for Toxic Substances and Disease Registry. 2007. Toxicological Profile for Benzene. Available at http://www.atsdr.cdc.gov/toxprofiles/tp.asp?id=40&tid=14. Accessed December 4, 2012.
11. Bradford County Government. Natural Gas Information. Chart of Permitted Gas Wells. Available athttp://bradfordcountypa.org/Natural-Gas.asp?specifTab=2. Accessed August 16, 2012.
12. Brathwaite, L. D. 2009. Shale-Deposited Natural Gas: A Review of Potential. Presented at California Energy Commission, Sacramento, California. Available at http://

www.energy.ca.gov/2009publications/CEC-200-2009-005/CEC-200-2009-005-SD. PDF. Accessed November 30, 2012.

13. Brown, D., Bridgeman, J. and West, J. R. 2011. Predicting chlorine decay and THM formation in water supply systems. Reviews in Environmental Science and Biotechnology 10 (1): 79-99.
14. Bruner, K. R. and Smosna, R. 2011. A Comparative Study of the Mississippi Barnett Shale, FortWorth Basin, and Devonian Marcellus Shale, Appalachian Basin. DOE/NETL-2011/1478. URS Corporation for US Department of Energy. Available at http://www.netl.doe.gov/technologies/oil-gas/publications/brochures/DOE-NETL-2011-1478%20MarcellusBarnett.pdf. Accessed December 5, 2012.
15. Cadmus Group Inc. 2012a. Quality Assurance Project Plan: Modeling the Impact of HydraulicFracturing on Water Resources Based on Water Acquisition Scenarios. Cadmus Group Inc. for US Environmental Protection Agency. Available at http://www.epa.gov/hfstudy/ qapps.html. Accessed November 30, 2012.
16. Cadmus Group Inc. 2012b. Supplemental Programmatic Quality Assurance Project Plan: National Hydraulic Fracturing Study Evaluation of Existing Production Well File Contents. Cadmus Group Inc. for US Environmental Protection Agency. Available at http://www.epa.gov/hfstudy/qapps.html. Accessed December 4, 2012.
17. Cao, J., Li, H., Chow, J. C., Watson, J. G., Lee, S., Rong, B., Dong, J. G. and Ho, K. F. 2011. Chemical composition of indoor and outdoor atmospheric particles at Emperor Qin's Terra-cotta Museum. Aerosol and Air Quality Research 11 (1): 70-79.
18. Cappa, F. and Rutqvist, J. 2011a. Impact of CO2 geological sequestration on the nucleation of earthquakes. Geophysical Research Letters 38 (17): L17313.
19. Cappa, F. and Rutqvist, J. 2011b. Modeling of coupled deformation and permeability evolutionduring fault reactivation induced by deep underground injection of CO2. International Journal of Greenhouse Gas Control 5 (2): 336-346.
20. Cappa, F. and Rutqvist, J. 2012. Seismic rupture and ground accelerations induced by CO2 injection in the shallow crust. Geophysical Journal International 190 (3): 1784-1789.
21. Cappa, F., Rutqvist, J. and Yamamoto, K. 2009. Modeling crustal deformation and rupture processes related to upwelling of deep CO2-rich fluids during the 1965–1967 Matsushiro earthquake swarm in Japan. Journal of Geophysical Research 114 (10): B10304.
22. Carlson, C. G. 1985. Geology of McKenzie County, North Dakota. North Dakota Geological Survey Bulletin 80 Part 1. North Dakota Geological Survey in cooperation with the US Geological Survey, North Dakota State Water Commission, and the McKenzie County WaterManagement District. 54 p. Available at http://www.swc.state.nd.us/4dlink9/4dcgi/ GetSubContentPDF/PB-285/McKenzie_Part_1.pdf. Accessed December 5, 2012.
23. Carter, K. M. and Harper, J. A. 2002. Oil and Gas Prospects in Northeastern Pennsylvania. In JD Inners and DD Braun, editors, From Tunkhannock to Starrucca: Bluestone, Glacial Lakes, and Great Bridges in the "Endless Mountains" of Northeastern Pennsylvania. Guidebook for the 67th Annual Field Conference of Pennsylvania Geologists. Field Conference of Pennsylvania Geologists, Tunkhannock, Pennsylvania. pp. 15-31.
24. CDM and GBSM. 2004. Statewide Water Supply Initiative Report. CDM and GBSM for ColoradoWater Conservation Board. Available at http://cospl.coalliance.org/fedora/repository/ co:3490. Accessed December 3, 2012.

25. Colorado Department of Public Health and Environment. 2012. Colorado's Section 303(D) List of Impaired Water and Monitoring and Evaluation List. Available at http://www.colorado.gov/cs/Satellite?blobcol=urldata&blobheadername1=Content-Disposition&blobheadername2=ContentType&blobheadervalue1=inline%3B+filename%3D%22Section+303(d)+List+and+Colorad o+Monitoring+and+Evaluation+List+(Regulation+%2393).pdf%22&blobheadervalue2=application%2Fpdf&blobkey=id&blobtable=MungoBlobs&blobwhere=1251806966544&ssbina ry=true. Accessed November 27, 2012.
26. Chapman, E. C., Capo, R. C., Stewart, B. W., Kirby, C.S., , Hammack, R. W., Schroeder, K. T. andEdenborn, H. M. 2012. Geochemical and strontium isotope characterization of produced water from Marcellus shale natural gas extraction. Environmental Science & Technology 46 (6): 3545-3553.
27. Molecular Operating Environment (MOE) Linux version 2011.10. 2011. Chemical Computing Group, Montreal, Quebec.
28. Colorado Oil and Gas Conservation Commission. 2011. Oil and Gas Industry Spills and Releases. Available at http://dnr.state.co.us/SiteCollectionDocuments/SpillsAndReleases.pdf. Accessed December 4, 2012.
29. Colorado Oil and Gas Conservation Commission. 2012a. Fact Sheet: Water Sources and Demand for the Hydraulic Fracturing of Oil and Gas Wells in Colorado from 2010 through 2015. Available at http://cogcc.state.co.us/Library/Oil_and_Gas_Water_Sources_Fact_Sheet.pdf. Accessed December 12, 2012.
30. Colorado Oil and Gas Conservation Commission. 2012b. Production Inquiry. Colorado Oil and GasInformation System. Available at http://cogcc.state.co.us. Accessed October 17, 2012.
31. Colborn, T., Kwiatkowski, C., Schultz, K. and Bachran, M. 2011. Natural Gas Operations from a Public Health Perspective. Human and Ecological Risk Assessment 17 (5): 1039-1056.
32. Coleman, J., Milici, R., Cook, T., Charpentier, R., Kirschbaum, M., Klett, T., Pollastro, R. and Schenk, C. 2011. Assessment of Undiscovered Oil and Gas Resources of the Devonian Marcellus Shale of the Appalachian Basin Province. National Assessment of Oil and Gas Fact Sheet 20113092. US Geological Survey. 2 p. Available at http://pubs.usgs.gov/fs/2011/3092/. Accessed November 30, 2012.
33. Cooper, J. R., Crelling, J. C., Rimmer, S. M. and Whittington, A. G. 2007. Coal Metamorphism by Igneous Intrusion in the Raton Basin, Colorado, and New Mexico: Implications forGeneration of Volatiles. International Journal of Coal Geology 71 (1): 15-27.
34. Davies, R. J. 2011. Methane contamination of drinking water caused by hydraulic fracturing remains unproven. Proceedings of the National Academy of Sciences 108 (43): E871.
35. Deng, Z.-Q., Bengtsson, L., Singh, V. P. and Adrian, D. D. 2002. Longitudinal dispersion coefficient insingle-channel streams. Journal of Hydraulic Engineering 128 (10): 901-916.
36. DiToro, D. M., Fitzpatrick, J. J. and Thomann, R. V. 1981. Documentation for Water Quality Analysis Simulation Program (WASP) and Model Verification Program (MVP). EPA 600/3-81-044. USEnvironmental Protection Agency, Duluth, Minnesota.

37. Dolly, E. D. and Meissner, F. F. 1977. Geology and Gas Exploration Potential, Upper Cretaceous and Lower Tertiary Strata, Northern Raton Basin, Colorado. In H.K. Veal, editors, ExplorationFrontiers of the Central and Southern Rockies: Rocky Mountain Association of Geologists Guidebook. Rocky Mountain Association of Geologists, Denver, Colorado. pp. 247-270.
38. Donigian Jr., A. S. Bibliography for HSPF and Related References. Available athttp://www.aquaterra.com/resources/hspfsupport/hspfbib.php. Accessed
39. Donigian Jr., A. S., Bicknell, B. and Bandurraga, M. 2011. Watershed Modeling for the Santa Clara River in Southern California. ASCE EWRI Conference Proceedings on CD-ROM: Bearing Knowledge for Sustainability, World Environmental and Water Resources Congress, PalmSprings, California. Available at http://www.aquaterra.com/resources/pubs/index.php. Accessed November 21, 2012.
40. Eastern Research Group Inc. 2011. Final Quality Assurance Project Plan for the Evaluation of Information on Hydraulic Fracturing. Eastern Research Group Inc. for US Environmental Protection Agency. Available at http://www.epa.gov/hfstudy/qapps.html. AccessedDecember 11, 2012.
41. Engelder, T. 2012. Peer review letter for Warner, N.R., Jackson, R.B., Darrah, T.H., Osborn, S.G., Down, A., Zhao, K., White, A., and Vengosh, A. (2012), Geochemical Evidence for Possible Natural Migration of Marcellus Formation Brine to Shallow Aquifers in Pennsylvania. Proceedings of the National Academy of Sciences 109 (30): 11961-11966.
42. ESRI. 2010a. US Counties Shapefile. ESRI Data & Maps Series. ESRI, Redlands, California.
43. ESRI. 2010b. US States Shapefile. ESRI Data & Maps Series. ESRI, Redlands, California.
44. ESRI. 2012. US Major Water Shapefile. ESRI Data & Maps Series. ESRI, Redlands, California.
45. Farhadian, M., Vachelard, C., Duchez, D. and Larroche, C. 2008. In situ bioremediation of monoaromatic pollutants in groundwater: A review. Bioresource Technology 99 (13): 5296– 5308.
46. Flores, R. M. 1993. Geologic and geomorphic controls of coal developments in some tertiary rocky-mountain basins, USA. International Journal of Coal Geology 23 (1-4): 43-73.
47. Flores, R. M. and Bader, L. R. 1999. A Summary of Tertiary Coal Resources of the Raton Basin,Colorado and New Mexico. In Resource Assessment of Tertiary Coalbeds and Zones in the Northern Rocky Mountains and Great Plains Region. Professional Paper 1625-A. US Geological Survey. 35 p. Available at http://pubs.usgs.gov/pp/p1625a/. Accessed December 5, 2012.
48. Freeman, C. M. 2010. Study of Flow Regimes in Multiply-Fractured Horizontal Wells in Tight Gasand Shale Gas Reservoir Systems. MS Thesis. Texas A & M University, College Station, Texas.
49. Freeman, C. M., Moridis, G. J. and Blasingame, T. A. 2011. A numerical study of microscale flow behavior in tight gas and shale gas reservoir systems. Transport in Porous Media 90 (1): 253-268.
50. Freeman, C. M., Moridis, G. J., Ilk, D. and Blasingame, T. A. 2009a. A Numerical Study of Microscale Flow Behavior in Tight Gas and Shale Gas Reservoir Systems. Proceedings ofthe 2009 TOUGH Symposium, Berkeley, California. Available at http://

www.netl.doe.gov/kmd/RPSEA_Project_Outreach/07122-23%20-%20TOUGH%20 2009%20Symposium%20-%20914-09.pdf. Accessed November 30, 2012.

51. Freeman, C. M., Moridis, G. J., Ilk, D. and Blasingame, T. A. 2009b. A Numerical Study of Performance for Tight Gas and Shale Gas Reservoir Systems. Presented at SPE Annual Technical Conference and Exhibition, New Orleans, Louisiana. Available at http://www.onepetro.org/mslib/servlet/onepetropreview?id=SPE-124961-MS. Accessed November 30, 2012.
52. Freeman, C. M., Moridis, G. J., Michael, G. E. and Blasingame, T. A. 2012. Measurement, Modeling, and Diagnostics of Flowing Gas Composition Changes in Shale Gas Wells. Presented at SPE LatinAmerican and Caribbean Petroleum Engineering Conference, Mexico City, Mexico. Available at http://www.onepetro.org/mslib/servlet/onepetropreview?id=SPE-153391-MS. Accessed November 30, 2012.
53. Gassman, P. W., Reyes, M. R., Green, C. H. and Arnold, J. G. 2007. The soil and water assessment tool:historical development, applications, and future research directions. Transactions of the American Society of Aricultural and Biological Engineers 50 (4): 1211-1250.
54. Greg Lewicki & Associates. 2001. Raton Basin Coal Mine Feature Inventory. Greg Lewicki &Associates for Colorado Oil and Gas Conservation Commission. Available at http://cospl.coalliance.org/fedora/repository/co:3284. Accessed December 3, 2012.
55. Ground Water Protection Council. 2012a. Disclosures Reported by Month in Frac Focus. Presentation. Ground Water Protection Council, Oklahoma City, Oklahoma.
56. Ground Water Protection Council. 2012b. FracFocus well records: January 1, 2011, through February 27, 2012. Available at http://www.fracfocus.org/.
57. Ground Water Protection Council and ALL Consulting. 2009. Modern Shale Gas Development in the US: A Primer. Ground Water Protection Council and ALL Consulting for US Department ofEnergy. Available at http://www.netl.doe.gov/technologies/oilgas/publications/epreports/shale_gas_primer_2009.pdf. Accessed December 12, 2012.
58. Hairer, E., Norsett, S. P. and Wanner, G. 1991. Solving Ordinary Differential Equations I: NonstiffProblems. Springer-Verlag, Berlin, Germany.
59. Haritash, A. K. and Kaushik, C. P. 2009. Biodegradation aspects of polycyclic aromatic hydrocarbons (PAHs): A review. Journal of Hazardous Materials 169 (1-3): 1-15.
60. Harrison, S. S. 1983. Evaluating system for ground-water contamination hazards due to gas-welldrilling on the glaciated Appalachian Plateau. Ground Water 21 (6): 689-700.
61. Harrison, S. S. 1985. Contamination of aquifers by overpressurizing the annulus of oil and gas wells. Ground Water 23 (7): 317-324.
62. Hayes, T. 2009. Sampling and Analysis of Water Streams Associated with the Development of Marcellus Shale Gas. Gas Technology Institute for Marcellus Shale Coalition. Available at http://eidmarcellus.org/wp-content/uploads/2012/11/MSCommission-Report.pdf. Accessed November 30, 2012.
63. Healy, R. W., Bartos, T. T., Rice, C. A., McKinley, M. P. and Smith, B. D. 2011. Groundwater chemistry near an impoundment for produced water, Powder River Basin, Wyoming USA. Journal of Hydrology 403 (1-2): 37-48.

64. Healy, R. W., Rice, C. A., Bartos, T. T. and McKinley, M. P. 2008. Infiltration from an impoundment for coal-bed natural gas, Powder River Basin, Wyoming: Evolution of water and sediment chemistry. Water Resources Research 44 (6): W06424.
65. Holditch, S. A. 1993. Completion methods in coal-seam reservoirs. Journal of Petroleum Technology 45 (3): 270-276.
66. Homer, C., Dewitz, J., Fry, J., Coan, M., Hossain, N., Larson, C., Herold, N., McKerrow, A., VanDriel, J. N. and Wickham, J. 2007. Completion of the 2001 national land cover database for theconterminous United States. Photogrammetric Engineering & Remote Sensing 73 (4): 337-341.
67. Howard, P. H. 1989. Handbook of Environmental Fate and Exposure Data for Organic Chemicals. CRC Press, Syracuse, New York.
68. Howard, P. H., Boethling, R. S., Jarvis, W. F., Meylan, W. M. and Michalenko, E. M. 1991. Handbook of Environmental Degradation Rates. CRC Press, Boca Raton, Florida.
69. Ivahnenko, T. and Flynn, J. L. 2010. Estimated Withdrawals and Use of Water in Colorado, 2005. Scientific Investigations Report 2010-5002. US Geological Survey. 61 p. Available at http://pubs.usgs.gov/sir/2010/5002/. Accessed November 30, 2012.
70. Jackson, R. B., Osborn, S. G., Vengosh, A. and Warner, N. 2011. Reply to Davies: Hydraulic fracturing remains a possible mechanism for observed methane contamination of drinking water. Proceedings of the National Academy of Sciences 108 (43): E872.
71. Jacob, R. 2011. Incident Action Plan, Franchuk 44-20 SWH Incident. Denbury Onshore, LLC, Plano, Texas.
72. Jeu, S. J., Logan, T. L. and McBane, R. A. 1988. Exploitation of Deeply Buried Coalbed Methane UsingDifferent Hydraulic Fracturing Techniques in the Piceance Basin, Colorado, and San Juan Basin, New Mexico. Presented at SPE Annual Technical Conference and Exhibition, Houston, Texas. Available at http://www.onepetro.org/mslib/app/ Preview.do?paperNumber=00018253&societyCode=SPE. Accessed November 30, 2012.
73. Jobson, H. E. 1996. Prediction of Traveltime and Longitudinal Dispersion in Rivers and Streams.Water-Resources Investigations Report 96-4013. US Geological Survey. 72 p. Available at http://water.usgs.gov/osw/pubs/wrir964013header.html. Accessed November 30, 2012.
74. Johnson, N., Gagnolet, T., Ralls, R., Zimmerman, E., Eichelberger, B., Tracey, C., Kreitler, G., Orndorff, S., Tomlinson, J., Bearer, S. and Sargent, S. 2010. Marcellus Shale Natural Gas and Wind. Pennsylvania Energy Impacts Assessment Report 1. The Nature Conservancy. 47 p. Available at http://www.nature.org/media/pa/tnc_energy_analysis.pdf. AccessedNovember 30, 2012.
75. Johnson, R. C. and Finn, T. M. 2001. Potential for a Basin-Centered Gas Accumulation in the Raton Basin, Colorado and New Mexico. Bulletin 2184-B. US Geological Survey. 18 p. Available athttp://pubs.usgs.gov/bul/b2184-b/b2184-b.pdf. Accessed December 3, 2012.
76. Johnson, T. E., Butcher, J. B., Parker, A. and Weaver, C. P. 2011. Investigating the sensitivity of U.S. streamflow and water quality to climate change: the U.S. EPA global change research program's "20 watersheds" project. Journal of Water Resources Planning and Management 138 (5): 453-464.

77. Jurich, A. and Adams, M. A. 1984. Geologic Overview, Coal, and Coalbed Methane Resources of Raton Mesa Region – Colorado and New Mexico. In C.T. Rightmire, G.E. Eddy and J.N. Kirr, editors, Studies in Geology Series 17: Coalbed Methane Resources of the United States. American Association of Petroleum Geologists, Tulsa, Oklahoma. pp. 163–184.
78. Keighin, C. W. 1995. Raton Basin-Sierra Grande Uplift Province (041). National Assessment of United States Oil and Gas Resources-Results, Methodology, and Supporting Data. Digital Data Series DDS-30. US Geological Survey. 1-7 p. Available athttp://certmapper.cr.usgs.gov/data/noga95/prov41/text/prov41.pdf. Accessed December 3, 2012.
79. Kenny, J. F., Barter, N. L., Hutson, S. S., Linsey, K. S., Lovelace, J. K. and Maupin, M. A. 2009. Estimated Use of Water in the United States in 2005. Circular 1344. US Geological Survey. p. Available at http://pubs.usgs.gov/circ/1344/. Accessed November 30, 2012.
80. Kim, J. and Moridis, G. J. 2012a. Analysis of fracture propagation during hydraulic fracturing operations in tight/shale gas systems. In preparation for journal submission. LawrenceBerkeley National Laboratory, Berkeley, California.
81. Kim, J. and Moridis, G. J. 2012b. Development of the T+M Coupled Flow-Geomechanical Simulator to Describe Fracture Propagation and Coupled Flow-Thermal-Geomechanical Processes inTight/Shale Gas Systems. Proceedings of the 2012 TOUGH Symposium Berkeley, California. Available at http://esd.lbl.gov/files/research/projects/tough/events/symposia/toughsymposium12/Kim_Jihoon-T+M.pdf. Accessed November 30, 2012.
82. Kim, J. and Moridis, G. J. 2012c. Gas Flow Tightly Coupled to Elastoplastic Geomechanics for Tight and Shale Gas Reservoirs: Material Failure and Enhanced Permeability. Presented at SPE Americas Unconventional Resources Conference, Pittsburgh, Pennsylvania. Available at http://www.onepetro.org/mslib/app/Preview.do?paperNumber=SPE-155640MS&societyCode=SPE. Accessed November 30, 2012.
83. Kim, J. and Moridis, G. J. 2012d. Numerical Geomechanical Analyses on Hydraulic Fracturing in TightGas and Shale Gas Reservoirs. Presented at 2012 SPE Annual Technical Conference and Exhibition, San Antonio, Texas.
84. Kim, J. and Moridis, G. J. 2012e. Numerical Studies for Naturally Fractured Shale Gas Reservoirs:Coupled Flow and Geomechanics in Multiple Porosity/Permeability Materials. Presented at 46th U.S. Rock Mechanics/Geomechanics Symposium, Chicago, Illinois.
85. Kish, L. 1965. Survey Sampling. John Wiley & Sons, New York, New York.
86. Klausing, R. L. 1979. Ground-Water Resources of Dunn County, North Dakota. US Geological Surveyfor North Dakota State Water Commission and North Dakota Geologic Survey. Available at http://www.swc.state.nd.us/4dlink9/4dcgi/GetSubContentPDF/PB-225/Dunn_Part_3.pdf. Accessed December 3, 2012.
87. Kuuskraa, V. A. and Ammer, J. 2004. Tight gas sands development–how to dramatically improve recovery efficiency. GasTIPS 10 (1): 15-20.
88. Lawrence Berkeley National Laboratory. 2011. Quality Assurance Project Plan: Analysis of Environmental Hazards Related to Hydrofracturing. Lawrence Berkeley National Laboratory for US Environmental Protection Agency. Available at http://www.epa.gov/ hfstudy/qapps.html. Accessed November 27, 2012.

89. Leadscope version 3.0.6-3. 2012. Leadscope Inc., Columbus, Ohio.
90. Lee, J. J. 2011. Hydraulic Fracturing and Safe Drinking Water. Proceedings of the US Environmental Protection Agency Technical Workshops for the Hydraulic Fracturing Study: Water Resources Management, Arlington, Virginia. Available at http://epa.gov/hfstudy/waterworkshop.html. Accessed November 30, 2012.
91. Leonard, B. P. 2002. Stability of explicit advection schemes. The balance point location rule. International Journal for Numerical Methods in Fluids 38 (5): 471-514.
92. Lide, D. R. 2008. CRC Handbook of Chemistry and Physics. 89th Edition. CRC Press, Boca Raton,Florida.
93. Macartney, H. 2011. Hydraulic Fracturing in Coalbed Methane Development, Raton Basin, Southern Colorado. Presented at US Environmental Protection Agency Technical Workshop for the Hydraulic Fracturing Study: Well Construction and Operations, Arlington, Virginia. Available at http://www.epa.gov/hfstudy/wellconstructworkshop.html. Accessed December 5, 2012.
94. Marcellus Center for Outreach and Research. 2012a. Depth of Marcellus Shale Base. PennsylvaniaState University. Available at http://www.marcellus.psu.edu/images/Marcellus_Depth.gif. Accessed December 6, 2012.
95. Marcellus Center for Outreach and Research. 2012b. Extent and Thickness of Marcellus Shale. Pennsylvania State University. Available at http://www.marcellus.psu.edu/images/ Marcellus_thickness.gif. Accessed December 6, 2012.
96. MarcellusGas.Org. 2012a. Estimated Anticipated Water Usage (for Hydraulic Fracturing) forPermitted Horizontal Wells, by Township, for Washington County, Pennsylvania. Available at http://www.marcellusgas.org/graphs/PA-Washington#waterstat. Accessed September 27, 2012.
97. MarcellusGas.Org. 2012b. Marcellus Well Permits, by Year, for Washington County, Pennsylvania. Available at http://www.marcellusgas.org/graphs/PA-Washington#pyear. Accessed September 25, 2012.
98. MarcellusGas.Org. 2012c. Marcellus Well Starts, by Year, for Washington County, Pennsylvania. Available at http://www.marcellusgas.org/graphs/PA-Washington#wstarts. Accessed September 25, 2012.
99. Material Safety Data Sheets. (a) Encana/Halliburton Energy Services, Inc.: Duncan, Oklahoma. Provided by Halliburton Energy Services during an onsite visit by the EPA on May 10, 2010; (b) Encana Oil and Gas (USA), Inc.: Denver, Colorado. Provided to US EPA Region 8.
100. Mazzoldi, A., Rinaldi, A. P., Borgia, A. and Rutqvist, J. 2012. Induced seismicity within geological carbon sequestration projects: Maximum earthquake magnitude and leakage potential from undetected faults. International Journal of Greenhouse Gas Control 10 (1): 434-442.
101. McCarthy. 2009. Travel Times, Streamflow Velocities, and Dispersion Rates in the Yellowstone River, Montana. Scientific Investigations Report 2009-5261. US Geological Survey. 25 p. Available at http://pubs.usgs.gov/sir/2009/5261/. Accessed November 30, 2012.
102. McGonigal, K. H. 2005. Nutrients and Suspended Sediment Transported in the Susquehanna River Basin, 2004, and Trends, January 1985 Through December 2004. Publication 241.

103. Susquehanna River Basin Commission. 61 p. Available at http://www.srbc.net/pubinfo/techdocs/Publication_241/NutrientReport.pdf. Accessed November 30, 2012.
104. Milici, R. C. and Swezey, C. S. 2006. Assessment of Appalachian Basin Oil and Gas Resources:Devonian Shale–Middle and Upper Paleozoic Total Petroleum System. Open-File Report Series 2006-1237. US Geological Survey. 70 p. Available at http://pubs.usgs.gov/of/2006/1237/of2006-1237.pdf. Accessed December 4, 2012.
105. Montgomery, J. H. 2000. Groundwater Chemicals Desk Reference. 3rd Edition. CRC Press, Boca Raton, Florida.
106. Moridis, F. and Webb, F. 2012. The RealGas and RealGasH2O Options of the TOUGH+ Code for the Simulation of Coupled Fluid and Heat Flow in Tight/Shale Gas Systems Proceedings of the 2012 TOUGH Symposium, Berkeley, California. Available at http://esd.lbl.gov/files/research/projects/tough/events/symposia/toughsymposium12/Moridis_GeorgeRealgasH2O.pdf. Accessed November 30, 2012.
107. Moridis, G. J., Blasingame, T. and Freeman, C. M. 2010. Analysis of Mechanisms of Flow in Fractured Tight-Gas and Shale-Gas Reservoirs. Presented at 2010 SPE Latin American & CaribbeanPetroleum Engineering Conference, Lima, Peru. Available at http://www.onepetro.org/ mslib/servlet/onepetropreview?id=SPE-139250-MS. Accessed November 30, 2012.
108. Moridis, G. J. and Freeman, C. M. 2012. The RGasH2OCont module of the TOUGH+ code for simulation of coupled fluid and heat flow, and contaminant transport, in tight/shale gas systems. In preparation for journal submission. Lawrence Berkeley National Laboratory, Berkely, California.
109. Moridis, G. J., Kowalsky, M. B. and Pruess, K. 2008. TOUGH+HYDRATE v1.0 User Manual: A Code for the Simulation of System Behavior in Hydrate-Bearing Geologic Media. Report 149E. Lawrence Berkley National Laboratory. 279 p. Available at http://esd.lbl.gov/files/ research/projects/tough/documentation/TplusH_Manual_v1.pdf. Accessed November 30, 2012.
110. Murphy, E. C. 2001. Geology of Dunn County. Bulletin 68 Part 1. North Dakota Geological Survey. 43p. Available at http://www.swc.state.nd.us/4dlink9/4dcgi/GetSubContentPDF/PB221/Dunn_Part_1.pdf. Accessed December 4, 2012.
111. Myers, T. 2012a. Author's Reply. Ground Water 50 (6): 828-830.
112. Myers, T. 2012b. Potential contaminant pathways from hydraulically fractured shale to aquifers. Ground Water 50 (6): 872-882.
113. National Academy of Sciences. 2012. Induced Seismicity Potential in Energy Technologies. Available at http://dels.nas.edu/besr. Accessed November 30, 2012.
114. North Dakota Industrial Commission. 2003. January 2003 Oil and Gas Production Report. Availableat https://www.dmr.nd.gov/oilgas/mpr/2003_01.pdf. Accessed November 27, 2012.
115. North Dakota Industrial Commission. 2012a. North Dakota Monthly Bakken Oil Production Statistics Available at https://www.dmr.nd.gov/oilgas/stats/historicalbakkenoilstats.pdf. Accessed November 27, 2012.
116. North Dakota Industrial Commission. 2012b. Oil and Gas Production Report. Available at https://www.dmr.nd.gov/oilgas/mpr/2012_07.pdf. Accessed December 5, 2012.
117. Newport, T. G. 1973. Summary Ground-Water Resources of Washington County, Pennsylvania.Water Resource Report 38. Pennsylvania Geological Survey. 32 p.

Available at http://www.dcnr.state.pa.us/topogeo/pub/water/pdfs/w038_text.pdf. Accessed December4, 2012.

118. Nicot, J., Hebel, A., Ritter, S., Walden, S., Baier, R., Galusky, P., Beach, J., Kyle, R., Symank, L. and Breton, C. 2011. Current and Projected Water Use in the Texas Mining and Oil and GasIndustry. The University of Texas at Austin Bureau of Economic Geology for Texas Water Development Board. Available at http://www.twdb.texas.gov/publications/reports/ contracted_reports/doc/0904830939_MiningWaterUse.pdf. Accessed November 10, 2012.
119. Noorishad, J. and Tsang, C.-F. 1997. Coupled Thermohydroelasticity Phenomena in VariablySaturated Fractured Porous Rocks -Formulation and Numerical Solution. In O. Stephanson,
120. L. Jing and C.-F. Tsang, editors, Coupled Thermo-Hydro-Mechanical Processes in FracturedMedia: Mathematical and Experimental Studies. Elsevier Science B.V., Amsterdam, The Netherlands. pp. 575.
121. Nordeng, SH. 2010. The Bakken Source System: Emphasis on the Three Forks Formation. NorthDakota Geological Survey. Available at http://www.ndoil.org/image/cache/ Stephan_Nordeng_-_NDGS.pdf. Accessed December 5, 2012.
122. Nordin, C. F. and Sabol, G. V. 1974. Empirical Data on Longitudinal Dispersion in Rivers. Water-Resources Investigations 20-74. US Geological Survey, p. 332.
123. Nordstrom, P. L. 1982. Occurrence, Availiability and Chemical Quality of Ground Water in the Cretaceous Aquifers of North-Central Texas. Volume 1. Report 269. Texas Department of Water Resources. 66 p. Available at http://www.twdb.state.tx.us/publications/reports/numbered_reports/doc/R269/R269v1/R269v1.pdf. Accessed December 3, 2012.
124. New York State Department of Environmental Conservation. 2011. Supplemental Generic Environmental Impact Statement on the Oil, Gas and Solution Mining Regulatory Program(Revised Draft). Well Permit Issuance for Horizontal Drilling and High-Volume Hydraulic Fracturing to Develop the Marcellus Shale and Other Low-Permeability Gas Reservoirs.Available at http://www.dec.ny.gov/energy/75370.html. Accessed September 1, 2011.
125. Olorode, O. M. 2011. Numerical modeling of fractured shale-gas and tight-gas reservoirs using unstructured grids. MS Degree. Texas A&M University, College Station, Texas.
126. Osborn, S. G., Vengosh, A., Warner, N. R. and Jackson, R. B. 2011. Methane contamination of drinking water accompanying gas-well drilling and hydraulic fracturing. Proceedings of the National Academy of Sciences 108 (20): 8172-8176.
127. Pennsylvania Department of Environmental Protection. 2010. Chemicals Used by Hydraulic Fracturing Companies in Pennsylvania for Surface and Hydraulic Fracturing Activities.Available at http://files.dep.state.pa.us/OilGas/BOGM/BOGMPortalFiles/MarcellusShale/ Frac%20list%206-30-2010.pdf. Accessed November 27, 2012.
128. Pennsylvania Department of the Environment. 2011. Letter from Pennsylvania Department of the Environment to US EPA Region 3 Administrator Shawn Garvin. Available at http://www.epa.gov/region3/marcellus_shale?#inforeqsbypadep. Accessed April 6, 2011.
129. Pennsylvania Department of Environmental Protection. 2012. State Water Plan Digital Water Atlas. Available at http://www.pawaterplan.dep.state.pa.us/statewaterplan/DWA/ DWAMain.aspx?theme=6&cacheId=227881113. Accessed August 13, 2012.

130. Palmer, I. D., Fryar, R. T., Tumino, K. A. and Puri, R. 1991. Water fracs outperform gel fracs in coalbed pilot. Oil & Gas Journal 89 (32): 71-76.
131. Palmer, I. D., Lambert, S. W. and Spitler, J. L. 1993. Coalbed Methane Well Completions and Stimulations: Chapter 14. In Ben E. Law and Dudley D. Rice, editors, SG 38: Hydrocarbons from Coal. American Association of Petroleum Geologists, Tulsa, Oklahoma. pp. 303-341.
132. Pancras, J. P., Vedantham, R., Landis, M. S., Norris, G. A. and Ondov, J. M. 2011. Application of EPA unmix and non-parametric wind regression on high time resolution trace elements and speciated mercury in Tampa, Florida, aerosol. Environmental Science & Technology 45 (8): 3511-3518.
133. Plewa, M. J., Muellner, M. G., Richardson, S. D., Fasano, F., Buettner, K. M., Woo, Y.-T., McKague, B. and Wagner, E. D. 2008. Occurrence, synthesis and mammalian cell cytotoxicity and genotoxicityof haloacetamides: an emerging class of nitrogenous drinking water disinfection byproducts. Environmental Science & Technology 42 (3): 955-961.
134. Plewa, M. J. and Wagner, E. D. 2009. Quantitative Comparative Mammalian Cell Cytotoxicity andGenotoxicity of Selected Classes of Drinking Water Disinfection By-Products. Presented at Water Research Foundation, Denver, Colorado. Available at http://prod2010.waterrf.org/ ExecutiveSummaryLibrary/91249_3089_profile.pdf. Accessed November 30, 2012.
135. Plewa, M. J., Wagner, E. D., Jazwierska, P., Richardson, S. D., Chen, P. H. and McKague, A. B. 2004.Chemical and biological characterization of newly discovered iodoacid drinking water disinfection byproducts. Environmental Science & Technology 38 (18): 4713-4722.
136. Puder, M. and Veil, J. 2006. Offsite Commercial Disposal of Oil and Gas Exploration and ProductionWaste: Availability, Options, and Costs. ANL/EVS/R-06/5. Argonne National Laboratory for US Department of Energy. Available at http://www.evs.anl.gov/pub/doc/ANL-EVS-R-065.pdf. Accessed October 18, 2012.
137. Reed, L. A. and Stuckey, M. A. 2002. Prediction of Velocities for a Range of Streamflow Conditions in Pennsylvania. Water-Resources Investigations Report 01-4214. US Geological Survey. 13 p. Available at http://pubs.er.usgs.gov/publication/wri014214. Accessed November 21, 2012.
138. Reutter, D. C. and Dunn, D. D. 2000. Water-Quality Assessment of the Trinity River Basin, Texas: Ground-Water Quality of the Trinity, Carrizo-Wilcox, and Gulf Coast Aquifers, February-August 1994. Water-Resources Investigations Report 99-4233. US Geological Survey. 64 p. Available at http://pubs.usgs.gov/wri/wri99-4233/. Accessed December 5, 2012.
139. Richardson, S. 2003. Disinfection by-products and other emerging contaminants in drinking water. Trends in Analytical Chemistry 22 (10): 666-684.
140. Richardson, S. D., Plewa, M. J., Wagner, E. D., Schoeny, R. and DeMarini, D. M. 2007. Occurrence, genotoxicity, and carcinogenicity of regulated and emerging disinfection by-products indrinking water: a review and roadmap for research. Mutatation Research 636 (1-3): 178-242.
141. Richenderfer, J. 2011. Natural Gas Industry Effects on Water Consumption and Management. Susquehanna River Basin Commission. 40 p. Available at http://www.mde.state.md.us/programs/Land/mining/marcellus/Pages/surfacewater.aspx. Accessed November 30, 2012.

142. Rogers, S. W., Ong, S. K., Kjartanson, B. H., Golchin, J. and Stenback, G. A. 2002. Natural attenuation of polycyclic aromatic hydrocarbon-contaminated sites: review. Practice Periodical of Hazardous, Toxic, and Radioactive Waste Management 6 (3): 141-155.
143. Rossenfoss, S. 2011. From flowback to fracturing: water recycling grows in the marcellus shale. Journal of Petroleum Technology 63 (7): 48-51.
144. Railroad Commission of Texas. 2012. Barnett Shale Information. Available at http://www.rrc.state.tx.us/barnettshale/index.php. Accessed December 5, 2012.
145. Rupp, B., Appel, K. E. and Gundert-Remy, U. 2010. Chronic oral LOAEL prediction by using a commercially available computational QSAR tool. Archives of Toxicology 84 (9): 681-688.
146. Rutqvist, J. 2012. The geomechanics of CO2 storage in deep sedimentary formations. Geotechnical and Geological Engineering 30 (3): 525-551.
147. Rutqvist, J., Birkholzer, J., Cappa, F. and Tsang, C.-F. 2007. Estimating maximum sustainableinjection pressure during geological sequestration of CO2 using coupled flow and geomechanical fault-slip analysis. Energy Conversion and Management 48 (6): 1798-1807.
148. Rutqvist, J., Borgesson, L., Chijimatsu, M., Kobayashi, A., Jing, L., Nguyen, T. S., Noorishad, J. andTsang, C.-F. 2001. Thermohydromechanics of partially saturated geological media: governing equations and formulation of four finite element models. International Journal of Rock Mechanics and Mining Sciences 38 (1): 105-127.
149. Rutqvist, J., Rinaldi, A. P., Cappa, F. and Moridis, G. J. 2012. Modeling of fault reactivation and induced seismicity during hydraulic fracturing of shale-gas reservoirs. In preparation forjournal submission. Lawrence Berkeley National Laboratory, Berkeley, California.
150. Papadopulos & Associates Inc. 2008. Coalbed Methane Stream Depletion Assessment Study – Raton Basin, Colorado. S.S. Papadopulos & Associates Inc. prepared in conjunction with the Colorado Geological Survey for Colorado Department of Natural Resources and the Colorado Oil and Gas Conservation Commission. Available at http://geosurvey.state.co.us/water/CBM%20Water%20Depletion/Documents/RatonCBMdepletion_FINAL.pdf. Accessed December 3, 2012.
151. Papadopulos & Associates Inc. 2007a. Coalbed Methane Stream Depletion Assessment Study – Piceance Basin, Colorado. S.S. Papadopulos & Associates Inc. for Prepared in conjunctionwith the Colorado Geological Survey for the State of Colorado Department of Natural Resources and the Colorado Oil and Gas Conservation Commission. Available at http://water.state.co.us/groundwater/GWAdmin/NontribGW/Archive/Pages/CBMStream DepletionStudies.aspx. Accessed November 30, 2012.
152. Papadopulos & Associates Inc. 2007b. Coalbed Methane Stream Depletion Assessment Study –Raton Basin, Colorado, Draft Final Report. S.S. Papadopulos & Associates Inc. for Prepared in conjunction with the Colorado Geological Survey for the State of Colorado Department of Natural Resources and the Colorado Oil and Gas Conservation Commission. Available at http://water.state.co.us/groundwater/GWAdmin/NontribGW/Archive/Pages/CBMStream DepletionStudies.aspx. Accessed November 30, 2012.
153. SAIC Energy Environment & Infrastructure LLC and Groundwater & Environmental Services Inc.2011. ATGAS Investigation Initial Site Characterization and Response, April 19, 2011 to May 2, 2011, ATGAS2H Well Pad, Permit No. 37-015-21237, Leroy

Township, Bradford County, PA. SAIC and GES for Chesapeake Appalachia, LLC. Available at http://www.chk.com/news/ articles/documents/atgas_initial_site_characterization_report_final_08292011.pdf. Accessed December 4, 2012.
154. Satterfield, J., Kathol, D., Mantell, M., Hiebert, F., Lee, R. and Patterson, K. 2008. Managing Water Resource Challenges in Select Natural Gas Shale Plays. Presented at Ground Water Protection Council Annual Forum, Oklahoma City, Oklahoma.
155. Schiesser, W. E. 1991. The Numerical Method-of-Lines: Integration of Partial Differential Equations. Academic Press, San Diego, California. Qikprop version 3.4. 2012. Schrodinger, LLC, New York, New York.
156. Schwarzenbach, R. P., Gschwend, P. M. and Imboden, D. M. 2002. Environmental Organic Chemistry. 2nd Edition. John Wiley & Sons, Inc, Hoboken, New Jersey.
157. Scott, G. and Armstrong, J. M. 1932. The Geology of Wise County, Texas. The University of TexasBulletin No. 3224. The University of Texas. 80 p. Available at http://www.lib.utexas.edu/books/landscapes/publications/txu-oclc-983185/txu-oclc-983185.pdf. Accessed December 5, 2012.
158. Scott, M. J. and Jones, M. N. 2000. The biodegradation of surfactants in the environment. Biochimica et Biophysica Acta 1508 (1-2): 235-251.
159. Seagren, E. A. and Becker, J. G. 2002. Review of natural attenuation of BTEX and MTBE in groundwater. Practice Periodical of Hazardous, Toxic, and Radioactive Waste Management 6 (3): 156-172.
160. Seo, J. S., Keum, Y. S. and Qing, X. L. 2009. Bacterial degradation of aromatic compounds. International Journal Environmental Research and Public Health 6 (1): 278-309.
161. Sharma, V. K., Anquandah, G. A. K., Yngard, R. A., Kim, H., Fekete J., Bouzek, K., Ray, A. K. and Golovko, D. 2009. Nonylphenol, octylphenol, and bisphenol-A in the aquatic environment: a review on occurrence, fate, and treatment. Journal of Environmental Science and Health, Part A: Toxic/Hazardous Substances and Environmental Engineering 44 (5): 423-442.
162. Shultz, C. H. 1999. The Geology of Pennsylvania. Pennsylvania Geological Survey and Pittsburgh Geological Society, Harrisburg, Pennsylvania.
163. Soares A., Guieysse, B., Jeffereson, B., Cartmell, E. and J.N., L. 2008. Nonylphenol in the environment:a critical review on occurrence, fate, toxicity and treatment in wastewaters. Environment International 34 (7): 1033-1049.
164. Soonthornnonda, P. and Christensen, E. 2008. Source apportionment of pollutants and flows of combined sewer wastewater. Water Research 42 (8-9): 1989-1998
165. Spahr, N. E., Apodaca, L. E., Deacon, J. R., Bails, J. B., Bauch, N. J., Smiath, C. M. and Driver, N. E. 2000. Water Quality in the Upper Colorado River Basin, Colorado, 1996-98. Circular 1214. US Geological Survey. 33 p. Available at http://pubs.water.usgs.gov/circ1214/. AccessedNovember 30, 2012.
166. Susquehanna River Basin Commission. Susquehanna River Basin Commission Information Sheet. Available at http://www.srbc.net/pubinfo/docs/Susq%20River%20Basin%20General%20%2811_06%29.PDF. Accessed November 27, 2012.
167. Susquehanna River Basin Commission. 2012. Flowback and Produced Water Volume. Data Providedto EPA Upon Request. Susquehanna River Basin Commission, Harrisburg, Pennsylvania.

168. Stangroom, S. J., Collins C.D. and Lester, J. N. 2010. Abiotic behavior of organic micropollutants in soils and the aquatic environment. A review: II. Transformations. Environmental Technology 21 (8): 865-882.
169. Staples, C. A., Williams, J. B., Craig, G. R. and Roberts., K. M. 2001. Fate, effects and potential environmental risks of ethylene glycol: a review. Chemosphere 43 (3): 377-383.
170. Talmage, S. S. 1994. Environmental and Human Safety of Major Surfactants—Alcohol Ethoxylates and Alkylphenol Ethoxylates. CRC Press, Boca Raton, Florida.
171. Taylor, L. E. 1984. Groundwater Resources of the Upper Susquehanna River Basin, Pennsylvania. Pennsylvania Bureau of Topographic and Geologic Survey, Harrisburg, Pennsylvania.
172. Topper, R., Scott, K. and Watterson, N. 2011. Geologic Model of the Purgatoire River Watershed within the Raton Basin, Colorado. Colorado Geological Survey for Colorado WaterConservation Board. Available at http://geosurvey.state.co.us/water/CBM%20Water%20Depletion/Documents/Raton%20Geology_CGS%20final%20report.pdf.Accessed December 3, 2012.
173. Tremain, C. M. 1980. The Coalbed Methane Potential of the Raton Basin, Colorado. Presented at SPE Unconventional Gas Recovery Symposium, Pittsburgh, Pennsylvania. Available athttp://www.onepetro.org/mslib/servlet/onepetropreview?id=00008927. Accessed December 3, 2012.
174. Tyler, R. 1995. Geologic and Hydrologic Assessment of Natural Gas from Coal: Greater Green River, Piceance, Powder River, and Raton Bains, Western United States. The University of Texas atAustin, Bureau of Economic Geology, Austin, Texas.
175. Tyler, R. and McMurry, R. G. 1995. Genetic Stratigraphy, Coal Occurrence, and Regional Cross Section of the Williams Fork Formation. Open File Report 95-2. Colorado Geological Survey.42 p. Available at http://geosurveystore.state.co.us/p-764-genetic-stratigraphy-coaloccurrence-and-regional-cross-section-of-the-coal-bearing-williams-fork-formation.aspx. Accessed November 30, 2012.
176. US Energy Information Administration. 2011a. Annual Energy Review 2010. Available at http://www.eia.gov/aer. Accessed September 26, 2012.
177. US Energy Information Administration. 2011b. Distribution and Production of Oil and Gas Wells by State, 1995 to 2009. Available at http://www.eia.gov/pub/oil_gas/petrosystem/ petrosysog.html. Accessed November 27, 2012.
178. US Energy Information Administration. 2011c. Shale Gas and Oil Plays, Lower 48 States. Available at http://www.eia.gov/pub/oil_gas/natural_gas/analysis_publications/maps/maps.htm. Accessed December 12, 2012.
179. US Energy Information Administration. 2011d. Shapefiles for Basin boundaries. Data for the US Shale Plays Map. May 9, 2011. US Energy Information Administration, Washington, DC. Available at http://www.eia.gov/pub/oil_gas/natural_gas/analysis_publications/ maps/maps.htm. Accessed November 30, 2012.
180. US Energy Information Administration. 2011e. Shapefiles for Play Boundaries. Data for the US Shale Plays Map. May 9, 2011. US Energy Information Administration, Washington, DC. Available at http://www.eia.gov/pub/oil_gas/natural_gas/analysis_publications/maps/maps.htm. Accessed November 30, 2012.
181. US Energy Information Administration. 2012. Annual Energy Outlook 2012. Available at http://www.eia.gov/forecasts/aeo/index.cfm. Accessed November 27, 2012.

182. US Environmental Protection Agency. 1995. Guidance for Methods Development and MethodsValidation for the Resource Conservation and Recovery Act (RCRA) Program. Available at http://www.epa.gov/sw-846/pdfs/methdev.pdf. Accessed December 12, 2012.
183. US Environmental Protection Agency. 2000a. EPA Quality Manual for Environmental Programs. Available at http://www.epa.gov/irmpoli8/policies/2105P010.pdf. Accessed December 3, 2012.
184. US Environmental Protection Agency. 2000b. Guidance on Technical Audits and RelatedAssessments for Environmental Data Operations. Available at http://www.epa.gov/ quality/qa_docs.html. Accessed December 3, 2012.
185. US Environmental Protection Agency. 2000c. Policy and Program Requirements for the MandatoryAgency-Wide Quality System. Available at http://www.epa.gov/irmpoli8/policies/ 21050.pdf. Accessed December 3, 2012.
186. US Environmental Protection Agency. 2002. Guidelines for Ensuring and Maximizing the Quality,Objectivity, Utility, and Integrity of Information Disseminated by EPA. US Environmental Protection Agency, Washington, DC.
187. US Environmental Protection Agency. 2003a. A Summary of General Assessment Factors for Evaluating the Quality of Scientific and Technical Information. Available athttp://www.epa.gov/spc/pdfs/assess2.pdf. Accessed November 27, 2012.
188. US Environmental Protection Agency. 2003b. Toxic Substances Control Act (TSCA) Confidential Business Information (CBI) Protection Manual. Available at http://www.epa.gov/oppt/pubs/tsca-cbi-protection-manual.pdf. Accessed November 27, 2012.
189. US Environmental Protection Agency. 2004a. Assuring the Competency of Environmental Protection Agency Laboratories. Available at http://epa.gov/osa/fem/pdfs/labdirective.pdf. Accessed December 3, 2012.
190. US Environmental Protection Agency, Office of Water. 2004b. Evaluation of Impacts to Underground Sources of Drinking Water by Hydraulic Fracturing of Coalbed MethaneReservoirs. EPA 816-R-04-003. Available at http://water.epa.gov/type/groundwater/uic/ class2/hydraulicfracturing/wells_coalbedmethanestudy.cfm. Accessed November 27, 2012.
191. US Environmental Protection Agency. 2006. Office of Research and Development Policies andProcedures Manual. US Environmental Protection Agency, Washington, D.C.
192. US Environmental Protection Agency. 2007. U.S. EPA Reach File 1 (RF1) for the Conterminous United States. Better Assessment Science Integrating point & Nonpoint Sources (BASINS). Washington, DC. Available at http://water.epa.gov/scitech/datait/models/basins/b3webdwn.cfm. Accessed 6/11/2007.
193. US Environmental Protection Agency. 2008. EPA Procedure for Quality Policy. Available at http://www.epa.gov/irmpoli8/policies/2106p01.pdf. Accessed December 3, 2012.
194. US Environmental Protection Agency. Chesapeake Bay Phase 5.3 Community Watershed Model. Available at http://www.chesapeakebay.net/about/programs/modeling/53/. Accessed November 27, 2012.
195. US Environmental Protection Agency. 2010b. Hydraulic Fracturing Study Consultation with TribalGovernments, August 5 and 30, 2010. Available at http://www.epa.gov/hfstudy/ publicoutreach.html. Accessed September 5, 2012.

196. US Environmental Protection Agency. 2010c. Integrated Climate and Land-Use Scenarios (ICLUS)version 1.3 User's Manual: ArcGIS Tools and Datasets for Modeling US Housing Density Growth. Available at http://www.epa.gov/ncea/global/iclus/. Accessed September 2010.
197. US Environmental Protection Agency. 2010d. Summary of Public Comments, Binghamton, NewYork, September 13 and 15, 2010. Hydraulic Fracturing Public Informational Meeting. Available at http://www.epa.gov/hfstudy/publicoutreach.html. Accessed September 5, 2012.
198. US Environmental Protection Agency. 2010e. Summary of Public Comments, Canonsburg, Pennsylvania, July 22, 2010. Hydraulic Fracturing Public Informational Meeting. Available at http://www.epa.gov/hfstudy/publicoutreach.html. Accessed September 5, 2012.
199. US Environmental Protection Agency. 2010f. Summary of Public Comments, Denver, Colorado, July13, 2010. Hydraulic Fracturing Public Informational Meeting. Available at http://www.epa.gov/hfstudy/publicoutreach.html. Accessed September 5, 2012.
200. US Environmental Protection Agency. 2010g. Summary of Public Comments, Fort Worth, Texas, July8, 2010. Hydraulic Fracturing Public Information Meeting. Available at http://www.epa.gov/hfstudy/publicoutreach.html. Accessed September 5, 2012.
201. US Environmental Protection Agency. 2010h. Technical Basis for the Lowest ConcentrationMinimum Reporting Level (LCMRL) Calculator. EPA 815-R-11-001. Available at http://water.epa.gov/scitech/drinkingwater/labcert/upload/LCMRLTechRpt.pdf. AccessedNovember 27, 2012.
202. US Environmental Protection Agency. 2011a. Counties with Oil and Gas Production Wells Hydraulically Fractured from September 2009 through October 2010 Shapefile. Data received from nine hydraulic fracturing service companies. US Environmental ProtectionAgency, Washington, DC.
203. US Environmental Protection Agency. 2011b. Data Received from Hydraulic Fracturing Service Companies. Non-confidential business information is available athttp://www.regulations.gov/#!docketDetail;rpp=100;so=DESC;sb=docId;po=0;D=EPA-HQORD-2010-0674. Accessed November 27, 2012.
204. US Environmental Protection Agency. 2011c. EPA Records Schedule 501. Applied and Directed Scientific Research. Available at http://www.epa.gov/records/policy/schedule/sched/501.htm. Accessed November 27, 2012.
205. US Environmental Protection Agency. 2011d. Location of Oil and Gas Production Wells Selected for Review Shapefile. US Environmental Protection Agency, Washington, DC.
206. US Environmental Protection Agency. 2011e. Plan to Study the Potential Impacts of HydraulicFracturing on Drinking Water Resources. EPA/600/R-11/122. Available at http://www.epa.gov/hfstudy/. Accessed November 27, 2012.
207. US Environmental Protection Agency. 2011f. Proceedings of the US Environmental ProtectionAgency Technical Workshops for the Hydraulic Fracturing Study: Water Resources Management, Arlington, Virginia. Available at http://www.epa.gov/hfstudy/waterworkshop.html. Accessed August 24, 2012.

208. US Environmental Protection Agency. 2011g. Quality Assurance Project Plan: Chemical Characterization of Select Constituents Relevant to Hydraulic Fracturing. Available at http://www.epa.gov/hfstudy/qapps.html. Accessed November 27, 2012.
209. US Environmental Protection Agency. 2011h. Quality Assurance Project Plan: Formation of Disinfection By-products from Hydraulic Fracturing Fluid Constituents. Available at http://www.epa.gov/hfstudy/qapps.html. Accessed November 27, 2012.
210. US Environmental Protection Agency. 2011i. Quality Assurance Project Plan: Hydraulic FracturingRetrospective Case Study, Bakken Shale, Killdeer and Dunn County, North Dakota. Available at http://www.epa.gov/hfstudy/bakken-qapp.pdf. Accessed November 27, 2012.
211. US Environmental Protection Agency. 2011j. Safe Drinking Water Information System. USEnvironmental Protection Agency, Washington, DC.
212. US Environmental Protection Agency. 2011k. Sampling Data for Flowback and Produced Water Provided to EPA by Nine Oil and Gas Well Operators (Non-Confidential Business Information). Available at http://www.regulations.gov/ #!docketDetail;rpp=100;so=DESC;sb=docId;po=0;D=EPA-HQ-ORD-2010-0674. AccessedNovember 27, 2012.
213. Estimation Programs Interface Suite for Microsoft® Windows (EPI Suite) version 4.10. 2012a. US Environmental Protection Agency, Washington DC.
214. US Environmental Protection Agency. 2012b. Exposure Assessment Tools and Models: Estimation Program Interface (EPI) Suite. Available at http://www.epa.gov/oppt/exposure/pubs/ episuite.htm. Accessed November 21, 2012.
215. US Environmental Protection Agency. 2012c. Interim EJSCREEN Common User Guidelines. Draft for Environmental Justice Committee Presentation. US Environmental Protection Agency, Washington, DC.
216. US Environmental Protection Agency. Key Documents About Mid-Atlantic Oil and Gas Extraction. Available at http://www.epa.gov/region3/marcellus_shale/. Accessed December 5, 2012.
217. US Environmental Protection Agency. 2012e. Peer Review Handbook. 3rd Edition. Available athttp://www.epa.gov/peerreview/pdfs/peer_review_handbook_2012.pdf. Accessed December 3, 2012.
218. US Environmental Protection Agency. 2012f. Quality Assurance Project Plan: Data and Literature Evaluation for the EPA's Study of the Potential Impacts of Hydraulic Fracturing (HF) on Drinking Water Resources. Available at http://www.epa.gov/hfstudy/qapps.html. Accessed December 4, 2012.
219. US Environmental Protection Agency. 2012g. Quality Assurance Project Plan: Analysis of Data Extracted from FracFocus. Available at http://www.epa.gov/hfstudy/qapps.html. AccessedDecember 4, 2012.
220. US Environmental Protection Agency. 2012h. Quality Assurance Project Plan: Analysis of Data Received from Nine Hydraulic Fracturing (HF) Service Companies. Available at http://www.epa.gov/hfstudy/qapps.html. Accessed December 4, 2012.
221. US Environmental Protection Agency. 2012i. Quality Assurance Project Plan: Chemical Information Quality Review and Physicochemical Property Calculations for Hydraulic Fracturing Chemical Lists. Available at http://www.epa.gov/hfstudy/qapps.html. Accessed November30, 2012.

222. US Environmental Protection Agency. 2012j. Quality Assurance Project Plan: Evaluation of Existing Production Well File Contents. Available at http://www.epa.gov/hfstudy/qapps.html. Accessed January 4, 2012.
223. US Environmental Protection Agency. 2012k. Quality Assurance Project Plan: Health and Toxicity Theme Hydraulic Fracturing Study. Available at http://www.epa.gov/hfstudy/qapps.html. Accessed November 30, 2012.
224. US Environmental Protection Agency. 2012l. Quality Assurance Project Plan: Hydraulic Fracturing (HF) Surface Spills Data Analysis. Available at http://www.epa.gov/hfstudy/qapps.html. Accessed December 4, 2012.
225. US Environmental Protection Agency. 2012m. Quality Assurance Project Plan: Hydraulic Fracturing Retrospective Case Study, Bradford-Susquehanna Counties, Pennsylvania. Available at http://www.epa.gov/hfstudy/pdfs/bradford-review-casestudy.pdf. Accessed November 27,2012.
226. US Environmental Protection Agency. 2012n. Quality Assurance Project Plan: Hydraulic Fracturing Retrospective Case Study, Marcellus Shale, Washington County, Pennsylvania. Available athttp://www.epa.gov/hfstudy/washingtonco-qapp.pdf. Accessed November 27, 2012.
227. US Environmental Protection Agency. 2012o. Quality Assurance Project Plan: Hydraulic Fracturing Retrospective Case Study, Raton Basin, Colorado. Available at http://www.epa.gov/hfstudy/qapps.html. Accessed November 27, 2012.
228. US Environmental Protection Agency. 2012p. Quality Assurance Project Plan: Hydraulic Fracturing Retrospective Case Study, Wise, Texas. Available at http://www.epa.gov/hfstudy/barnettqapp.pdf. Accessed November 27, 2012.
229. US Environmental Protection Agency. 2012q. Quality Assurance Project Plan: Hydraulic Fracturing Wastewater Source Apportionment. Available at http://www.epa.gov/hfstudy/qapps.html. Accessed November 27, 2012.
230. US Environmental Protection Agency. 2012r. Quality Assurance Project Plan: Inter-Laboratory Verification and Validation of Diethylene Glycol, Triethylene Glycol, Tetraethylene Glycol, 2Butoxyethanol and 2-Methoxyethanol in Ground and Surface Waters by Liquid Chromatography/Tandem Mass Spectrometry. Available at http://www.epa.gov/hfstudy/ qapps.html. Accessed November 27, 2012.
231. US Environmental Protection Agency. 2012s. Quality Assurance Project Plan: Surface WaterTransport of Hydraulic Fracturing-Derived Waste Water. Available at http://www.epa.gov/ hfstudy/qapps.html. Accessed November 27, 2012.
232. US Environmental Protection Agency. 2012t. Quality Management Plan: Plan to Study the Impactsof Hydraulic Fracturing on Drinking Water Resources. Available at http://www.epa.gov/ hfstudy/qapps.html. Accessed December 3, 2012.
233. US Environmental Protection Agency. 2012u. Request for Information to Inform HydraulicFracturing Research Related to Drinking Water Resources. Federal Register 77:218 p. 67361.
234. US Environmental Protection Agency. 2012v. Science Advisory Board Staff Office Request forNominations of Experts for the SAB Hydraulic Fracturing Advisory Panel. Federal Register 77:162. p. 50505.

235. US Environmental Protection Agency. 2012w. Unconventional Gas Drilling Projections for the US:2011-2040 AEO2012 Reference Case. National Energy Modeling System. US Environmental Protection Agency, Washington, DC.
236. US Fish and Wildlife Service. 2008. Programmatic Biological Opinion for Water Depletions Associated with Bureau of Land Management's Fluid Mineral Program within the UpperColorado River Basin in Colorado. Memorandum. US Fish and Wildlife Service, Grand Junction, Colorado.
237. US Government Accountability Office. 2012. Information on the Quantity, Quality, and Managementof Water Produced during Oil and Gas Production. Energy-Water Nexus GAO-12-156. Available at http://www.gao.gov/products/GAO-12-156. Accessed December 12, 2012.
238. US Office of Management and Budget. 2004. Final Information Quality Bulletin for Peer Review. Available at http://www.whitehouse.gov/sites/default/files/omb/assets/omb/memoranda/fy2005/m05-03.pdf. Accessed August 23, 2012.
239. US Census Bureau. 2010a. Bradford County, Pennsylvania. State and County QuickFacts. Available at http://quickfacts.census.gov/qfd/states/42/42015.html. Accessed November 27, 2012.
240. US Census Bureau. 2010b. Dunn County, North Dakota. State and County QuickFacts. Available athttp://quickfacts.census.gov/qfd/states/38/38025.html. Accessed November 27, 2012.
241. US Census Bureau. 2010c. Huerfano County, Colorado. State and County QuickFacts. Available at http://quickfacts.census.gov/qfd/states/08/08055.html. Accessed November 27, 2012.
242. US Census Bureau. 2010d. Las Animas County, Colorado. State and County QuickFacts. Available athttp://quickfacts.census.gov/qfd/states/08/08071.html. Accessed November 27, 2012.
243. US Census Bureau. 2010e. Washington County, Pennsylvania. State and County Quick Facts. US Census Bureau, Available at http://quickfacts.census.gov/qfd/states/42/42125.html. Accessed December 5, 2012.
244. US Census Bureau. 2010f. Wise County, Texas. State and County QuickFacts. Available at http://quickfacts.census.gov/qfd/states/48/48497.html. Accessed November 27, 2012.
245. US Census Bureau. 2012a. Counties (and Equivalent) Shapefile. 2010 TIGER/Line Shapefiles. USCensus Bureau, Washington, DC. Available at http://www.census.gov/cgibin/geo/shapefiles2010/main. Accessed December 4, 2012.
246. US Census Bureau. 2012b. Places Shapefile. 2010 TIGER/Line Shapefiles. US Census Bureau, Washington, DC. Available at http://www.census.gov/cgi-bin/geo/shapefiles2010/main. Accessed December 5, 2012.
247. US Census Bureau. 2012c. States (and Equivalent) Shapefile. 2010 TIGER/Line Shapefiles. USCensus Bureau, Washington, DC. Available at http://www.census.gov/cgibin/geo/shapefiles2010/main. Accessed December 4, 2012.
248. US Department of Agriculture. 2012. Information on Hydrologic Units and the Watershed Boundary Dataset. Available at http://www.ncgc.nrcs.usda.gov/wps/portal/nrcs/detail/national/technical/nra/dma/?cid=nrcs143_0216_16. Accessed November 27, 2012.

249. US Geological Survey. 2003. Petroleum Systems and Geologic Assessment of Oil and Gas in the Uinta-Piceance Province, Utah and Colorado. Digital Data Series DDS-69-B. Available athttp://pubs.usgs.gov/dds/dds-069/dds-069-b/chapters.html. Accessed November 30, 2012.
250. US Geological Survey. 2011a. Allegheny River at Franklin, Pennsylvania (03025500). Surface-WaterMonthly Statistics for the Nation. May 2006-September 2011. Available at http://waterdata.usgs.gov/nwis/inventory/?site_no=03025500&agency_cd=USGS&amp. Accessed October 10, 2012.
251. US Geological Survey. 2011b. Blacklick Creek at Josephine, Pennsylvania (03042000). Surface-Water Monthly Statistics for the Nation. May 2006-September 2011. Available at http://waterdata.usgs.gov/nwis/inventory/?site_no=03042000. Accessed November 27, 2012.
252. US Geological Survey. 2011c. Information Relevant to the U.S. Geological Survey Assessment of the Middle Devonian Shale of the Appalacian Basin Province. Open-File Report 2011-1298. US Geological Survey. 22 p. Available at http://pubs.usgs.gov/of/2011/1298/. Accessed
253. US Geological Survey. 2012a. Allegheny River at Franklin, Pennsylvania (03025500). Current Conditions for Selected Sites (Provisional). Available at http://nwis.waterdata.usgs.gov/nwis/uv?site_no=03025500. Accessed November 27, 2012.
254. US Geological Survey. 2012b. Blacklick Creek at Josephine, Pennsylvania (03042000). Current Conditions for Selected Sites (Provisional). US Geological Survey, Available at http://nwis.waterdata.usgs.gov/nwis/uv?site_no=03042000. Accessed November 27, 2012.
255. US House of Representatives. 2009. Appropriations Committee Report for the Department of the Interior, Environment, and Related Agencies Appropriations Bill, HR 2996. Available at http://www.gpo.gov/fdsys/pkg/CRPT-111hrpt180/pdf/CRPT-111hrpt180.pdf. AccessedDecember 6, 2012.
256. US House of Representatives 2011. Chemicals Used in Hydraulic Fracturing. Available at http://democrats.energycommerce.house.gov/sites/default/files/documents/Hydraulic%20Fracturing%20Report%204.18.11.pdf. Accessed November 27, 2012.
257. Van Ginkel, C. G. 1996. Complete degradation of xenobiotic surfactants by consortia of aerobic microorganisms. Biodegradation 7 (2): 151-164.
258. Van Schie, P. M. and Young L.Y. 2000. Biodegradation of phenol: mechanisms and applications. Bioremediation Journal 4 (1): 1-18.
259. Wallis, S. 2007. The numerical solution of the advection-dispersion equation: a review of some basic principles. Acta Geophysica 55 (1): 85-94.
260. Warner, N. R., Jackson, R. B., Darrah, T. H., Osborn, S. G., Down, A., Zhao, K., White, A. and Vengosh, A.2012. Geochemical evidence for possible natural migration of marcellus formation brine to shallow aquifers in Pennsylvania. Proceedings of the National Academy of Sciences of the United States of America 109 (30): 11961-11966.
261. Watts, K. R. 2006a. Hydrostratigraphic Framework of the Raton, Vermejo, and Trinidad Aquifers in the Raton Basin, Las Animas County, Colorado. Scientific Investigations Report 2006-5129. US Geological Survey 37 p. Available at http://pubs.usgs.gov/sir/2006/5129/pdf/SIR065129_508.pdf. Accessed December 3, 2012.

262. Watts, K. R. 2006b. A Preliminary Evaluation of Vertical Separation between Production Intervals of Coalbed-Methane Wells and Water-Supply Wells in the Raton Basin, Huerfano and LasAnimas Counties, Colorado, 1999-2004. Scientific Investigations Report 2006-5109. US Geological Survey. 15 p. Available at http://pubs.usgs.gov/sir/2006/5109/pdf/sir2006_5109.pdf. Accessed December 3, 2012.
263. Westat. 2011. Quality Assurance Project Plan v. 1.1 for Hydraulic Fracturing. Westat for US Environmental Protection Agency. Available at http://www.epa.gov/hfstudy/qapps.html. Accessed December 11, 2012.
264. Williams, D. R., Felbinger, J. K. and Squillace, P. J. 1993. Water Resources and the Hydrologic Effects of Coal Mining in Washington County, Pennsylvania. Open-File Report 89-620. US Geological Survey, Lemoyne, Pennsylvania.
265. Williams, J. E., Taylor, L. E. and Low, D. J. 1998. Hydrogeology and Groundwater Quality of the Glaciated Valleys of Bradford, Tioga, and Potter Counties, Pennsylvania. Water Resources Report 68. US Geological Survey and Pennsylvania Geological Survey. 98 p. Available athttp://eidmarcellus.org/wp-content/uploads/2011/06/USGS-Bradford-Report.pdf. Accessed December 5, 2012.
266. [267] Ziemkiewicz, P. 2011. Wastewater from Gas Development: Chemical Signatures in the Monongahela River Basin. Proceedings of the US Environmental Protection Agency Technical Workshops for the Hydraulic Fracturing Study: Water Resources Management, Arlington, Virginia. Available at http://epa.gov/hfstudy/waterworkshop.html. Accessed November 30, 2012.

U.S. Environmental Protection Agency. Study of the Potential Impacts of Hydraulic Fracturing on Drinking Water Resources: Progress Report. Washington, D.C.: U.S. EPA, Office of Research and Development; December, 2012. EPA/601/R-12/011.

PART II

HYDRAULIC FRACKING AND WILDLIFE

CHAPTER 9

HABITAT SELECTION BY MULE DEER DURING MIGRATION: EFFECTS OF LANDSCAPE STRUCTURE AND NATURAL-GAS DEVELOPMENT

PATRICK E. LENDRUM, CHARLES R. ANDERSON, JR., RYAN A. LONG, JOHN G. KIE, and R. TERRY BOWYER

9.1 INTRODUCTION

Migration is a remarkable life-history strategy that represents an essential component of the ecological niche of a variety of taxa, including mammals (Dingle and Drake 2007). This seasonal movement between ranges allows animals to increase access to important forage resources (Baker 1978) and reduce risk of predation (Fryxell and Sinclair 1988), both of which affect survival and reproduction (Nicholson et al. 1997). For example, large herbivores living in temperate regions often move from low elevations in winter to higher elevations in spring and summer, which provides release from a restricted food supply and access to newly available forage in spring (Garrott et al. 1987, Fryxell and Sinclair 1988, Mysterud 1999, Hebblewhite et al. 2008, Monteith et al. 2011). Such strategies ostensibly are favored by natural selection because the development of scale-dependent behaviors, which increase access to high-quality forage, improves both survival and reproductive success (Senft et al. 1987). Access to high-quality forage during spring migration is of particular importance for migratory ungulates living in temperate regions, because migration closely coincides with the timing of parturition (Singh and Milner-Gulland 2011).

Ungulate migrations generally occur along traditional routes (Baker 1978, McCullough 1985, Andersen 1991), many of which have been disrupted, especially over the past 4 decades, because of human activities (e.g., anthropogenic barriers and habitat loss; Sawyer et al. 2005, Harris et al. 2009, Beckmann et al. 2012). Bolger et al. (2008) observed that for many mammalian species (e.g., wildebeest, *Connochaetes taurinus,* Ottichilo et al. 2001; zebra, *Equis burchelli,* Williamson and Williamson 1985; and mule deer, *Odocoileus hemionus*, Bertram and Rempel 1977), the disruption of migratory routes has caused rapid population collapses. Indeed, populations of mule deer have declined throughout much of the Intermountain West, USA (Unsworth et al. 1999, Johnson et al. 2000, Stewart et al. 2002, Bishop et al. 2009, Hurley et al. 2011), and unprecedented levels of energy development throughout the region represent a critical threat to traditional migration routes for mule deer (Copeland et al. 2009). In particular, effects of natural-gas development can include the direct loss of habitat around well pads, access roads, and pipeline constructions, as well as indirect losses caused by increased human disturbance (e.g., traffic, noise) associated with infrastructure. These disturbances may displace mule deer or alter their patterns of habitat use along migration routes (Hayes and Krausman 1993, Sawyer et al. 2006). As a result, large-scale migration corridors that are protected or managed specifically to mitigate energy development may be critical for protecting the life-history strategy of long-distance migration by mule deer (Sawyer et al. 2005).

Effective conservation and planning must account for the inherent dynamics of ecological processes and effects of anthropogenic disturbance on habitat use and availability (Morrison 2001, Pressey et al. 2007). Land-use activities such as recreation, agriculture, and infrastructure development can influence spatial and temporal patterns of animal occurrence and demographics (Rost and Bailey 1979, Sawyer et al. 2009a, Dzialak et al. 2011). As landscape fragmentation increases because of human development and land-use practices, understanding effects of development and landscape characteristics on variation in range use, fidelity to migration paths, and demographics of animals will be of increasing importance for the conservation of these large, vagile mammals (Webb et al. 2011). For example, anthropogenic disturbances associated with energy development have been related to changes in resource selection by ungulates (Sawyer et al. 2006, Dzialak et al. 2011), as well as to reduced range fidelity (Webb et al. 2011).

Estimating resource-selection functions (RSFs) for mammals can be a valuable research tool for mitigating influences of human activities (Sawyer et al. 2006, Long et al. 2008, Harju et al. 2011). Conservation actions may be misguided, however, when resource selection is only evaluated for animals that already are influenced by high levels of human activity, because selection for some resources may be partially or largely a function of avoidance of human disturbance (Harju et al. 2011). We used resource-selection functions (Compton et al. 2002, Boyce 2006) to determine if varying levels of natural-gas development altered selection for landscape characteristics by migratory mule deer. We used a novel approach to define availability based on movement parameters collected from GPS collars, which included step length and turning angles between successive deer locations during spring migration.

We conducted our study in the Piceance Basin of northwestern Colorado, USA, from 2008 to 2010. The Piceance Basin supports one of the largest populations of migratory mule deer in North America, estimated at 21,000–27,000 animals over the past several decades (White and Lubow 2002). This region also includes one of the largest natural-gas reserves in North America, with projections of energy development throughout northwestern Colorado over the next 20 years to increase from approximately 500 to 15,000 wells. Within the Piceance Basin, we monitored patterns of spring migration for mule deer that were exposed to differing levels of natural-gas development.

We hypothesized that, during spring migration, step lengths and turning angles of mule deer would vary with the level of natural-gas development and time of day because of disturbances associated with human activity. We predicted that mule deer in the most-developed areas would have longer step lengths with a more straight-forward direction of travel compared with deer in less-developed study areas. We further predicted that movement rates would be higher at night in highly developed areas compared with those of less development to compensate for deer avoiding human activities during the day. We also hypothesized that landscape characteristics and human disturbances associated with natural-gas development would influence resource selection by adult female mule deer along their migration routes in spring. We predicted that mule deer would select travel routes along south-facing aspects, with moderate slopes and low levels of ruggedness, at low elevations in all areas, regardless of levels of development. In contrast, we predicted that deer in the most highly developed areas would select for habitat types with greater concealment cover further from areas of human activity, such as roads and well pads, whereas in areas with low levels of development, selection by mule deer would be influenced more by habitat types that provide high forage availability and less by the presence of development.

9.2 STUDY AREA

We monitored four populations of mule deer that wintered in different areas of the Piceance Basin: North Ridge (53 km^2) in the northeastern portion of the Basin; Ryan Gulch (141 km^2) in the southwestern portion of the Basin; and North Magnilia (79 km^2) and South Magnolia (83 km^2) in the central portion of the Basin (Fig. 1). During spring migration, mule deer from North Ridge and North Magnolia moved in an easterly direction across US Highway 13 towards the Flat Top Mountain Range, where they resided at high elevations during summer (Fig. 1). Mule deer from Ryan Gulch and South Magnolia migrated in a southerly direction to high elevations along the Roan Plateau (Fig. 1).

The climate of the region was typified by warm dry summers (28°C average high) and cold winters (−12°C average low); most annual moisture was from snow (144.0 cm; Western Regional Climate Center, 2008–2010). The primary winter habitat for mule deer ranged from 1,675 to 2,285 m in elevation, and summer habitat ranged from 2,000 to 2,800 m. The Piceance Basin varied topographically with numerous ridges and draws. The area contained other large herbivores including North American elk (*Cervus elaphus)*, wild horses (*Equus caballus*), and moose (*Alces alces*), the latter of which occurred infrequently on summer range. Common species of large carnivores included coyotes (*Canis latrans*),

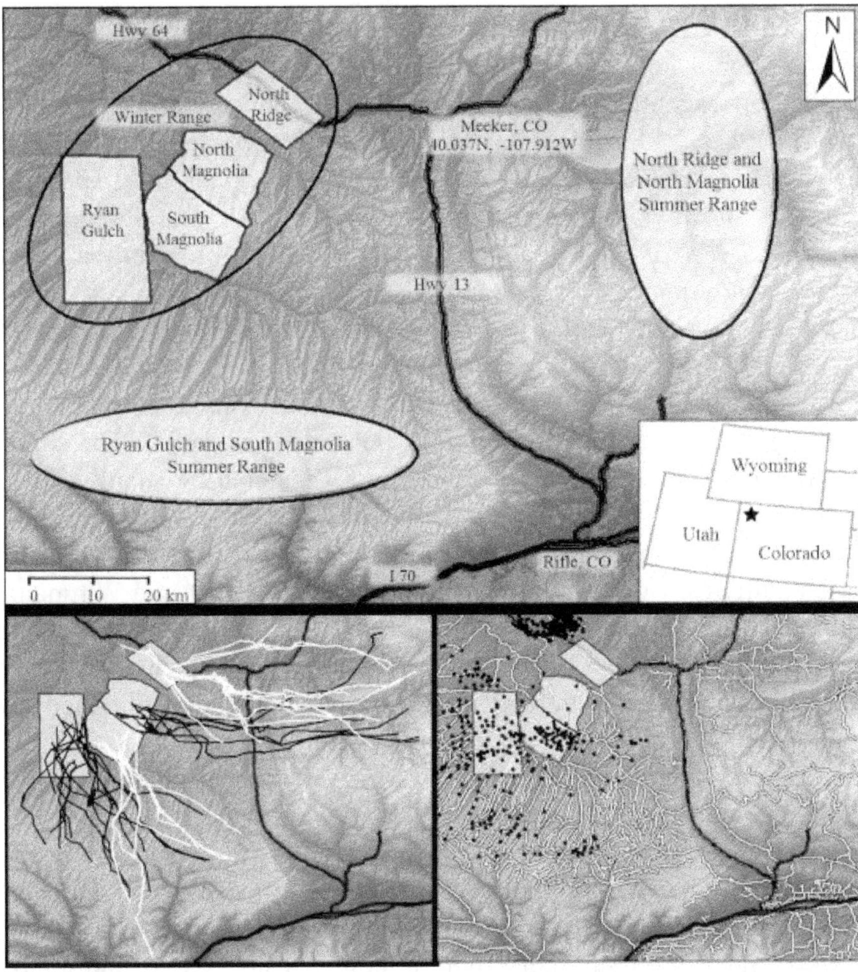

FIGURE 1: (Top) Study areas in the Piceance Basin of northwestern Colorado, USA, (lower left) spring 2009 migration routes of adult female mule deer (n = 52), and (lower right) active natural-gas well pads (black dots) and roads (state, county, and natural-gas; white lines) from May 2009.

mountain lions (*Puma concolor)*, bobcats (*Lynx rufus*), and black bears (*Ursus americana*).

Pinion pine (*Pinus edulis*) and Utah juniper (*Juniperus osteosperma*) were the dominant overstory species on winter range; common shrubs included big sagebrush (*Artemisia tridentata*), Utah serviceberry (*Amelanchier utahensis)*, mountain mahogany *(Cercocarpus montanus*), bitterbrush (*Purshia tridentate*), Gamble's oak (*Quercus gambelii*), mountain snowberry (*Symphoricarpos oreophilus*), and rabbitbrush (*Crysothamnus spp*.; Bartmann et al. 1992). Primary vegetation communities on summer range included a Gambel's oak-mountain shrub complex at lower elevations. This community was mixed with quaking aspen (*Populus tremuloides*)-Douglas-fir (*Pseudotsuga menziesii*) forest, and Engelmann spruce (*Picea engelmannii*)-subalpine fir (*Abies lasiocarpa*) forest at higher elevations (Garrott et al. 1987). The study area was dissected by numerous drainages vegetated by stands of big sagebrush, saltbrush (*Atriplex spp*.), and black greasewood (*Sarcobatus vermiculatus*), with most of the primary drainages converted to mixed-grass hay fields. Botanical nomenclature follows Weber and Wittmann (2001).

Within the Piceance Basin, levels of natural-gas development varied markedly (Fig. 1). North Ridge (low development) contained no development on either winter or summer range; however, the transition between those ranges included increased levels of human activity from vehicle traffic and housing infrastructure because of proximity to the town of Meeker, Colorado. North Magnolia (medium-low development) exhibited a low density of active well pads on winter range ($\leq$0.05 pads/km^2) and along migration paths (0.17 pads/km^2), and no active well pads on summer range, although deer crossed one major highway with scattered ranch holdings along their migration path. Ryan Gulch (medium-high development) exhibited moderate development on winter range (0.37 pads/km^2), and throughout the transition range (1.54 pads/km^2), with a decreased density of development on summer range as deer spread across the landscape (0.06 pads/km^2). South Magnolia (high development) had the highest level of development activity on winter range (0.70 pads/km^2), and along migration corridors (1.99 pads/km^2), with low levels of development on summer range (0.04 pads/km^2).

9.3 METHODS

9.3.1 ANIMAL CAPTURE

We net-gunned mule deer from a helicopter (Krausman et al. 1985) to obtain a sample of adult (≥1.5 years old) females (n = 205; North Ridge = 60, North Magnolia = 43, South Magnolia = 42, and Ryan Gulch = 60) during 2008–2010. Total number of deer captured was 45 during January 2008, 60 during March 2009, and 100 during March 2010. All deer were fitted with GPS collars programmed to attempt a fix once every 5 h during spring migration. We only retained 3D fixes or fixes with a positional dilution of precision <10 m (D'eon and Delparte 2005); 90% of fixes had <20 m accuracy. Fourteen females received remotely downloadable GPS collars during the first year of study (GPS-4400S; Lotek Wireless, Newmarket, Ontario, Canada), and remaining females received store-on-board GPS collars (G2110B; Advanced Telemetry Systems, Isanti, Minnesota, USA). All collars were equipped with timed drop-off mechanisms, scheduled to release during April of the year following deployment, and mortality sensors that increased pulse rate following 4–8 h of inactivity. All aspects of animal handling and research complied with methods adopted by the American Society of Mammalogists for research on wild mammals (Sikes et al. 2011), and were approved by an Animal Care and Use Committee at Idaho State University (protocol # 670 0410).

9.3.2 PATTERNS OF RESOURCE SELECTION

We retrieved GPS collars from the field each spring or following mortality events. Deer locations were plotted in ArcGIS 9.3 (ESRI, Redlands, California, USA) and spring migration routes identified. Only deer that completed spring migration were included in analyses (n = 167). We used

Hawth's Analysis Tools to derive 95% kernel-density estimates of seasonal ranges for each individual. We determined the initiation of spring migration based on the day a particular deer left the winter range on a trajectory path (i.e., three successive locations leading away from winter range), and arrival on summer range was determined as the first location inside the summer range for that same deer (Garrott et al. 1987).

We estimated resource-selection functions based on a matched-case design (Manly et al. 2002, Boyce 2006) to model resource selection by mule deer along migration paths. Locations from individual deer represented used points. Random locations were generated based on step lengths and turning angles derived from deer locations to define available habitat for each sub-population (i.e., study area) along migratory paths. Hence, we used a critical life-history characteristic of mule deer to determine the scale at which to measure availability of habitat characteristics during migration (sensu Bowyer and Kie 2006). Constructed on principles of a correlated random walk (Turchin 1998), "steps" were characterized by the straight-line distance between animal locations and turning angles determined as the angle from a previous step length to the next. Steps that had >5 h between locations were not included in analyses. We used Hawth's Analysis Tools (Beyer 2004) to determine a distribution of average step lengths and turning angles from known locations of mule deer (n = 6,433). We then grouped turning angles into 20° bins to create an average frequency distribution for each study area (Fortin et al. 2005). Step-length distributions were separated into 1-km bins, with a maximum step-length of 8 km. We chose 8 km as the maximum distance because <1% of the steps for each area were greater than that distance. An 8-km buffer was then placed around the outermost set of used locations as a boundary for the distribution of random locations. Each observed step was then paired with 10 random steps derived from the distribution of known steps and turning angles with a correlated random-walk simulation (Geospatial Modelling Environment, Beyer et al. 2010). We chose 10 random locations because those points provided a relatively uniform distribution across the area of interest. We did not allow used and random locations to overlap to avoid a loss of statistical power (Bowyer and Kie 2006). We analyzed steps separately for day (n = 3,651) and nighttime (n = 2,782) use, because of fluctuation in human activities associated with time of day. We used sunrise (5:59 h) and sunset (20:17 h) of the median departure date (5 May) of deer from winter range to divide diel patterns into day and night.

9.3.3 LANDSCAPE VARIABLES

Sawyer et al. (2006) identified five landscape variables as potentially important predictors of winter distributions of mule deer in areas with oil and gas development: elevation; slope; aspect; road density; and distance to wells. We expanded that list by including additional variables for modeling resource selection by female mule deer along migration routes in spring. We examined five characteristics of landscape structure including elevation (m), slope (%), aspect (transformed into categories of North, East, South, West), terrain ruggedness (vector ruggedness measure; VRM), and vegetation type. In addition, we included two anthropogenic disturbances associated with natural-gas development: distance to nearest well pad; and distance to nearest road. Within the least-developed areas, well pads were present at low density (<0.05 pads/km^2) only on winter range, and not throughout the migration corridor. Consequently, we did not include distance to well pad as a predictor variable for those two study sites. Prior to analyses, we transformed elevation, distance to nearest well pad, and distance to nearest road so that a 1-unit change in elevation represented 50 m, and a 1-unit change in distance to well pads and roads represented 100 m. Accordingly, odds ratios for those variables indicate the predicted change in odds of selection by mule deer for every 50-m change in elevation, and 100-m change in distance to roads.

We estimated elevation using a digital-elevation model (DEM) at a resolution of 30 m (http://datagateway.nrcs.usda.gov/). We then used ArcGIS 9.3 Spatial Analyst Tools to derive values of slope and aspect from the DEM. A vector ruggedness measure also was derived from the DEM following the method of Sappington et al. (2007). Ruggedness values ranged between 0 (flat) and 1 (most rugged). A map of vegetation types was obtained from the Colorado Vegetation Classification Project (http://ndis.nrel.colostate.edu/coveg/), which provided a landscape-level vegetation dataset for the state of Colorado at a resolution of 25 m. The vegetation map included 87 habitat classes, which we reclassified into six categorical habitats based on similarity of vegetation types: (1) forbs and grasslands (herbaceous); (2) sagebrush-steppe; (3) pinion-juniper dominated (PJ); (4) aspen and conifer stands (forest); (5) riparian; and (6) bare rocky ground (barren).

Locations of well pads were obtained from the Colorado Oil and Gas Conservation Commission (http://cogcc.state.co.us/). We selected well-pad records from June during 2008–2010 and designated each well as either producing or in development. Datasets for roads were obtained from the TIGER/Line shape files of the U.S. Census Bureau (http://www.census.gov/geo/www/tiger/shp.html) and the Colorado Department of Transportation (http://apps.coloradodot.info/dataaccess/). We included county roads, highways, and roads used for purposes of natural-gas extraction in our models. We did not differentiate levels of vehicle use among roads because there was insufficient information to do so.

9.3.4 INDEPENDENCE OF LOCATIONS

We used association matrices to investigate the spatiotemporal association among individual radio-collared deer (ASSOC1; Weber et al. 2001, Long et al. 2008). We considered deer to be associated (i.e., part of the same herd) if a pair of individuals were within 500 m of each other during >50% of the total number of days during migration. No patterns of association were detected; therefore, all deer remained in the analyses as separate sample units.

9.3.5 DATA ANALYSES

To estimate resource-selection functions, we compared used and random locations along spring migration routes with conditional logistic regression (PROC LOGISTIC; SAS Institute Inc. 1990, Compton et al. 2002, Boyce 2006). Each individual mule deer was considered as a stratified variable to control for variation among individuals (i.e., individuals were sampling units), and the logistic model for each study area was conditioned upon that variable (Long et al. 2009a). Prior to modeling, we used a correlation matrix to evaluate collinearity ($|r| > 0.7$) among predictor variables (PROC CORR; SAS Institute, Cary, North Carolina). Distance to developing well pads was highly correlated with distance to producing well pads ($|r| > 0.9$), so we combined categories to create a single variable for distance to well pads. No other predictor variables were correlated (all $|r| < 0.50$) and therefore, they remained in the modeling process. To initially evaluate potential differences in patterns of resource selection between two broad categories of development (least developed = North Ridge and North Magnolia, most developed = South Magnolia and Ryan Gulch), we fit a global model that included all main effects along with possible main effect × development-level interactions for both night and daytime locations (Long et al. 2008). Statistical significance ($P \leq 0.05$) of interaction terms indicated a difference in selection for that variable between development levels. After evaluation of the global model, we then modeled each study area separately for night and day, with all possible combinations of the seven predictor variables (Long et al. 2008). Whenever distance to well pads entered the model, we also included distance to roads, and vice versa, because we considered these two variables as indicative of levels of development. For categorical variables, we used southerly aspect, which we predicted to receive high use, and sagebrush-steppe, because of its importance to mule deer (Stewart et al. 2010, Anderson et al. 2012), as reference categories.

We calculated Akaike's Information Criterion adjusted for small sample size (AICc), ΔAICc, and Akaike weights (wi) for each model (Burnham and Anderson 2002). We intended to use model-averaged parameter estimates and unconditional standard errors (SE) to assess the influence of each predictor variable on resource selection (Burnham and Anderson 2002). Our global models, however, contained >95% of the Akaike weights, with the next-best models having a ΔAICc of >30 for all study areas. Consequently, we used the global model for interpreting patterns of resource selection for each study area and time of day. We converted parameter estimates to odds ratios by exponentiation for simplicity of interpretation. If the 95% confidence interval around an odds ratio contained 1, then that variable was considered not significant. We considered odds ratios for each predictor variable to differ significantly among study areas if their 95% confidence intervals did not overlap (Long et al. 2008, Anderson et al. 2012). We tested descriptive statistics among study areas and years using analysis of variance (ANOVA) with Bonferroni pairwise comparisons in Minitab 16.1.0 (State College, Pennsylvania, USA, 2010). Additionally, we calculated a Pearson correlation coefficient to determine if step lengths were correlated with the distance traveled along migratory routes among study areas.

We used k-fold cross validation (Boyce et al. 2003, Anderson et al. 2005, Long et al. 2009b) to evaluate predictive strength of the resource-selection functions for adult female mule deer within study areas, for both day and nighttime models. We withheld 1 year at a time as test data and used the remaining 2 years as training data, which resulted in three total iterations for each model. During each iteration of the procedure, we used the model derived from the training data to obtain predicted RSF values for the random locations for each deer. Next, we sorted random locations from lowest to highest based on their predicted values and binned them into 10 groups of equal size (Boyce et al. 2003, Anderson et al. 2005, Long et al. 2009b). We then obtained predicted RSF values for test data using the same model, and placed locations from the test dataset into the bins we created with the random data based on their associated RSF values (Anderson et al. 2005, Long et al. 2009b). Finally, we regressed the number of locations from the test dataset in each bin against the median RSF value of the random locations, and recorded the coefficient of determination (r^2) and its slope. We averaged these statistics across the three iterations for each model, and considered the combination of a high coefficient of determination and a positive slope to be indicative of a model that predicted well (Long et al. 2009b). In addition, we calculated a Spearman's rank correlation (r_s), which makes no assumptions concerning line shape, for each iteration of the procedure and used the mean value as an additional metric of predictive strength.

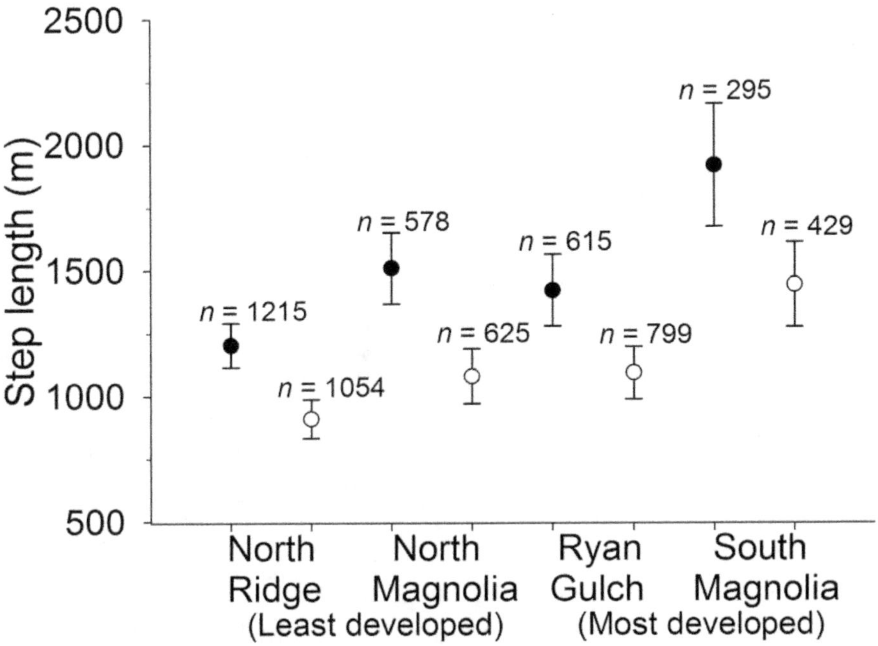

FIGURE 2: Step lengths (m) for adult female mule deer (n = 167) during spring 2008–2010 migration in four study areas of the Piceance Basin, northwestern Colorado, USA for night (solid circle) and day (open circle) locations. Error bars = 95% CI and n = number of step lengths for each category.

9.4 RESULTS

9.4.1 STEP LENGTHS AND TURNING ANGLES

Step lengths were significantly longer ($P < 0.05$) at night than during the day for all study areas (Fig. 2). Step lengths also were significantly greater in South Magnolia (high development) compared with all other study areas (Fig. 2). Furthermore, within broad development categories (least

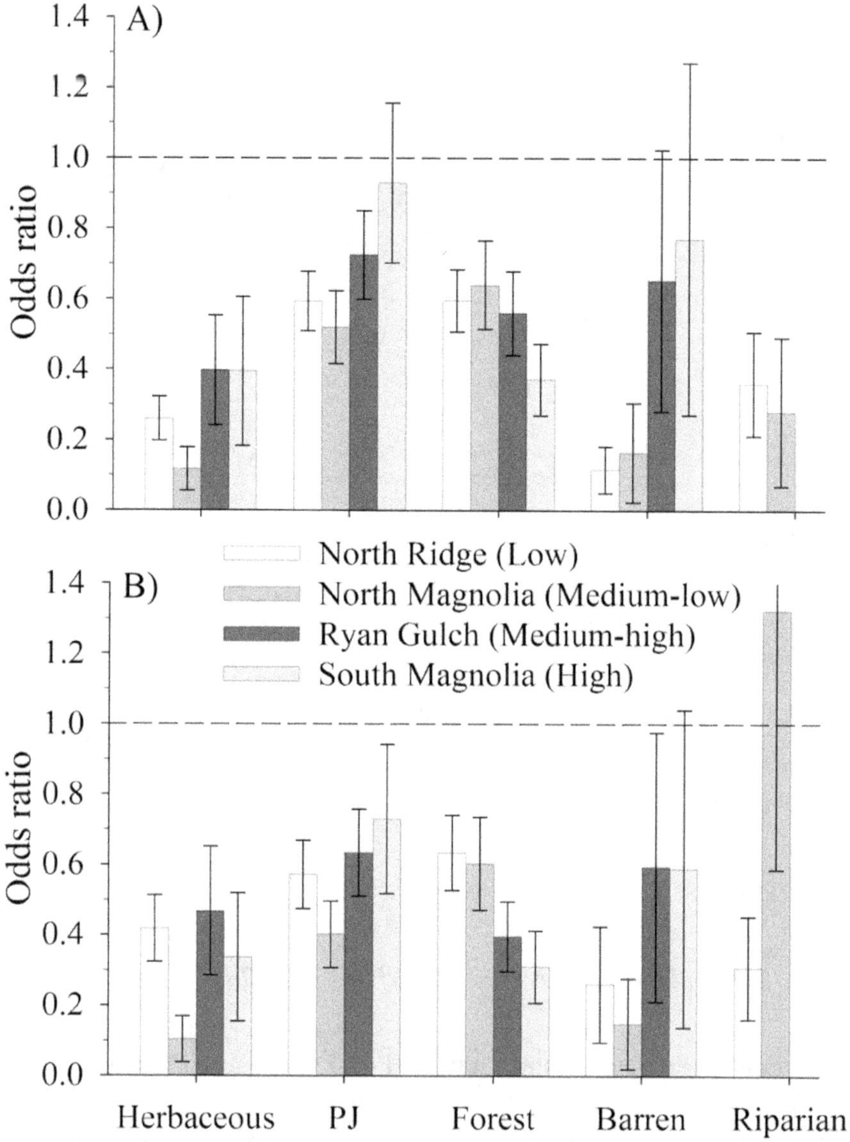

FIGURE 3: Odds ratios for grasses (herbaceous), pinyon-juniper (PJ), aspen and conifer (Forest), bare-rocky ground (Barren), and riparian habitat (Riparian) obtained from resource-selection functions for adult female mule deer (n = 167) during spring 2008–2010 migration from 4 study areas in the Piceance Basin, northwestern Colorado, USA for day (A) and night (B). Odds ratios indicate the percent change (1 = no change) in odds of use by mule deer for each habitat type relative to sagebrush steppe. Error bars = 95% CI.

developed, most developed), deer from the study areas with higher development had longer step lengths than their counterparts experiencing lower development. Deer from North Magnolia had longer step lengths than deer from North Ridge, and deer from South Magnolia had longer step lengths than deer from Ryan Gulch (Fig. 2). Step lengths were not correlated with distance traveled ($r = -0.002$, $P = 0.93$). In addition, deer from all study areas exhibited a strong tendency to travel in a generally forward direction during both day and night (= −0.08°, SD ± 90.05°, n = 5,610).

9.4.2 RESOURCE SELECTION

Global models of resource selection for mule deer that contained interactions between each main effect (i.e., slope, elevation, habitat type) and level of development (least developed vs. most developed) indicated notable differences in patterns of selection between levels of development during both day and night. The daytime model indicated significant differences in selection for distance to roads ($P = 0.0002$), elevation ($P < 0.0001$), slope ($P = 0.0005$), and aspect ($P = 0.0057$) between the least developed and most developed sites. Similarly, the nighttime model indicated significant differences in selection for distance to roads ($P = 0.0021$), elevation ($P < 0.0001$), and slope ($P = 0.0004$) between the least developed and most developed study areas; therefore, we produced separate model sets for each study area.

Percentage of vegetation types occurring along the migratory paths of mule deer was similar among study areas (Friedman's 2-way ANOVA for goodness of fit, $\chi 25 = 96.4$, $P < 0.001$), which allowed for meaningful comparisons. Models of resource selection indicated that female deer generally selected sagebrush-dominated communities significantly more than other vegetation types across study areas (odds ratios for most habitat types <1.0; Fig. 3). No significant difference occurred in selection of sagebrush versus pinyon juniper habitat by deer from South Magnolia (high development) during the day, or in selection of sagebrush habitat versus barren ground in that study area regardless of time of day (CI's overlapping 1.0; Fig. 3). Similarly, no difference was observed in selection of sagebrush habitat versus barren ground in Ryan Gulch (medium-high development) during the day, and in selection of sagebrush versus riparian habitats in North Magnolia (medium-low development) during the night.

Deer from South Magnolia (high development) selected habitat types that provided a greater degree of concealment cover (e.g., pinyon-juniper) more strongly than deer from either of the least developed study areas (North Ridge, low development, during the day, and North Magnolia, medium-low development, during both day and night; Fig. 3). In contrast, deer from North Ridge and North Magnolia (least developed) selected habitat types that increased access to both forage and cover (e.g., aspen-conifer forests) more strongly than did deer from South Magnolia (high development) regardless of time of day (Fig. 3). Additionally, deer from North Ridge (low development) selected aspen-conifer habitats more strongly at night than deer from Ryan Gulch (medium-high development; Fig. 3).

South-facing slopes were selected significantly less than other aspects in all study areas during both day and night (all odds ratios >1.0, Fig. 4). Models also indicated that deer selected gentle slopes and gentle terrain (low VRM) across study areas (all odds ratios <1.0). Selection for elevation, however, varied with respect to study area (Figs. 5, 6). Deer from South Magnolia (high development) showed the greatest response to elevation, with a 13.4% increase in the odds of selection for every 50-m increase in elevation during the day, and a 13.3% increase in odds of selection for every 50-m increase in elevation at night (Figs. 5, 6). Conversely, deer from North Ridge (low development) showed selection for low elevations, with a 7.8% increase in odds of selection for every 50-m decrease in elevation during the day, and a 7.7% increase in odds of selection for every 50-m decrease in elevation at night.

In the most developed study areas (Ryan Gulch, South Magnolia), female deer selected areas closer to well pads, regardless of time of day (Figs. 5, 6). Deer from Ryan Gulch (medium-high development) selected areas farther from roads (4.5% increase in odds of selection for every 100-m increase in distance to roads during both day and night), whereas deer from South Magnolia (high development) showed the opposite pattern (≤3.7% decrease in odds of selection for every 100-m increase in distance to roads; Figs. 5, 6). In the least developed study areas (North Ridge, North Magnolia), the only significant effect of roads was for deer

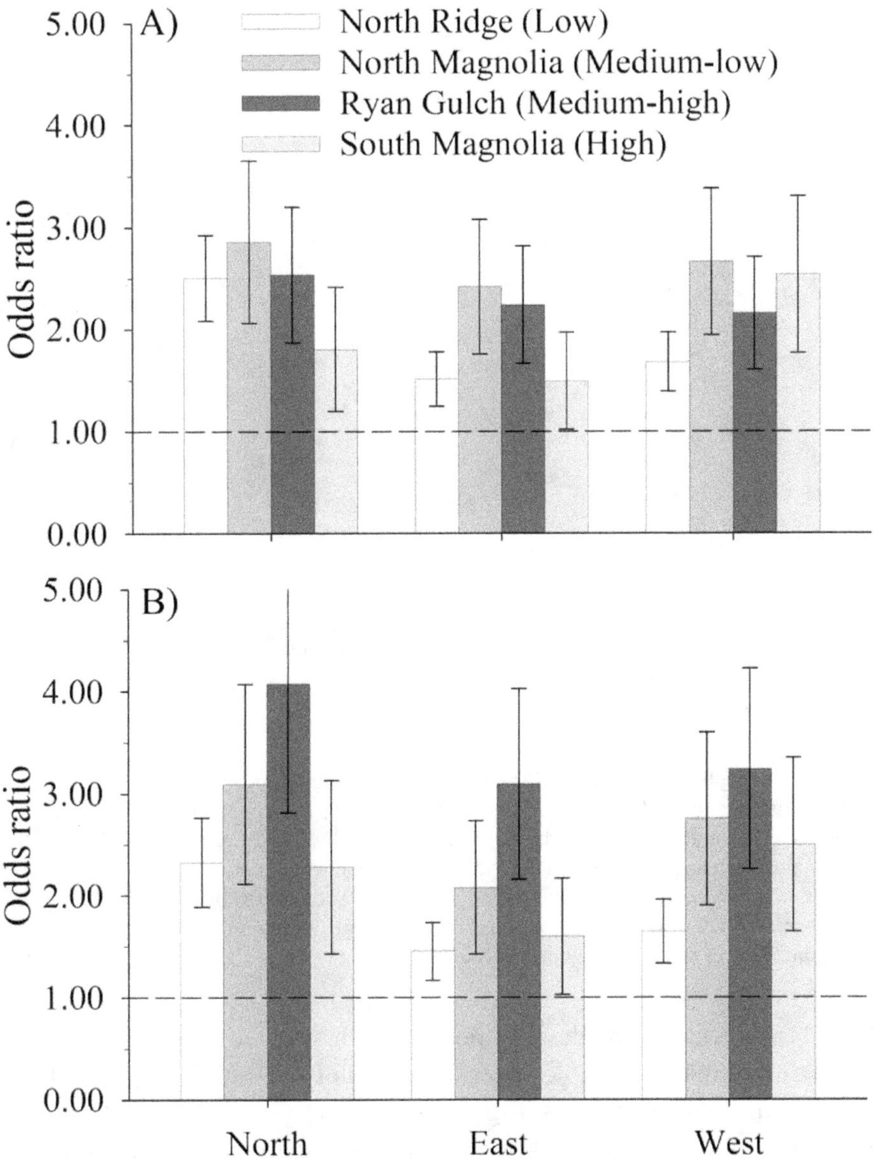

FIGURE 4: Odds ratios for north-, east-, and west-facing aspects obtained from resource-selection functions for adult female mule deer (n = 167) during spring 2008–2010 migrating from four study areas in the Piceance Basin, northwestern Colorado, USA, for daytime (A) and nighttime (B). Odds ratios indicate the percent change (1 = no change) in odds of use by mule deer for each habitat type relative to south-facing aspects. Error bars = 95% CI.

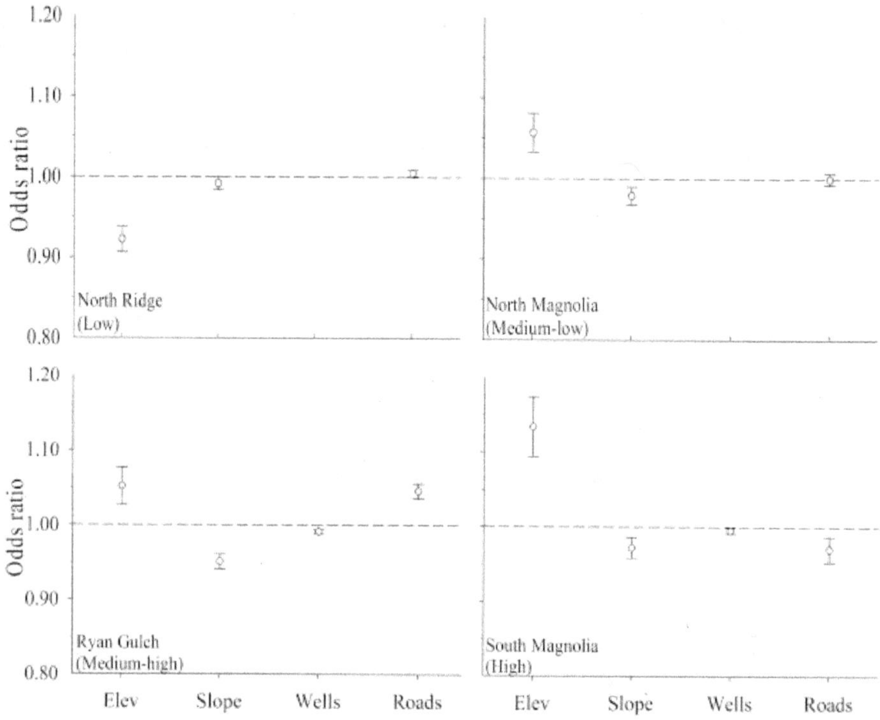

FIGURE 5: Daytime odds ratios for elevation (m; Elev), slope (%), terrain ruggedness (VRM), distance to well pads (m; Wells), and distance to roads (m; Roads) obtained from resource-selection functions for adult female mule deer (n = 167) during spring 2008–2010 migrating from four study areas in the Piceance Basin, northwestern Colorado, USA. Odds ratios indicate the percent change (1 = no change) in odds of use by mule deer for every 50-m increase in elevation, 1 % increase in slope, change in terrain ruggedness (VRM), and 100-m increase in distance to wells and roads. Error bars = 95% CI.

from North Ridge (low development), which selected areas further from roads during the day (0.5% decrease in odds of selection for every 100-m increase in distance to roads; Fig. 5).

Cross-validation analyses indicated that resource-selection functions were highly predictive for all study areas except Ryan Gulch (Table 1). Mean slopes of the regression lines were positive for all models, and mean coefficients of determination and Spearman rank correlations were high for all models other than those for Ryan Gulch (Table 1). Lower predictive strength indicated that there was more variability in patterns of selection during spring migration among individual deer, among years, or both, in Ryan Gulch than in the other study areas.

9.5 DISCUSSION

We obtained several critical tests of our hypotheses related to patterns of spring migration of adult female mule deer in an area strongly influenced by anthropogenic disturbances. Patterns of resource selection and movement differed between deer that migrated through areas of highest well-

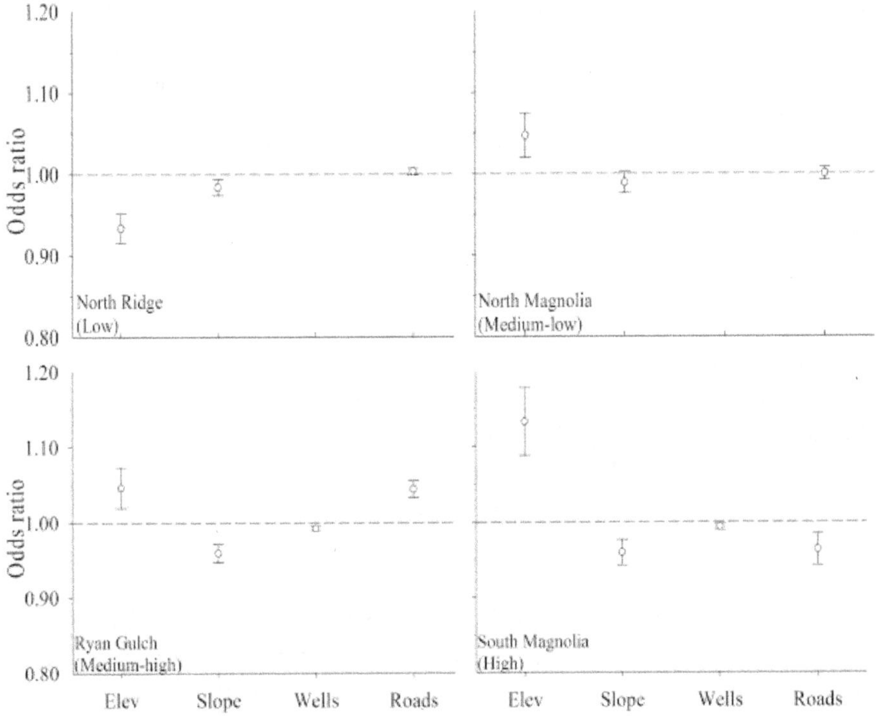

FIGURE 6: Nighttime odds ratios for elevation (m; Elev), slope (%), terrain ruggedness (VRM), distance to well pads (m; Wells), and distance to roads (m; Roads) obtained from resource-selection functions for adult female mule deer (n = 167) during spring 2008–2010 migration from four study areas in the Piceance Basin, northwestern Colorado, USA. Odds ratios indicate the percent change (1 = no change) in odds of use by mule deer for every 50-m increase in elevation, 1 % increase in slope, change in terrain ruggedness (VRM), and 100-m increase in distance to wells and roads. Error bars = 95% CI.

pad density and those that migrated through the least-developed areas. Patterns of behavior exhibited by deer that migrated through the sites of intermediate development did not differ from those of deer that migrated through either the highly developed or the least-developed study areas. Consequently, we hypothesize that mule deer may exhibit a threshold response to natural-gas development in which behavior is altered only after a relatively high degree of development occurs on the landscape. Additionally, we hypothesize that the low predictive strength of the resource-selection functions for deer occurring in Ryan Gulch, compared with high predictive strength of models for other areas, likely occurred because some deer captured from Ryan Gulch during the first 2 years of the study ultimately migrated outside of the boundary of the study area.

TABLE 1: Cross-validation results for resource-selection functions for spring migration of adult female mule deer (n = 167), indicating slope, coefficient of determination (r^2), and Spearman rank correlation (r_s), Piceance Basin, northwestern Colorado, USA, 2008–2010.

	Mean		
Model	Slope	r^2	r_s
North Ridge day	99.71	0.91	0.94
North Ridge Night	76.53	0.90	0.92
North Magnolia day	6.36	0.77	0.81
North Magnolia night	3.52	0.79	0.87
Ryan Gulch day	5.11	0.38	0.44
Ryan Gulch night	4.09	0.31	0.52
South Magnolia day	1.49	0.70	0.80
South Magnolia night	3.71	0.75	0.81

We hypothesized that step lengths and turning angles would vary with respect to levels of natural-gas development and time of day during which mule deer migrated. As predicted, mule deer migrating thorough the study area of greatest well-pad density had longer step lengths compared with deer migrating through the least-developed areas. Again, however, deer did not exhibit a difference in behavior when migrating through areas of moderate development (i.e., medium-low and medium-high). Avoidance of disturbed areas, such as those associated with natural-gas development, may affect patterns of migration by causing mule deer to alter rates of movement, potentially increasing energetic costs (sensu Parker et al. 1984). Increased energetic costs at a time when deer are physically stressed (i.e., after sever winters, increased development levels) could potentially lead to decreased survivorship (Parker and Robbins 1984). In addition, Hayes and Krausman (1993) observed that in areas with high levels of human disturbance, patterns of habitat use by female mule deer varied with respect to time of day. We observed longer step lengths for mule deer at night compared with daytime, although contrary to our prediction, this pattern was observed regardless of development level. Movement rates of cervids often are highest during crepuscular hours and greatly diminished during midday (Bowyer 1981, Beier and McCullough 1990, Ager et al. 2003), which may explain the generality of our results across development levels. Similarly, turning angles did not vary among levels of development or with time of day in our study, resulting in rejection of our third prediction that mule deer in developed areas would travel in a more straightforward direction compared with deer in the least-developed study areas.

We also hypothesized that landscape characteristics and human disturbances associated with natural-gas development would influence resource selection by female mule deer along migration routes during spring. Although sagebrush-steppe was the primary habitat selected, we observed that during the day, deer migrating through the highly developed landscapes selected pinyon-juniper habitats more often than deer migrating through the least-developed landscapes. In a shrub-steppe community, scattered trees may provide improved microclimates for deer (Parker and Gillingham 1990), or serve as concealment cover from perceived risks (Bowyer 1986). Similarly, McClure et al. (2005) demonstrated that risk, or risk avoidance, was more pronounced with deer living in urban environments, compare with their counterparts living in rural environments. Deer inhabiting urban areas attempted to minimize risks associated with harassment by concentrating in areas with greater concealment cover. Deer were more likely to select for aspen and conifer stands along the least-developed migration corridors compared with the area of greatest development. Thomas and Irby (1990) observed that deer selected aspen patches that provided cover and nutritious forage during migration. The difference in habitat selection we observed between development levels also could have resulted from the long, continuous forest stands along migration corridors in the least-developed areas, which contrasted with the most-developed areas, where a patch-work mosaic of forest stands resulted in pinyon-juniper being the more accessible cover type.

Mule deer selected for moderate slopes with less-rugged terrain, but avoided south-facing slopes, across development levels. This result was contrary to our prediction that deer would select south-facing slopes. The use of south and west aspects by mule deer is thought to be associated with higher solar radiation and higher primary production (Bowyer et al. 1998, D'Eon and Serrouya 2005). Nicholson et al. (1997), however, observed that migratory deer selected for north-facing slopes, which had a greater proportion of available water and were further from human disturbance than south-facing slopes. Similarly, Garrott et al. (1987) observed a shift in use from southerly to northerly aspects in April, prior to migration. These observations are comparable to our results, which may be an effect of the relatively dry climate of the Piceance Basin. Ager et al. (2003) observed a change in habitats selected by mule deer towards flatter slopes during the onset of spring, which also was consistent with our observations. Mule deer also selected for higher elevations along migratory routes in all study areas except North Ridge. We suspect this outcome may be a result of increased availability of agriculture fields at lower elevations and the use of a natural travel corridor created by the White River and associated tributaries, which are not present in the migration paths of deer on the other study areas. Agricultural lands consisting of forbs and grain crops can be beneficial to mule deer as newly available vegetation emerges in spring (Garrott et al. 1987, Stewart et al. 2010, Anderson et al. 2012). Although agricultural fields are present in South Magnolia, the rapid movements observed through that area may preclude deer from using those fields in a similar manner as deer from North Ridge.

Deer selected areas closer to well pads in the most developed areas, which was contrary to our prediction. Our results differ from previous studies in which ungulates have been observed to avoid anthropogenic disturbances (Nicholson et al. 1997, Dyer et al. 2001, Cameron et al. 2005, Sawyer et al. 2006, Singh et al. 2010). For example, Sawyer et al. (2006) observed that, during winter months, mule deer were less likely to occupy areas in close proximity to well pads than those farther away. Additionally, Singh et al. (2010) noted that the location and density of calving aggregations of Saiga antelope (*Saiga tatarica*) also have been affected by human disturbance on spring and summer ranges. During migration, however, ungulates demonstrate a strong fidelity to particular routes (Garrott et al. 1987, Thomas and Irby 1990, Andersen 1991, Sawyer et al. 2009b, Sawyer and Kauffman 2011). In addition, although migration routes often include stopover sites (Sawyer et al. 2009b, Sawyer and Kauffman 2011), in the Piceance Basin migration was rapid and traditional stopovers did not occur. Perhaps such fidelity and the rapid rate at which migrations occurred in the Piceance Basin (median spring migration periods = 3-8 days), overrode the behavioral response to avoid anthropogenic disturbances. The unprecedented rates of natural-gas development in the Piceance Basin and other areas throughout the Intermountain West may not allow deer sufficient time to adapt and alter their behaviors.

Deer often are observed thriving in areas of human development, such as residential environments. Natural-gas development may disturb deer more than residential development, because once residential neighborhoods are established they are relatively permanent, which may allow deer sufficient time to acclimate. Natural-gas developments, however, are constantly changing in nature and intensity. In the Piceance Basin, development first began in what is now the highly developed study area, and while acclimation is a possible explanation for the lack of avoidance, without pre-development data this possibility cannot be addressed directly. Sawyer et al. (2006) noted that changes in habitat selection appeared to be immediate, with no evidence of well-pad acclimation occurring over the 3 years during which their study took place. Furthermore, mule deer selected areas further from well pads as development progressed (Sawyer et al. 2006). The Pinedale Anticline, Wyoming, USA, however, is a very different landscape than the Piceance Basin. Deer do not have concealment cover on the Anticline because of wide open, flat, sagebrush winter range versus the topographic and vegetative diverse conditions present in the Piceance Basin, and these conditions may have minimized deer behavioral responses as development progressed.

Within the most developed study areas, mule deer avoided roads along migration routes in the moderately developed sectors, while selecting areas closer to roads in sectors of higher development. We hypothesize that deer may avoid high-traffic areas if they can do so without greatly altering their migration routes. If development levels are too great, however, deer may not have the option of avoiding roads (Wagner et al. 2011). Storm et al. (2007) suggested that in areas with high levels of human development, such as suburban environments, the ability for deer to exhibit an avoidance of anthropogenic disturbances may be diminished because of such uniform and wide-spread disturbance. Our hypothesis is further supported by deer also avoiding roads along migration routes that passed nearest the town of Meeker, where vehicle traffic was moderate, and slight alterations in patterns of movement allowed for avoidance of traffic without greatly altering migration routes. Disturbance caused by humans may be analogous to predation risk (Berger et al. 1983, Frid and Dill 2002). The risk-disturbance hypothesis postulates that animal responses should track disturbance stimuli, with responses being stronger when perceived risk is greater (Frid and Dill 2002). Our results support that hypothesis, including deer selecting concealment cover in the most developed area.

Interspecific competition with North American elk also might explain behavioral responses of mule deer during migration. Mule deer demonstrate strong avoidance of elk (Johnson et al. 2000, Stewart et al. 2002, Long et al. 2008, Stewart et al. 2010), and elk occur in large herds throughout the Piceance Basin (White et al. 2001). In addition, elk tend to avoid roads and other human activities (Johnson et al. 2000, Stewart et al. 2002, Ager et al. 2003, Long et al. 2008). If mule deer are displaced because of interference or exploitive competition, mule deer would be expected to distribute themselves into lower-quality habitats, which might result in deer using areas closer to roads to avoid elk.

As the intensity of human land use increases, so does the potential for disruption of important migration routes (Thomas and Irby 1990). Anthropogenic disturbances may affect wildlife via direct and indirect mortality, habitat loss, or by altering behavior (Trombulak and Frissell 2000, Sawyer et al. 2006). The need for effective conservation of migration routes of mammals necessitates a more complete understanding of the biology of this complex behavior (Bolger et al. 2008, Wilcove 2008, Monteith et al. 2011). Results of this study have broad implications for the conservation of mammals that make long-distance migratory movements. For example, the ability of long-distance migrators to avoid anthropogenic disturbance may depend, in part, on the degree to which they would be required to alter their traditional migratory paths to do so. Furthermore, faster rates of movement through more highly developed areas may impose additional energetic costs if, for example, the ability of females to take advantage of ideal forage conditions along migratory routes is reduced (Sawyer et al. 2009b, Sawyer and Kauffman 2011). The ability of herbivores to track phenological progression of newly emergent vegetation across the landscape is of particular importance to pregnant females attempting to supporting the increased demands of late gestation (Parker et al. 2009). Understanding factors affecting movements between seasonal ranges can be critical for biologists to sustain viable populations of these large migratory ungulates (Berger 2004, Sawyer et al. 2005). As anthropogenic development increases, biologists must balance the need for human expansion with maintenance of healthy populations of mammals. Improved understanding of responses of mammals to development activities could allow landscape alterations to be manipulated to maintain necessary behaviors (i.e., migration, foraging, parturition), while still allowing infrastructure development for human benefits.

REFERENCES

1. Ager, A. A., B. K. Johnson, J. W. Kern, and J. G. Kie. 2003. Daily and seasonal movements and habitat use by female Rocky Mountain elk and mule deer. Journal of Mammalogy 84:1076–1088.
2. Andersen, R. 1991. Habitat deterioration and the migratory behavior of moose in Norway. Journal of Applied Ecology 28:102–108.
3. Anderson, D. P., M. G. Turner, J. D. Forester, J. Zhu, M. S. Boyce, H. Beyer, and L. Stowell. 2005. Scale-dependent summer resource selection by reintroduced elk in Wisconsin, USA. Journal of Wildlife Management 69:298–310.
4. Anderson, E. D., R. A. Long, M. P. Atwood, J. G. Kie, T. R. Thomas, P. Zager, and R. T. Bowyer. 2012. Winter resource selection by female mule deer Odocoileus hemi-

nous: functional response to spatiotemporal changes in habitat. Wildlife Biology 18:153–163.

5. Baker, R. R. 1978. The evolutionary ecology of animal migration. Hooder and Stoughton, London, UK.
6. Bartmann, R. M., G. C. White, and L. H. Carpenter. 1992. Compensatory mortality in a Colorado mule deer population. Wildlife Monographs 121:1–39.
7. Beckmann, J. P., K. Murray, R. G. Seidler, and J. Berger. 2012. Human-mediated shifts in animal habitat use: sequential changes in pronghorn use of a natural gas field in Greater Yellowstone. Biological Conservation 147:222–233.
8. Beier, P., and D. R. McCullough. 1990. Factors influencing white-tailed deer activitypatterns and habitat use. Wildlife Monographs 109:1–51.
9. Berger, J. 2004. The last mile: how to sustain long-distance migration in mammals. Conservation Biology 18:320–331.
10. Berger, J., D. Daneke, J. Johnson, and S. H. Berwick. 1983. Pronghorn foraging economy and predator avoidance in a desert ecosystem: implications for the conservation of large mammalian herbivores. Biological Conservation 25:193–208.
11. Bertram, R. C., and R. D. Rempel. 1977. Migration of the North Kings deer herd. California Fish and Game 63:157–179.
12. Beyer, H. L. 2004. Hawth's analysis tools for ArcGIS. http://www.spatialecology.com/htools
13. Beyer, H. L., J. Jenness, and S. A. Cushman. 2010. Components of spatial informationmanagement in wildlife ecology: software for statistical and modeling analysis. Pages 245–253 in S. A. Cushman and F. Huettmann, editors. Spatial complexity, informatics and wildlife conservation. New York, New York, USA.
14. Bishop, C. J., G. C. White, D. J. Freddy, B. E. Watkins, and T. R. Stephenson. 2009. Effect of enhanced nutrition on mule deer population rate of change. Wildlife Monographs 172:1–28.
15. Bolger, D. T., W. D. Newmark, T. A. Morrison, and D. F. Doak. 2008. The need for integrative approaches to understand and conserve migratory ungulates. Ecology Letters 11:63–77.
16. Bowyer, R. T. 1981. Activity, movement, and distribution of Roosevelt elk during rut. Journal of Mammalogy 62:574–582.
17. Bowyer, R. T. 1986. Habitat selection by southern mule deer. California Fish and Game 72:153–169.
18. Bowyer, R. T., and J. G. Kie. 2006. Effects of scale on interpreting life-history characteristics of ungulates and carnivores. Diversity and Distributions 12:244–257.
19. Bowyer, R. T., J. G. Kie, and V. Ballenberghe. 1998. Habitat selection by neonatal black tailed deer: climate, forage, or risk of predation? Journal of Mammalogy 79:415–425.
20. Boyce, M. S. 2006. Scale for resource selection functions. Diversity and Distributions 12:269–276.
21. Boyce, M. S., J. S. Mao, E. H. Merrill, D. Fortin, M. G. Turner, J. M. Fryxell, and P. Turchin. 2003. Scale and heterogeneity in habitat selection by elk in Yellowstone National Park. Écoscience 10:421–431.
22. Burnham, K. P., and D. R. Anderson. 2002. Model selection and multimodel inference: a practical information theoretic approach. Second edition. Springer-Verlag, New York, New York, USA.

23. Cameron, R. D., W. T. Smith, R. G. White, and B. Griffith. 2005. Central arctic caribou and petroleum development: distributional, nutritional, and reproductive implications. Arctic 58:1–9.
24. Compton, B. W., J. M. Rhymer, and M. McCollough. 2002. Habitat selection by wood turtles (Clemmys insculpta): an application of paired logistic regression. Ecology 83:833–843.
25. Copeland, H. E., K. E. Doherty, D. E. Naugle, A. Pocewicz, and J. M. Kiesecker. 2009. Mapping oil and gas development potential in the US Intermountain West and estimating impacts to species. PLoS ONE 4:e7400.
26. D'Eon, R. G., and D. Delparte. 2005. Effects of radio-collar position and orientation on GPS radio-collar performance, and the implications of PDOP in data screening. Journal of Applied Ecology 42:383–388.
27. D'Eon, R. G., and R. Serrouya. 2005. Mule deer seasonal movements and multiscale resource selection using global positioning system radiotelemetry. Journal of Mammalogy 86:736–744.
28. Dingle, H., and V. A. Drake. 2007. What is migration? BioScience 57:113–121.
29. Dyer, S. J., J. P. O'Neill, S. M. Wasel, and S. Boutin. 2001. Avoidance of industrial development by woodland caribou. Journal of Wildlife Management 65:531–542.
30. Dzialak, M. R., S. M. Harju, R. G. Osborn, J. J. Wondzell, L. D. Hayden-Wing, J. B. Winstead, and S. L. Webb. 2011. Prioritizing conservation of ungulate calving resources in multiple-use landscapes. PLoS ONE 6:e14597.
31. Fortin, D., H. L. Beyer, M. S. Boyce, D. W. Smith, T. Duchesne, and J. S. Mao. 2005. Wolves influence elk movements: behavior shapes a trophic cascade in Yellowstone National Park. Ecology 86:1320–1330. [Abstract]
32. Frid, A., and L. Dill. 2002. Human-caused disturbance stimuli as a form of predation risk. Conservation Ecology 6:art11.
33. Fryxell, J. M., and A. R. E. Sinclair. 1988. Causes and consequences of migration by large herbivores. Trends in Ecology and Evolution 3:237–241.
34. Garrott, R. A., G. C. White, R. M. Bartmann, L. H. Carpenter, and A. W. Alldredge. 1987. Movements of female mule deer in northwest Colorado. Journal of Wildlife Management 51:634–643.
35. Harju, S. M., M. R. Dzialak, R. G. Osborn, L. D. Hayden-Wing, and J. B. Winstead. 2011. Conservation planning using resource selection models: altered selection in the presence of human activity changes spatial prediction of resource use. Animal Conservation 14:502–511.
36. Harris, G., S. Thirgood, J. G. C. Hopcraft, J. P. G. M. Cromsigt, and J. Berger. 2009. Global decline in aggregated migrations of large terrestrial mammals. Endangered Species Research 7:55–76.
37. Hayes, C. L., and P. R. Krausman. 1993. Nocturnal activity of female desert mule deer. Journal of Wildlife Management 57:897–904.
38. Hebblewhite, M., E. Merrill, and G. McDermid. 2008. A multi-scale test of the forage maturation hypothesis in a partially migratory ungulate population. Ecological Monographs 78:141–166.
39. Hurley, M. A., J. W. Unsworth, P. Zager, M. Hebblewhite, E. O. Garton, D. M. Montgomery, J. R. Skalski, and C. L. Maycock. 2011. Demographic response of mule deer to experimental reduction of coyotes and mountain lions in southeastern Idaho. Wildlife Monographs 178:1–33.

40. Johnson, B. K., J. W. Kern, M. J. Wisdom, S. L. Findholt, and J. G. Kie. 2000. Resource selection and spatial separation of mule deer and elk during spring. Journal of Wildlife Management 64:685–697.
41. Krausman, P. R., J. J. Hervert, and L. L. Ordway. 1985. Capturing deer and mountain sheep with a net-gun. Wildlife Society Bulletin 13:71–73.
42. Long, R. A., J. G. Kie, R. T. Bowyer, and M. A. Hurley. 2009a. Resource selection and movements by female mule deer Odocoileus hemionus: effects of reproductive stage. Wildlife Biology 15:288–298.
43. Long, R. A., J. D. Muir, J. L. Rachlow, and J. G. Kie. 2009b. A comparison of two modeling approaches for evaluating wildlife-habitat relationships. Journal of Wildlife Management 73:294–302.
44. Long, R. A., J. L. Rachlow, and J. G. Kie. 2008. Effects of season and scale on response of elk and mule deer to habitat manipulation. Journal of Wildlife Management 72:1133–1142.
45. Manly, B. F. J., L. L. McDonald, D. A. Thomas, T. L. McDonald, and W. E. Erickson. 2002. Resource selection by animals, statistical design and analysis for field studies. Second edition. Kluwer Academic, Dordrecht, Netherlands.
46. McClure, M. F., J. A. Bissonette, and M. R. Conover. 2005. Migratory strategies, fawn recruitment, and winter habitat use by urban and rural mule deer (Odocoileus hemionus). European Journal of Wildlife Research 51:170–177.
47. McCullough, D. R. 1985. Long range movements of large terrestrial animals. Contributions in Marine Science 27:444–465.
48. Monteith, K. L., V. C. Bleich, T. R. Stephenson, B. M. Pierce, M. M. Conner, R. W. Klaver, and R. T. Bowyer. 2011. Timing of seasonal migration in mule deer: effects of climate, plant phenology, and life-history characteristics. Ecosphere 2:art47.
49. Morrison, M. L. 2001. A proposed research emphasis to overcome the limits of wildlife-habitat relationship studies. Journal of Wildlife Management 6:613–623.
50. Mysterud, A. 1999. Seasonal migration pattern and home range of roe deer (Capreolus capreolus) in an altitudinal gradient in southern Norway. Journal of Zoology 247:479–486.
51. Nicholson, M. C., R. T. Bowyer, and J. G. Kie. 1997. Habitat selection and survival of mule deer: tradeoffs associated with migration. Journal of Mammalogy 78:483–504.
52. Ottichilo, W. K., J. de Leeuw, and H. H. T. Prins. 2001. Population trends of resident wildebeest (Connochaetes taurinus hecki (Neumann)) and factors influencing them in the Masai Mara ecosystem, Kenya. Biological Conservation 97:271–282.
53. Parker, K. L., P. S. Barboza, and M. P. Gillingham. 2009. Nutrition integrates environmental responses of ungulates. Functional Ecology 23:57–69.
54. Parker, K. L., and M. P. Gillingham. 1990. Estimates of critical thermal environments for mule deer. Journal of Range Management 43:73–81.
55. Parker. K. L., and C. T. Robbins. 1984. Thermoregulation in mule deer and elk. Canadian Journal of Zoology 62:1409–1422.
56. Parker, K. L., C. T. Robbins, and T. A. Hanley. 1984. Energy expenditures for locomotion by mule deer and elk. Journal of Wildlife Management 48:474–488.
57. Pressey, R. L., M. Cabeza, M. E. Watts, R. M. Cowling, and K. E. Wilson. 2007. Conservation planning in a changing world. Trends in Ecology and Evolution 22:583–592.
58. Rost, G. R., and J. A. Bailey. 1979. Distribution of mule deer and elk in relation to roads. Journal of Wildlife Management 43:634–641.

59. Sappington, J. M., K. M. Longshore, and D. B. Thompson. 2007. Quantifying landscape ruggedness for animal habitat analysis: a case study using bighorn sheep in the Mojave Desert. Journal of Wildlife Management 71:1419–1426.
60. SAS Institute Inc. 1990. SAS/STAT user's guide. Release 6.03. SAS Institute, Cary, North Carolina, USA.
61. Sawyer, H., and M. J. Kauffman. 2011. Stopover ecology of a migratory ungulate. Journal of Animal Ecology 80:1078–1087.
62. Sawyer, H., M. J. Kauffman, and R. M. Nielson. 2009a. Influence of well pad activity on winter habitat selection patterns of mule deer. Journal of Wildlife Management 73:1052–1061.
63. Sawyer, H., M. J. Kauffman, R. M. Nielson, and J. S. Horne. 2009b. Identifying and prioritizing ungulate migration routes for landscape-level conservation. Ecological Applications 19:2016–2025.
64. Sawyer, H., F. Lindzey, and D. McWhirter. 2005. Mule deer and pronghorn migration in western Wyoming. Wildlife Society Bulletin 33:1266–1273.
65. Sawyer, H., R. M. Nielson, F. G. Lindzey, and L. L. McDonald. 2006. Winter habitat selection of mule deer before and during development of a natural gas field. Journal of Wildlife Management 70:396–403.
66. Senft, R. L., M. B. Coughenhour, D. W. Bailey, L. R. Rittenhouse, O. E. Sala, and D. M. Swift. 1987. Large herbivore foraging and ecological hierarchies. BioScience 37:789–799.
67. Sikes, R. S., and W. L. Gannon. and the Animal Care and Use Committee of the American Society of Mammalogists. 2011. Guidelines of the American Society of Mammalogists for the use of wild mammals in research. Journal of Mammalogy 92:235–253.
68. Singh, N. J., I. A. Grachev, A. B. Bekenov, and E. J. Milner-Gulland. 2010. Saiga antelope calving site selection is increasingly driven by human disturbance. Biological Conservation 143:1770–1779.
69. Singh, N. J., and E. J. Milner-Gulland. 2011. Conserving a moving target: planning protection for a migratory species as its distribution changes. Journal of Applied Ecology 48:35–46.
70. Stewart, K. M., R. T. Bowyer, J. G. Kie, N. J. Cimon, and B. K. Johnson. 2002. Temporospatial distributions of elk, mule deer, and cattle: resource partitioning and competitive displacement. Journal of Mammalogy 83:229–244.
71. Stewart, K. M., R. T. Bowyer, J. G. Kie, and M. A. Hurley. 2010. Spatial distributions of mule deer and North American elk: resource partitioning in a sage-steppe environment. American Midland Naturalist 163:400–412.
72. Storm, D. J., C. K. Nielsen, E. M. Schauber, and A. Woolf. 2007. Space use and survival of white-tailed deer in an exurban landscape. Journal of Wildlife Management 71:1170–1176.
73. Thomas, T. R., and L. R. Irby. 1990. Habitat use and movement patterns by migrating mule deer in southeastern Idaho. Northwest Science 64:19–27.
74. Trombulak, S. C., and A. Frissell. 2000. Review of ecological effect of roads on terrestrial and aquatic communities. Conservation Biology 14:18–30.
75. Turchin, P. 1998. Quantitative analysis of movement: measuring and modeling population redistribution in animals and plants. Sinauer Associates, Sunderland, Massachusetts, USA.

76. Unsworth, J. W., D. F. Pac, G. C. White, and R. M. Bartmann. 1999. Mule deer survival in Colorado, Idaho, and Montana. Journal of Wildlife Management 63:315–326.
77. Wagner, T., D. R. Diefenbach, A. S. Norton, and S. A. Christensen. 2011. Using multilevel models to quantify heterogeneity in resource selection. Journal of Wildlife Management 75:1788–1796.
78. Webb, S. L., M. R. Dzialak, S. M. Harju, L. D. Hayden-Wing, and J. B. Winstead. 2011. Influence of land development on home range use dynamics of female elk. Wildlife Research 38:163–167.
79. Weber, K. T., M. Burcham, and C. L. Marcum. 2001. Assessing independence of animal locations with association matrices. Journal of Range Management 54:21–24.
80. Weber, W. A., and R. C. Wittmann. 2001. Colorado flora western slope. Third edition. University Press of Colorado, Boulder, Colorado, USA.
81. White, G. C., D. J. Freddy, R. B. Gill, and J. H. Ellenberger. 2001. Effect of adult sex ratio on mule deer and elk productivity in Colorado. Journal of Wildlife Management 65:543–551.
82. White, G. C., and B. C. Lubow. 2002. Fitting population models to multiple sources of observed data. Journal of Wildlife Management 66:300–309.
83. Wilcove, D. S. 2008. No way home. Island Press, Washington, D.C., USA.
84. Williamson, D., and J. Williamson. 1985. Botswana's fences and the depletion of Kalahari wildlife. Parks 10:5–7.

CHAPTER 10

POTENTIAL ENVIRONMENTAL IMPACTS

NEW YORK STATE DEPARTMENT OF ENVIRONMENTAL CONSERVATION

The 1992 GEIS discusses the significant habitats known to exist at the time in or near then existing oil and gas fields (heronries, deer wintering areas, and uncommon, rare and endangered plants). Significant habitats are defined as areas that provide one or more of the key factors required for survival, variety, or abundance of wildlife, and/or for human recreation associated with such wildlife. This section considers the potential impact of high-volume hydraulic fracturing on all terrestrial habitat types, including forests, grasslands (including old fields managed for grasslands, and pasture and hay fields) and shrublands. Four areas of concern related to high-volume hydraulic fracturing are:

1) fragmentation of habitat;
2) potential transfer of invasive species;
3) potential impacts on endangered and threatened species; and
4) use of certain State-owned lands.

When the 1992 GEIS was developed, the scale and scope of the anticipated impact of oil and gas drilling in New York State was much different than it is today. Development of low-permeability reservoirs by high-volume hydraulic fracturing have the potential to draw substantial development into

New York, which is reasonably anticipated to result in potential impacts to habitats (fragmentation, loss of connectivity, degradation, etc.), species distributions and populations, and overall natural resource biodiversity.

The development of Marcellus Shale gas will have a large footprint (Environmental Law Clinic, 2010). In addition to direct loss of habitat, constant activity on each well pad from construction, drilling, and waste removal can be expected for 4 to 10 months, further affecting species. If a pad has multiple wells, it might be active for several years. More land is disturbed for multi-well pads, but fewer access roads, infrastructure, and total pads would be needed. Well pad sites are partially restored after drilling, but 1-3 acres is typically left open for the life of the well (as are access roads and pipelines), which is expected to be 20 to 40 years.

10.1 IMPACTS OF FRAGMENTATION TO TERRESTRIAL HABITATS AND WILDLIFE

Fragmentation is an alteration of habitats resulting in changes in area, configuration, or spatial patterns from a previous state of greater continuity, and usually includes the following:

- Reduction in the total area of the habitat;
- Decrease of the interior to edge ratio;
- Isolation of one habitat fragment from other areas of habitat;
- Breaking up of one patch of habitat into several smaller patches; and
- Decrease in the average size of each patch of habitat.

10.1.1 GENERAL DIRECT, INDIRECT, AND CUMULATIVE IMPACTS

Habitat loss, conversion, and fragmentation (both short-term and long-term) would result from land grading and clearing, and the construction of well pads, roads, pipelines, and other infrastructure associated with gas drilling (Environmental Law Clinic 2010).

Habitat loss is the direct conversion of surface area to uses not compatible with the needs of wildlife, and can be measured by calculating the physical dimensions of well pads, roads, and other infrastructure. In addition to loss of habitat, other potential direct impacts on wildlife from drilling in the Marcellus Shale include increased mortality, increase of edge habitats, altered microclimates, and increased traffic, noise, lighting, and well flares. Existing regulation of wellhead and compressor station noise levels is designed to protect human noise receptors. Little definitive work has been done on the effects of noise on wildlife (New Mexico Dept. of Game and Fish, 2007).

Habitat degradation is the diminishment of habitat value or functionality; its indirect and cumulative effects on wildlife are often assessed through analysis of landscape metrics. Indirect and cumulative impacts may include a loss of genetic diversity, species isolation, population declines in species that are sensitive to human noise and activity or dependent on large blocks of habitat, increased predation, and an increase of invasive species. Certain life-history characteristics, including typically long life spans, slow reproductive rates, and specific habitat requirements for nesting and foraging, make raptor (birds of prey) populations especially vulnerable to disturbances. Direct habitat loss has less impact than habitat degradation through fragmentation and loss of connectivity due to widespread activities like oil and gas development (New Mexico Dept. of Game and Fish, 2007).Biological systems are exceedingly complex, and there can be serious cascading ecological consequences when these systems are disturbed. Little baseline data are available with which comparisons can later be made in the attempt to document changes, or lack thereof, due to oil and gas development. In cases where serious adverse consequences may reasonably be expected, it is prudent to err on the side of caution (New Mexico Dept. of Game and Fish, 2007).

Habitat fragmentation from human infrastructure has been identified as one of the greatest threats to biological diversity. Research on habitat fragmentation impacts from oil and gas development specific to New York is lacking. However, the two following studies from the western United States are presented here to illustrate qualitatively the potential impacts to terrestrial habitats that could occur in New York. A quantitative comparison between these studies and potential impacts in New York is not possible because these studies were conducted under a regulatory structure that resulted in well spacing that differs from those anticipated for high-volume hydraulic fracturing in New York. Additional research would be necessary to determine the precise impacts to species and wildlife expected from such drilling in New York's Marcellus Shale.

While fragmentation of all habitats is of conservation concern, the fragmentation of grasslands and interior forest habitats are of utmost concern in New York. Some of the bird species that depend on these habitat types are declining. This decline is particularly dramatic for grasslands where 68% of the grassland-dependent birds in New York are declining (Post, 2006)

10.1.2 PROJECTED DIRECT IMPACTS

10.1.2.1 STUDY 1, GENERAL DISCUSSION

The Wilderness Society conducted a study in 2008 (Wilbert et al., 2008) that provided both an analytical framework for examining habitat fragmentation and results from a hypothetical GIS analysis simulating the incremental development of an oil and gas field to progressively higher well pad numbers over time. Results of the sample analysis gave a preliminary estimate of the minimum potential fragmentation impacts of oil and gas development on wildlife and their habitats; the results were not intended to be a substitute for site-specific analyses.

The study identified a method to measure fragmentation (landscape metrics), and a way to tie various degrees of fragmentation to their impacts on wildlife (from literature). Two fragmentation indicator values (road density and distance-to-nearest-road or well pad) were analyzed for impacts to a few important wildlife species present in oil and gas development areas across the western U.S.

10.1.2.2 STUDY 1, FINDINGS

The total area of direct disturbance from well pads and roads used in oil and gas development was identified for a hypothetical undeveloped 120-acre site, with seven separate well-pad densities - one pad per 640 acres, 320 acres, 160 acres, 80 acres, 40 acres, 20 acres, and 10 acres:

1. Well pads: the disturbance area increased approximately linearly as pad density increased;
2. Total road length: the disturbance area increased more rapidly in the early stages of development;
3. Mean road density: the rate of increase was higher at earlier stages of development. The size of the pre-development road system had an effect on the magnitude of change between subsequent development stages, but the effect decreased as development density increased;
4. Distance-to-nearest-road (or well pad): the rate of decrease was higher at earlier stages of development than at later stages; and
5. Significant negative effects on wildlife were predicted to occur over a substantial portion of a landscape, even at the lower well pad densities characteristic of the early stages of development in gas or oil fields.

This suggests that landscape-level planning for infrastructure development and analysis of wildlife impacts need to be done prior to initial development of a field. Where development has already occurred, the study authors recommend that existing impacts on local wildlife species be measured and acknowledged, and the cumulative impacts from additional development be assessed.

10.1.2.3 STUDY 1, IMPLICATIONS FOR NEW YORK:

The study results emphasize the importance of maintaining undeveloped areas. Note that the degree of habitat fragmentation and the associated impacts on wildlife from such development in real landscapes would be even greater than those found in the study, which used conservative estimates of road networks (no closed loops, shorter roads, and few roads pre-development) and did not include pipelines and other infrastructure.

10.1.3 PROJECTED INDIRECT AND CUMULATIVE IMPACTS

10.1.3.1 STUDY 2, GENERAL DISCUSSION

The Wilderness Society conducted a study in 2002 (Weller et al., 2002) that analyzed the landscape of an existing gas and oil field in Wyoming to identify habitat fragmentation impacts. As fragmentation of the habitat occurred over a wide area, cumulative and indirect impacts could not be adequately addressed at the individual well pad site level. Rather, analyzing the overall ecological impacts of fragmentation on the composition, structure, and function of the landscape required a GIS spatial analysis. A variety of metrics were developed to measure the condition of the landscape and its level of fragmentation, including: density of roads and linear features; acreage of habitat in close proximity to infrastructure; and acreage of continuous uniform blocks of habitat or core areas.

10.1.3.2 STUDY 2, FINDINGS

The study area covered 166 square miles, and contained 1864 wells, equaling a density of 11 wells per square mile (Note that this density is between that of single horizontal wells (9 per square mile) and vertical wells (16 per square mile) expected in New York). The direct physical footprint of oil and gas infrastructure was only 4% of the study area; however, the ecological impact of that infrastructure was much greater. The entire study area was within one-half mile of a road, pipeline corridor, well head, or other infrastructure, while 97% fell within one-quarter mile. Study results also showed the total number, total acreage, and the percent of study area remaining in core areas decreased as the width of the infrastructure impact increased. No core areas remained within one-half mile of infrastructure, and only 27% remained within 500 feet of infrastructure. These results, combined with a review of the scientific literature for fragmentation impacts to western focal species, indicated there was little to no place in the study area where wildlife would not be impacted.

10.1.3.3 STUDY 2, IMPLICATIONS FOR NEW YORK

This study demonstrated that impacts to wildlife extended beyond the direct effects from the land physically altered by oil and gas fields. Note that the overall impacts predicted in the study were likely conservative as the data were only assessed at the individual gas field scale, not the broader landscape. While well densities from multiple horizontal wells from a common pad (a minimum of 1 well pad per square mile) would be less than in this study, all three drilling scenarios might result in negative impacts to wildlife in New York, as the impacts predicted to the complement of species in Wyoming were so extreme.

10.1.4 IMPACTS OF GRASSLAND FRAGMENTATION

Grassland birds have been declining faster than any other habitat-species suite in the northeastern United States (Morgan and Burger, 2008). The primary cause of these declines is the fragmentation of habitat caused by the abandonment of agricultural lands, causing habitat loss due to reversion to later successional stages or due to sprawl development. Remaining potential habitat is also being lost or severely degraded by intensification of agricultural practices (e.g., conversion to row crops or early and frequent mowing of hayfields).

Stabilizing the declines of populations of grassland birds has been identified as a conservation priority by virtually all of the bird conservation initiatives, groups, and agencies in the northeastern US, as well as across the continent, due to concern over how precipitous their population declines have been across portions of their ranges (for the list of species of concern and their population trends, see Table 2). In New York, grassland bird population declines are linked strongly to the loss of agricultural grasslands, primarily hayfields and pastures; it is therefore critical to conserve priority grasslands in order to stabilize or reverse these declining trends.

TABLE 1: Grassland bird population trends at three scales from 1966 to 2005 (Morgan and Burger, 2008) (New July 2011)

	New York		USFWS Region 5		Survey-wide	
Species	trend (%/year)	population remaining (%)	trend (%/year)	population remaining (%)	trend (%/year)	population remaining (%)
Northern Harrier	-3.4	25.9	1.1	153.2	-1.7	51.2
Upland Sandpiper	-6.9	6.2	-0.7	76.0	0.5	121.5
Short-eared Owl	--	--	--	--	-4.6	15.9
Sedge Wren	-11.5	0.9	0.5	121.5	1.8	200.5
Henslow's Sparrow	-13.8	0.3	-12.6	0.5	-7.9	4.0
Grasshopper Sparrow	-9.4	2.1	-5.2	12.5	-3.8	22.1
Bobolink	-0.5	82.2	-0.3	88.9	-1.8	49.2
Loggerhead Shrike	--	--	-11.4	0.9	-3.7	23.0
Horned Lark	-4.7	15.3	-2.1	43.7	-2.1	43.7
Vesper Sparrow	-7.9	4.0	-5.4	11.5	-1.0	67.6
Eastern Meadowlark	-4.9	14.1	-4.3	18.0	-2.9	31.7
Savannah Sparrow	-2.6	35.8	-2.3	40.4	-0.9	70.3

Some of New York's grassland birds have experienced steeper declines than others, or have a smaller population size and/or distribution across the state or region, and are therefore included in the highest priority tier: northern harrier (*Circus cyaneus*), upland sandpiper (Bartramia longicauda), short-eared owl (*Asio flammeus*), sedge wren (*Cistothorus platensis*), Henslow's sparrow (*Ammodramus henslowii*), grasshopper sparrow (*Ammodramus savannarum*), bobolink (*Dolichonyx oryzivorus*), and loggerhead shrike (*Lanius ludovicianus*). Species included in the high priority tier are those that have been given relatively lower priority, but whose populations are also declining and are in need of conservation. The high priority tier in Table 1 includes: horned lark (*Eremophila alpestris*), vesper sparrow (*Pooecetes gramineus*), eastern meadowlark (*Sturnella magna)*, and savannah sparrow (*Passerculus sandwichensis*).

While these birds rely on grasslands in New York as breeding habitat (in general), two of these species (northern harrier and short-eared owl) and several other raptor species also rely on grasslands for wintering habitat. For this reason, a third target group of birds are those species that rely on grassland habitats while they over-winter (or are year-round residents) in New York, and include: snowy owl (*Bubo scandiacus*), rough-legged hawk (*Buteo lagopus*), red-tailed hawk (*Buteo jamaicensis*), American kestrel (*Falco sparverius*), and northern shrike (*Lanius excubitor*).

The specific effects of drilling for natural gas on nesting grassland birds are not well studied. However, the level of development expected for multi-pad horizontal drilling and minimum patch sizes of habitat necessary for bird reproduction, unless mitigated, will result in substantial impacts from the fragmentation of existing grassland habitats. Minimum patch sizes would vary by species and by surrounding land uses, but studies have shown that a minimum patch size of between 30-100 acres is necessary to protect a wide assemblage of grassland-dependent species (USFWS, Sample and Mossman 1997, Mitchell et al, 2000.)

10.1.5 IMPACTS OF FOREST FRAGMENTATION

Forest fragmentation issues were the subject of two assessments referenced below which are specific to the East and address multiple horizontal well drilling from common pads. These studies, therefore, are more directly applicable to New York than previously mentioned western studies of vertical drilling. The Multi-Resolution Land Characteristic Dataset (MRLCII) (2004) indicates the following ratios of habitat types in the area underlain by the Marcellus shale in New York: 57% forested; 28% grassland/agricultural lands; and 3% scrub/shrub. The other 12% is divided evenly between developed land and open water/wetlands. As forests are the most common cover type, it is reasonable to assume that development of the Marcellus Shale would have a substantial impact on forest habitats and species.

Today, New York is 63% (18.95 million acres) forested (NYSDEC, Forest Resource Assessment and Strategy, 2010) and is unlikely to substantially increase. Current forest parcelization and fragmentation trends will likely result in future losses of large, contiguous forested areas (NYSDEC, Forest Resource Assessment and Strategy, 2010). Therefore, protecting these remaining areas is very important for maintaining the diversity of wildlife in New York. The forest complex provides key ecosystem services that provide substantial ecological, economic, and social benefits (water quality protection, clean air, flood protection, pollination, pest predation, wildlife habitat and diversity, recreational opportunities, etc.) that extend far beyond the boundaries of any individual forested area.

Large contiguous forest patches are especially valuable because they sustain wide-ranging forest species, and provide more habitat for forest interior species. They are also more resistant to the spread of invasive species, suffer less tree damage from wind and ice storms, and provide more ecosystem services – from carbon storage to water filtration – than small patches (Johnson, 2010), Lands adjacent to well pads and infrastructure can also be affected, even if they are not directly cleared. This is most notable in forest settings where clearings fragment contiguous forest patches, create new edges, and change habitat conditions for sensitive wildlife and plant species that depend on interior forest conditions.

Forest ecologists call this the edge effect. While the effect is somewhat different for each species, research has shown measurable impacts often extend at least 330 feet (100 meters) into forest adjacent to an edge (Johnson, 2010). Interior forest species avoid edges for different reasons. Blackthroated blue warblers and other interior forest birds, for example, avoid areas near edges during nesting season because of the increased risk of predation. Tree frogs, flying squirrels and certain woodland flowers are sensitive to forest fragmentation because of changes in canopy cover, humidity and light levels. Some species, such as white-tailed deer and cowbirds, are attracted to forest edges – often resulting in increased competition, predation, parasitism, and herbivory.

Invasive plant species, such as tree of heaven, stilt grass, and Japanese barberry, often thrive on forest edges and can displace native forest species. As large forest patches become progressively cut into smaller patches, populations of forest interior species decline.

10.1.5.1 LESSONS LEARNED FROM PENNSYLVANIA

Assessment 1, General Discussion:

The Nature Conservancy (TNC) conducted an assessment in 2010 (Johnson, 2010) to develop credible energy development projections for horizontal hydraulic fracturing in Pennsylvania's Marcellus Shale by 2030, and how those projections might affect high priority conservation areas, including forests. The projections were informed scenarios, not predictions, for how much energy development might take place and where it was more and less probable. Project impacts, however, were based on measurements of actual spatial footprints for hundreds of well pads.

Potential Direct Impacts, Methodology and Assessment Findings:

Projections of future Marcellus gas development impacts depended on robust spatial measurements for existing Marcellus well pads and infrastructure. This assessment compared aerial photos of Pennsylvania Department of Environmental Protection (PADEP) Marcellus well permit locations taken before and after development and precisely documented the spatial foot print of 242 Marcellus well pads (totaling 435 drilling permits) in Pennsylvania (see Table 2).

Well pads in Pennsylvania occupy 3.1 acres on average while the associated infrastructure (roads, water impoundments, pipelines) takes up an additional 5.7 acres, or a total of nearly 9 acres per well pad (This is larger than the 7.4 acres predicted by IOGA to be disturbed in New York).

TABLE 2: Average Spatial Disturbance for Marcellus Shale Well Pads in Forests Context (acres)

<table>
<tr><td>Forest Cleared for Marcellus Shale well pad</td><td>3.1</td><td rowspan="2">8.8</td></tr>
<tr><td>Forest cleared for associated infrastructure (roads, pipelines, water impoundments, etc.)</td><td>5.7</td></tr>
<tr><td>Indirect forest impact from new edges</td><td colspan="2">21.2</td></tr>
<tr><td>TOTAL DIRECT AND INDIRECT IMPACTS</td><td colspan="2">30</td></tr>
</table>

Another key variable for determining land-use and habitat impacts in this assessment was the number of wells on each pad; more wells per pad translated to less disturbance and infrastructure on the landscape. It is technically possible to put a dozen or more Marcellus wells on one pad. For the 242 well pads assessed in this study, the average in Pennsylvania has been 2 wells per pad to date (IOGA estimates the same for New York) as companies quickly moved on to drill other leases to test productivity and to secure as many potentially productive leases as possible (leases typically expire after 5 years if there is no drilling activity). TNC assumed that in many cases, the gas company would return to these pads later and drill additional wells. This assumption may not be valid in New York where there is a three-year limit on well development (ECL 23-0501).

The TNC assessment developed low, medium, and high scenarios for the amount of energy development that might take place in Pennsylvania. The projections included a conservative estimate of 250 horizontal drilling rigs, each of which could drill one well per month, resulting in an estimated 3,000 wells drilled annually. Estimates in New York predict less activity than this, but activity could result in approximately 40,000 wells by 2040.

The low scenario (6,000 well pads) assumed that each pad on average would have 10 wells, or 1 well pad per 620 acres. Because many leases are irregularly shaped, in mixed ownership, or their topography and geology impose constraints, TNC concluded that it is unlikely this scenario would develop in Pennsylvania. It would take relatively consolidated leaseholds and few logistical constraints for this scenario to occur (Note that while no definitive number is provided in section 5.1.3.2, this is expected to be the most common spacing for horizontal drilling in New York's Marcellus Shale.)

The medium scenario for well pads assumed 6 wells on average would be drilled from each pad (10,000 well pads), or 1 pad per 386 acres. Industry generally agreed that 6 is the most likely number of wells they would be developing per pad for most of their leaseholds in Pennsylvania (Note that IOGA assumes that 6 horizontal wells would be drilled per pad in New York).

The high scenario assumed each pad would have 4 wells drilled on average (15,000 well pads), or 1 pad per 258 acres. This scenario is more likely if there is relatively little consolidation of lease holds between companies in the next several years. While this scenario would result in a loss of less than 1% of Pennsylvania's total forest acreage, areas with intensive Marcellus gas development could see a loss of 2-3% of local forest habitats.

In summary, 60,000 wells could be drilled by 2030 in the area underlain by the Marcellus Shale in Pennsylvania on between 6,000 and 15,000 new well pads (there are currently about 1,000),

depending on how many wells are placed on each pad. A majority (64%) of projected well locations were found in a forest setting for all three scenarios. By 2030, a range of between 34,000 and 82,000 acres of forest cover could be cleared by new Marcellus gas development in Pennsylvania. Some part of the cleared forest area would become reforested after drilling is completed, but there has not been enough time to establish a trend since the Marcellus development started.

Potential Direct Impacts, Implications for New York:

Direct land disturbance from horizontal hydraulic fracturing of Marcellus Shale in New York is expected to result in 7.4 acres of direct impacts from each well pad and associated infrastructure. This is different from the experiences in Pennsylvania where nearly 9 acres of habitat was removed for each well pad and its associated infrastructure. Under either scenario, the direct impacts are substantial.

The most likely drilling scenario in Pennsylvania would result in a density of 1 pad per 386 acres. However, given New York's regulatory structure, a spacing of 1 pad per 640 acres is anticipated. If spacing units are less than 640 acres, or if there are less than 6-8 horizontal wells per pad, the percentage of land disturbance could be greater. Again, using the set of currently pending applications as an example, the 47 proposed horizontal wells would be drilled on eleven separate well pads, with between 2 and 6 wells for each pad. Therefore, greater than 1.2% land disturbance per pad estimated by industry can be expected in New York.

Potential Indirect Impacts, Methodology and Assessment Findings:

To assess the potential interior forest habitat impact, a 100-meter buffer was created into forest patches from new edges created by well pad and associated infrastructure development (Figure 1). For those well sites developed in forest areas or along forest edges (about half of the assessed sites), TNC calculated an average of 21 acres of interior forest habitat was lost. Thus, the total combined loss of habitat was 30 acres per well pad due to direct and indirect impacts.

In addition to the direct clearing of between 34,000 to 82,000 acres of forest cover in Pennsylvania, forest interior species could be negatively impacted within an additional 85,000 to 190,000 forest acres adjacent to Marcellus development. Forest impacts would be concentrated where many of Pennsylvania's largest and most intact forest patches occur, resulting in fragmentation into smaller patches by well pads, roads, and other infrastructure. In contrast to overall forest loss, projected Marcellus gas development scenarios in Pennsylvania indicate a more pronounced impact on large forest patches. Impacts to forest interior species would vary depending on their geographic distribution and density. Some species, such as the black-throated blue warbler, could see widespread impacts to their relatively restricted breeding habitats in the state, while widely distributed species such as the scarlet tanager, would be relatively less affected.

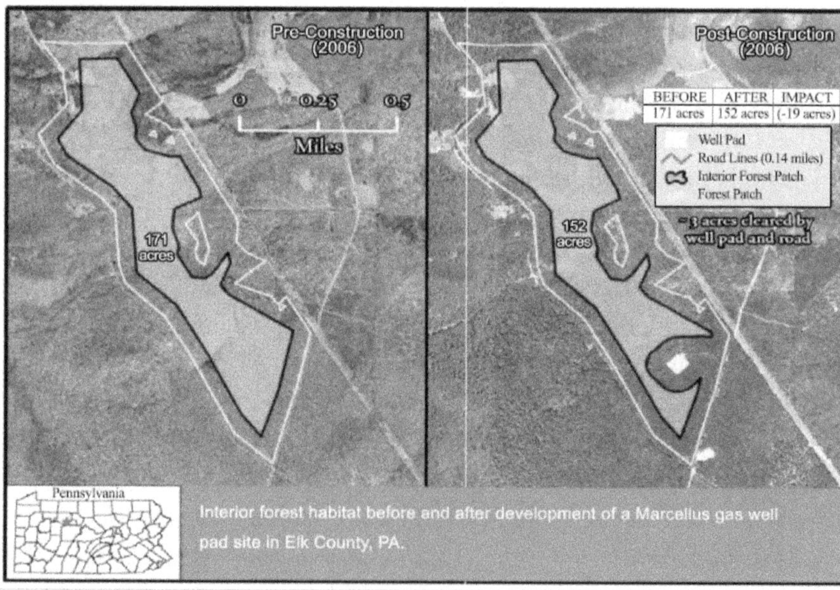

FIGURE 1: Interior Forest Habitat Before & After Development of a Marcellus Gas Well Pad, Elk County PA (New July 2011)

This study went on to find that locating energy infrastructure in open areas or toward the outer edges of large patches can significantly reduce impacts to important forest areas. To address this finding and explore potential ways in which conservation impacts could be minimized, TNC examined how projected Marcellus gas pads could be relocated to avoid forest patches in a specific region of Pennsylvania. To reduce the impacts to forest habitats, the wells were hypothetically relocated, where practicable, to nearby existing openings maintained by human activity (e.g., old fields, agricultural fields). If nearby open areas did not exist, the locations of the well pads were moved toward the edges of forest patches to minimize impacts to forest interior habitats. This exercise did not eliminate forest impacts in this heavily forested Pennsylvania landscape, but there was a significant reduction in impacts. Total forest loss declined almost 40% while impacts to interior forest habitats adjacent to new clearings declined by one-third (Figure 2). The study authors recommend that information about

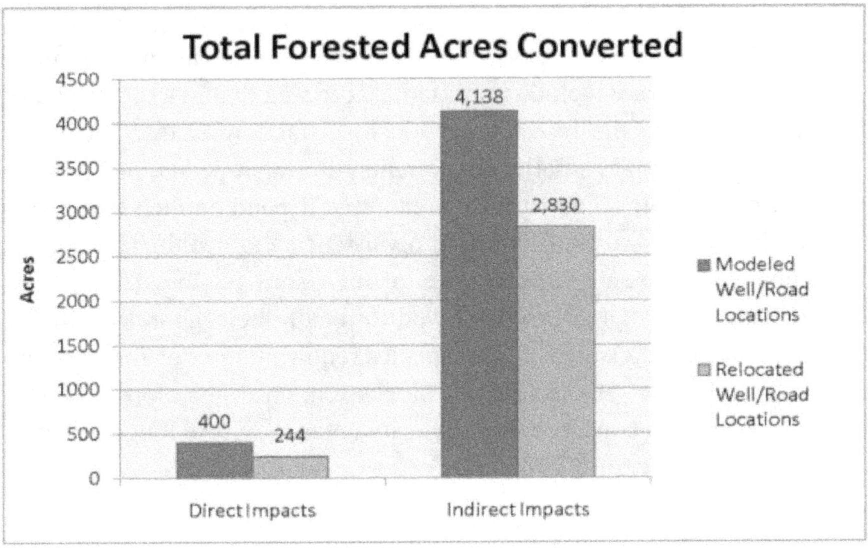

FIGURE 2: Total Forest Areas Converted (New July 2011)

Pennsylvania's important natural habitats be an important part of the calculus about trade-offs and optimization as energy development proceeds.

Potential Indirect Impacts, Implications for New York: For each acre of forest directly cleared for well pads and infrastructure in New York, an additional 2.5 acres can be expected to be indirectly impacted. Interior forest bird species with restricted breeding habitats, such as the black-throated blue and cerulean warblers, might be highly impacted.

Additional assessment work conducted for New York based on estimates and locations of well pad densities across the Marcellus landscape could better quantify expected impacts to forest interior habitats and wildlife.

10.1.5.2 NEW YORK FOREST MATRIX AND LANDSCAPE CONNECTIVITY

Forest matrix blocks contain mature forests with old trees, understories, and soils that guarantee increased structural diversity and habitat important to many species. They include important stabilizing features such as large, decaying trunks on the forest floor and big, standing snags. Set within these matrix forests are smaller ecosystems offering a wide range of habitat (wetlands, streams, and riparian areas) that depend on the surrounding forested landscape for their longterm persistence and health. These large, contiguous areas are viable examples of the dominant forest types that, if protected, and in some cases allowed to regain their natural condition, serve as critical source areas for all species requiring interior forest conditions. Few remnants of such matrix blocks remain in the Northeast; it is therefore critical to conserve these priority areas to ensure long-term conservation of biodiversity (TNC 2004).

10.1.5.3 ASSESSMENT 2, GENERAL DISCUSSION

The New York Natural Heritage program in 2010 (NYSDEC, Strategic Plan for State Forest Management, 2010) identified New York's forest matrix blocks and predicted corresponding forest connectivity areas. Securing connections between major forested landscapes and their imbedded matrix forest blocks is important for the maintenance of viable populations of species, especially those that are wide-ranging and highly mobile, and ecological processes such as dispersal and pollination over the long term. Identifying, maintaining, and enhancing these connections represents a critical adaptation strategy if species are to shift their ranges in response to climate change and other landscape changes.

10.1.5.4 ASSESSMENT 2, FINDINGS

Figure 3 depicts the large forested landscapes within New York and predicts the linkages between them, called least-cost path (LCP). A least-cost path corridor represents the most favorable dispersal path for forest species based on a combination of percent natural forest cover in a defined area, barriers to movement, and distance traveled. Thus, as many species that live in forests generally prefer to travel through a landscape with less human development (i.e., fewer impediments to transit) as well as in a relatively direct line, the predicted routes depict a balance of these sometimes opposing needs.

10.1.5.5 ASSESSMENT 2, IMPLICATIONS FOR NEW YORK

The area underlain by the Marcellus Shale in New York is 57% forested with about 7% of that forest cover occurring on State-owned lands. It is reasonable to assume high-volume horizontal hydraulic fracturing would have negative impacts to forest habitats similar to those predicted in Pennsylvania.

In order to minimize habitat fragmentation and resulting restrictions to species movement in the area underlain by the Marcellus, it is recommended that forest matrix blocks be managed to create, maintain, and enhance the forest cover characteristics that are most beneficial to the priority species that may use them.

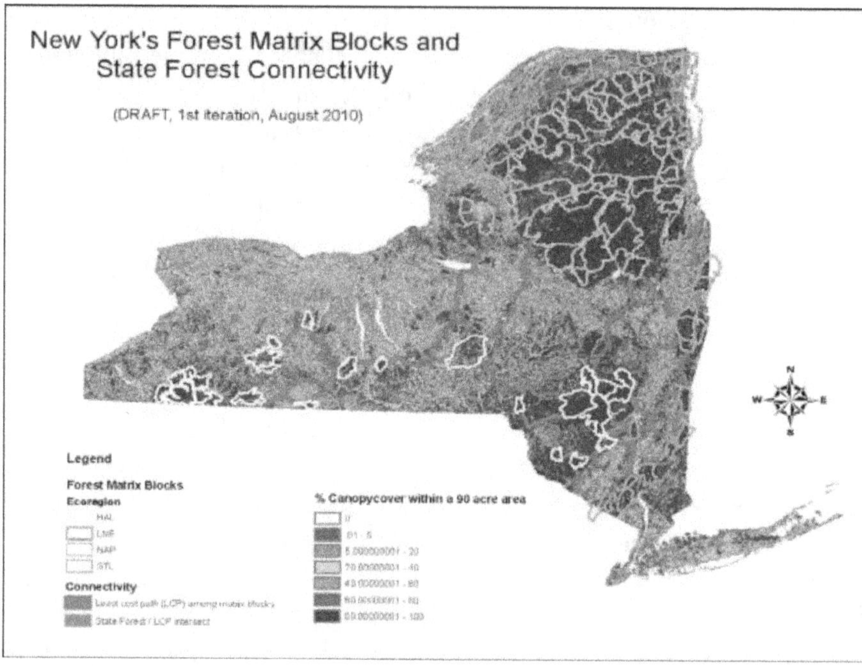

FIGURE 3: Areas of Concern for Endangered and Threatened Animal Species within the Area Underlain by the Marcellus Shale in New York, March 31, 2011 (New July 2011)

10.2 INVASIVE SPECIES

An invasive species, as defined by ECL §9-1703, is a species that is nonnative to the ecosystem under consideration and whose introduction causes or is likely to cause economic or environmental harm or harm to human health. Invasive species can be plants, animals, and other organisms such as microbes, and can impact both terrestrial and aquatic ecosystems.

While natural means such as water currents, weather patterns and migratory animals can transport invasive species, human actions - both intentional and accidental - are the primary means of invasive species introductions to new ecosystems. Once introduced, invasive species usually spread profusely because they often have no native predators or diseases to limit their reproduction and control their population size. As a result, invasive species out-compete native species that have these controls in place, thus diminishing biological diversity, altering natural community structure and, in some cases, changing ecosystem processes. These environmental

TABLE 3: Terrestrial invasive plant species in New York State (interim list)

Terrestrial - Herbaceous	
Common Name	**Scientific Name**
Garlic Mustard	*Alliaria petiolata*
Mugwort	*Artemisia vulgaris*
Brown Knapweed	*Centaurea jacea*
Black Knapweed	*Centaurea nigra*
Spotted Knapweed	*Centaurea stoebe ssp. micranthos*
Canada Thistle	*Cirsium arvense*
Bull Thistle	*Cirsium vulgare*
Crown Vetch	*Coronilla varia*
Black Swallow-Wort	*Cynanchum louiseae (nigrum)*
European Swallow-Wort	*Cynanchum rossicum*
Fuller's Teasel	*Dipsacus fullonum*
Cutleaf Teasel	*Dipsacus laciniatus*
Giant Hogweed	*Heracleum mantegazzianum*
Japanese Stilt Grass	*Microstegium vimineum*
Terrestrial - Vines	
Common Name	**Scientific Name**
Porcelain Berry	*Ampelopsis brevipedunculata*
Oriental Bittersweet	*Celastrus orbiculatus*
Japanese Honeysuckle	*Lonicera japonica*
Mile-a-Minute Weed	*Persicaria perfoliata*
Kudzu	*Pueraria montana var. lobata*
Terrestrial - Shrubs and Trees	
Common Name	**Scientific Name**
Norwa Maple	*Acer platanoides*
Tree of Heaven	*Ailanthus altissima*
Japenese Barberry	*Berberis thunbergii*
Russian Olive	*Elaeagnus angustifolia*
Autumn Olive	*Elaeagnus umbellata*
Glossy Buckthorn	*Frangula alnus*
Border Privet	*Ligustrum obtusifolium*
Amur Honeysuckle	*Lonicera maackii*
Shrub Honeysuckles	*Lonicera morrowii/tatarica/x bella*
Bradford Pear	*Pyrus calleryana*
Common Buckthorn	*Rhamnus cathartica*
Black Locust	*Robinia pseudoacacia*
Multiflora Rose	*Rosa multiflora*

impacts can further impose economic impacts as well, particularly in the water supply, agricultural and recreational sectors (ECL §9-1701).

The number of vehicle trips associated with high-volume hydraulic fracturing, particularly at multi-well sites, has been identified as an activity which presents the opportunity to transfer invasive terrestrial species. Surface water withdrawals also have the potential to transfer invasive aquatic species.

10.2.1 TERRESTRIAL

Terrestrial plant species which are widely recognized as invasive (as per ECL §9-1703) or potentially-invasive in New York State, and are therefore of concern, are listed in Table 3.

Operations involving land disturbance such as the construction of well pads, access roads, and engineered surface impoundments for fresh water storage have the potential to both introduce and transfer invasive species populations. Machinery and equipment used to remove vegetation and soil may come in contact with invasive plant species that exist at the site and may inadvertently transfer those species‘ seeds, roots, or other viable plant parts via tires, treads/tracks, buckets, etc. to another location on site, to a separate project site, or to any location
in between.

The top soil that is stripped from the surface of the site during construction and set aside for reuse during reclamation also presents an opportunity for the establishment of an invasive species population if it is left exposed. Additionally, fill sources (e.g., gravel, crushed stone) brought to the well site for construction purposes also have the potential to act as a pathway for invasive species transfer if the fill source itself contains viable plant parts, seeds, or roots.

10.2.2 AQUATIC

The presence of non-indigenous aquatic invasive species in New York State waters is recognized, and, therefore, operations associated with the withdrawal, transport, and use of water for horizontal well drilling and high volume hydraulic fracturing operations have the potential to transfer invasive species. Species of concern include, but are not necessarily limited to; zebra mussels, eurasian watermilfoil, alewife, water chestnut, fanwort, curly-leaf pondweed, round goby, white perch, didymo, and the spiny water flea. Other aquatic, wetland and littoral plant species that are of concern due to their status as invasive or potentially-invasive in New York State are listed in Table 4.

TABLE 4: Aquatic, wetland, and littoral invasive plant species in New York State (interim list)

Floating and Submerged Aquatic	
Common Name	**Scientific Name**
Carolina Fanwort	*Cabomba caroliniana*
Rock Snot (didymo)	*Didymosphenia geminata*
Brazilian Elodea	*Egeria densa*
Water Thyme	*Hydrilla verticillata*
European Frog's Bit	*Hydrocharis morus-ranae*
Floating Water Primrose	*Ludwigia peploides*
Parrot-Feather	*Myriophyllum aquaticum*
Variable Watermilfoil	*Myriophyllum heterophyllum*
Eurasian Watermilfoil	*Myriophyllum spicatum*
Brittle Naiad	*Najas minor*
Starry Stonewort (green alga)	*Nitellopsis obtusa*
Yellow Floating Heart	*Nymphoides peltata*
Water-Lettuce	*Pistia stratiotes*
Curly-Leaf Pondweed	*Potamogeton crispus*
Water Chestnut	*Trapa natans*
Emergent Wetland and Littoral	
Common Name	**Scientific Name**
Flowering Rush	*Butomus umbellatus*
Japanese Knotweed	*Fallopia japonica*
Giant Knotweed	*Fallopia sachalinensis*
Yellow Iris	*Iris pseudacorus*
Reed Canarygrass	*Phalaris arundinacea*
Common Reed - nonnative variety	*Phragmites australis var. australis*
Purple Loosestrife	*Lythrum salicaria*

Invasive species may be transported with the fresh water withdrawn for, but not used for drilling or hydraulic fracturing. Invasive species may potentially be transferred to a new area or watershed if unused water containing such species is later discharged at another location. Other potential mechanisms for the possible transfer of invasive aquatic species may include trucks, hoses, pipelines and other equipment used for water withdrawal and transport.

10.3 IMPACTS TO ENDANGERED AND THREATENED SPECIES

The area underlain by the Marcellus Shale includes both terrestrial and aquatic habitat for 18 animal species listed as endangered or threatened in New York State (Table 5 and Figure 4) protected under the State Endangered Species Law (ECL 11-0535) and associated regulations (6 NYCRR Part 182). Some species, such as the northern harrier and upland sandpiper, are dependent upon grassland habitat for breeding and foraging and can be found in many counties within the project area. Species such as the rayed bean mussel and mooneye fish are aquatic species limited to only two counties on the western edge of the project area. Other species are associated with woodlands, with bald eagles nesting in woodlands adjacent to lakes, rivers and ponds throughout many counties within the project area. The area also includes habitat for cerulean warblers and eastern hellbenders, two species currently under consideration for listing by both the State and the federal government.

Endangered and threatened wildlife may be adversely impacted through project actions such as clearing, grading and road building that occur within the habitats that they occupy. Certain species are unable to avoid direct impact due to their inherent poor mobility (e.g., Blanding's turtle, club shell mussel). Certain actions, such as clearing of vegetation or alteration of stream beds, can also result in the loss of nesting and spawning areas. If these actions occur during the time of year that species are breeding, there can be a direct loss of eggs and/or young. For species that are limited to specific habitat types for breeding, the loss of the breeding area can result in a loss of productivity in future years as adults are forced into less suitable habitat. Any road construction through streams or wetlands within habitats occupied by these species can result in the creation of impermeable barriers to movement for aquatic species and reduce dispersal for some terrestrial species. Other impacts from the project, such as increased vehicle traffic, can result in direct mortality of adult animals. In general, the loss of habitat in areas occupied by listed species can result in reduced numbers of breeding pairs and lowered productivity.

TABLE 5: Endangered & Threatened Animal Species within the Area Underlain by the Marcellus Shale (New July 2011)

Common Name	Scientific Name	NYS Listing	Primary Habitats
Henslow's Sparrow	*Ammodramus henslowii*	Threatened	Grassland
Short-eared Owl	*Asio flammeus*	Endangered	Grassland
Upland Sandpiper	*Bartramia longicauda*	Threatened	Grassland
Northern Harrier	*Circus cyaneus*	Threatened	Grassland, Wetlands
Sedge Wren	*Cistothorus platensis*	Threatened	Grassland
Peregrine Falcon	*Falco peregrinus*	Endangered	Cliff faces
Bald Eagle	*Haliaeetus leucocephalus*	Threatened	Forest, Open water
Least Bittern	*Ixobruchus exilis*	Threatened	Wetlands
Pie-billed Grebe	*Podilymbus podiceps*	Threatened	Wetlands
Eastern Sand Darter	*Ammocrypta pellucida*	Threatened	Streams
Mooneye	*Hiodon tergisus*	Endangered	Large Lakes, Rivers
Longhead Darter	*Percina macrocephala*	Threatened	Large Streams, Rivers
Brook Floater	*Alasmidonta varicosa*	Threatened	Streams and Rivers
Wavyrayed Lampmussel	*Lampsilis fasciola*	Threatened	Small, Medium Streams
Green Floater	*Lasmigona subviridis*	Threatened	Small, Medium Streams
Clubshell	*Pleurobema clava*	Endangered	Small, Medium Streams
Rayed Bean	*Villosa fabalis*	Endangered	Small Streams
Timber Rattlesnake	*Crotalus horridus*	Threatened	Forest

10.4 IMPACTS TO STATE-OWNED LANDS

State-owned lands play a unique role in New York's landscape because they are managed under public ownership to allow for sustainable use of natural resources, provide recreationalopportunities for all New Yorkers, and provide important wildlife habitat and open space. They represent the most significant portions of large contiguous forest patch in the study area.

Industrial development on these lands is, for the most part, prohibited, and any type of clearing and development on these lands is limited and managed. Given the level of development expected for multi-pad horizontal drilling, it is anticipated that there would be additional pressure for surface disturbance on state-owned lands. Surface disturbance associated with gas extraction could have a significant adverse impact on habitats contained on the state-owned lands, and recreational use of those lands.

10.4.1 FOREST HABITAT FRAGMENTATION

As described earlier, large contiguous forest patches are especially valuable because they sustain wide-ranging forest species, and provide more habitat for forest interior species. State-owned lands, by their very nature, consist of large contiguous forest patches. While some fragmentation has occurred, the level of activity associated with multi-well horizontal drilling (e.g., well pad construction, access roads, pipelines, etc.) would negatively impact the state's ability to maintain the existing large contiguous patches of forest.

The Department has stated that protecting these areas from further fragmentation is a high priority. One of the objectives stated in the Strategic Plan for State Forest Management is to emphasize closed canopy and interior forest conditions to maintain and enhance‖ forest matrix blocks. It is critical therefore, that any additional road, pipeline and well pad construction be carefully assessed in order to avoid further reducing this habitat. Given the State's responsibility to protect these lands as steward of the public trust, the State has a heightened responsibility, as compared to its role with respect to private lands, to ensure that any State permitted action does not adversely impact the ecosystems and habitat on these public lands so that they may be enjoyed by future generations.

10.4.2 PUBLIC RECREATION

State-owned lands have been acquired over the past century to provide compatible public recreation opportunities, protect watersheds, and provide sustainable timber harvesting. Drilling and trucking activities disturb the tranquility found on these lands and can cause significant visual impacts. Also, many State Forest roads serve as recreational trails for bicyclists, horseback riders, snowmobilers and others. The level of truck traffic associated with horizontal drilling and high-volume hydraulic fracturing presents safety issues, and would significantly degrade the experience for users of these roads, if not altogether during the drilling and construction phases of development.

10.4.3 LEGAL CONSIDERATIONS

State Forests have an identity that is distinct from private lands, prescribed by the NYS Constitution, the ECL and the Environmental Quality Bond Acts of 1972 and 1986, under the provisions of which they were acquired. New York State Constitution Article XIV, Section 3(1) states:

> "Forest and wild life conservation are hereby declared to be policies of the state. For the purposes of carrying out such policies the legislature may appropriate moneys for the acquisition by the state of land, outside of the Adirondack and Catskill parks as now fixed by law, for the practice of forest or wild life conservation."

ECL Section 9-0501(1), in keeping with the above constitutional provision, authorizes the state to acquire reforestation areas, —which are adapted for reforestation and the establishment and maintenance thereon of forests for watershed protection, the production of timber and other forests products, and for recreation and kindred purposes,. . .which shall be forever devoted to the planting, growth and harvesting of such trees...‖ Similarly, ECL Section 11-2103(1) authorizes the state to acquire —lands, waters or lands and waters…for the purpose of establishing and maintaining public hunting, trapping and fishing grounds.

ECL Section 9-0507 provides the Department discretionary authority to lease oil and gas rights on reforestation areas, provided that —such leasehold rights shall not interfere with the operation of such reforestation areas for the purposes for which they were acquired and as defined in Section 3 of Article XIV of the Constitution. The expected volume of truck traffic, the expected acreage that would be converted to non-forest use in the form of well pads, roads and pipelines, and noise and other impacts, raise serious questions as to how the surface activities anticipated with horizontal drilling and high-volume hydraulic fracturing could be viewed as consistent with this provision of the ECL.

For Wildlife Management Areas (WMAs) there are additional legal considerations stemming from the use of federal funds. Many WMAs were purchased using Federal Aid in Wildlife Restoration (Pittman-Robertson) funds and all are managed/maintained using Pittman-Robertson funds. Under these provisions, any surface use of the land must not be in conflict with the intended use as a WMA. These areas are managed for natural habitats to benefit wildlife, and disturbance associated with multi-pad wells raises questions about compatibility with essential wildlife behaviors such as breeding, raising young, and preparation for migration. Also, selling or leasing of minerals rights must be approved by the U.S. Fish and Wildlife Service, and may require reimbursement of the federal government for revenue generated. In addition, siting well pads on WMAs purchased with Conservation Fund monies may require additional mitigation under federal statutes and/or compensation.

From: Chapter 6: Potential Environmental Impacts, in NYSDEC, Revised Draft: Supplemental Generic Environmental Impact Statement on the Oil, Gas and Solution Mining Regulatory Program. Albany, NY: NYSDEC, Division of Water, Division of Air Resources, Division of Lands and Forests, and Division of Fish, Wildlife, and Marine Resources; September 2011. pp. 6-66–6-93.

PART III

HUMAN HEALTH PERSPECTIVES

CHAPTER 11

MISSING FROM THE TABLE: ROLE OF THE ENVIRONMENTAL PUBLIC HEALTH COMMUNITY IN GOVERNMENTAL ADVISORY COMMISSIONS RELATED TO MARCELLUS SHALE DRILLING

BERNARD D. GOLDSTEIN, JILL KRIESKY, and BARBARA PAVLIAKOVA

The development of hydrofracking technology has led to rapid growth in drilling for oil and natural gas in the United States and globally. Public concern about potential environmental and public health consequences has led to the formation of governmental advisory committees that are looking at the risks and consequences of the drilling activity. In 2011, President Barack Obama and the governors of Pennsylvania and Maryland independently established commissions to provide advice about a broad range of issues related to drilling for natural gas.

The Marcellus Shale is a rich natural gas field, said to be the second largest in the world, that extends under much of Pennsylvania, New York, and West Virginia and parts of Maryland, Ohio, Virginia, Kentucky, and Tennessee (Considine et al. 2010). The technology that permits access to natural gas in the Marcellus Shale includes drilling first vertically to the shale level and then horizontally within the shale (Arthur et al. 2009; Pennsylvania Department of Environmental Protection 2011). Holes are then cut in the horizontal pipe, followed by injection of high volumes of hydraulic fracturing fluids (primarily water and sand plus chemical additives) to break open the shale layers and maintain gas flow (for a compendium of links to research and data on the Marcellus Shale, see FracTracker 2011).

The amount of water used is in the range of a million gallons per well injected over perhaps days to a few weeks (New York City Department of Environmental Protection 2009), and the level of chemicals additives is in the range of 0.5–2.0% [U.S. Department of Energy (DOE) 2009]. About 30–70% of the fracking fluid returns to the surface and must be discharged somewhere (DOE 2009). Specific chemical and physical agents used in the fracking mixture to increase the release and flow of the fossil fuel and prevent microbial growth, corrosion, and scale formation vary by company and by location. Lists of these additives have been published (U.S. House of Representatives 2011), and some companies have been cooperative in revealing additives used (FracFocus Chemical Disclosure Registry 2011; Range Resources 2010), but the lack of complete information complicates interpretation of public complaints about health impacts. Secrecy itself may heighten public concern, as appeared to occur as a result of the high-volume use of a dispersant with an unknown component during the Deepwater Horizon disaster (Goldstein et al. 2011). Other unknowns include whether fracking chemicals react with other compounds to produce new chemicals, and whether naturally occurring agents such as arsenic, bromine, and radioactive compounds may be displaced into waste fracking fluids or groundwater (Goldstein 2011). Concerns about potential endocrinedisrupting effects of fracking chemicals have been raised (Finkel and Law 2011) and the potential health impacts of oil and gas development have been reviewed (Witter et al. 2008).

In the context of environmental and public health concerns, we reviewed the organizational aspects of three of these advisory committees related to drilling for natural gas in the Marcellus Shale and nationally. We also analyzed the extent to which human health issues are of concern to the public by reviewing presentations at the public meeting of the Secretary of Energy Advisory Board (SEAB) Natural Gas Subcommittee.

11.1 MATERIALS AND METHODS

We reviewed the charge and the composition of three advisory committees formally established in 2011 to review drilling for natural gas. Two were established by governors, one Republican (Pennsylvania Governor

Tom Corbett) and one Democrat (Maryland Governor Martin O'Malley), and the third by President Obama, also a Democrat. We classified individual committee members according to employment or participation in government, academia, environmental groups, civil society groups, or industry, based on their current positions. Possible health expertise was determined through review of biographies available as part of the official record of their appointment, review of information available on their organization's website, or a web search via Google.

For this study, we distinguished between environmental public health, which focuses on the human consequences of an activity affecting the environment, and environmental health, which we define more broadly as alterations of the environment. We recognize that this is an arbitrary distinction, and that there is a continuum between environmental and human health. By identifying a committee member as having environmental public health expertise, we mean that this individual has experience in evaluation of or response to the direct toxicological effects of chemical and physical agents on human health or indirect effects such as psychosocial stresses, highway safety, and drug and alcohol abuse. For the purposes of this study, we have extended this definition to include medical and health professionals who could be presumed to have some health background related to environmental health, however minimal (e.g., physicians, nurses, pharmacists, psychologists).

Public response was assessed by review of presentations to the 13 June 2011 federal SEAB Natural Gas Subcommittee meeting held in Washington, Pennsylvania, the one public meeting of the subcommittee in the Marcellus Shale area. The meeting was widely advertised and well attended both by supporters and by opponents of Marcellus Shale drilling, many of whom were from surrounding states. Because ofthe large number of speakers, each was limited to two minutes. Two of us (B.P.and J.K.) independently reviewed the video of these presentations (SEAB Natural Gas Subcommittee 2011) to develop a list of codes summarizing the main points covered by each speaker and categorize speakers as either supporters (51) or opponents (59) of shale gas drilling. The responses of the opponents were further categorized into a variety of subheadings. Differences between the two raters were reconciled before analysis.

11.2 RESULTS

11.2.1 REVIEW OF THE EXECUTIVE ORDERS

11.2.1.1 U.S. FEDERAL GOVERNMENT

The federal review of fracking issues was requested in President Obama's "Blueprint for a Secure Energy Future" (Obama 2011) that states:

> To provide recommendations from a range of independent experts, the Secretary of Energy, in consultation with the U.S. EPA Administrator and Secretary of Interior, should task the Secretary of Energy Advisory Board (SEAB) with establishing a subcommittee to examine fracking issues. The subcommittee will . . . include leaders from industry, the environmental community, and states. The subcommittee will work to identify . . . any immediate steps that can be taken to improve the safety and environmental performance of fracking and to develop . . . consensus recommended advice to the agencies on practices for shale extraction to ensure the protection of public health and the environment.

Note that this executive order gives leadership to the DOE in consultation with the Department of Interior and U.S. Environmental Protection Agency (EPA). Although this statement culminates with the charge to ensure the protection of public health and the environment, the U.S. Department of Health and Human Services (DHHS), despite its environmental health components [National Institute of Environmental Health Sciences (NIEHS), Centers for Disease Control and Prevention/Agency for Toxic Substances and Disease Registry], is not included.

11.2.1.2 PENNSYLVANIA

In Pennsylvania, the newly elected Governor, Tom Corbett, had stressed the importance of the Marcellus Shale to economic development during his campaign.Early in his administration, in March 2011, he appointed a 31-member Governor's Marcellus Shale Advisory Commission. The executive order establishing the commission states:

> The Commonwealth takes seriously its responsibility to ensure the development of gas in a manner that protects the environment and safeguards the health and welfare of its citizens. (Corbett 2011)

Four work groups were designated by the commission, including one on public health, safety, and environmental protection. They were charged with the following task:

> Consideration of additional measures necessary to ensure the protection of the Commonwealth's environment and natural resources and the enhancement of public health and safety. (Governor's Marcellus Shale Advisory Commission 2011b)

11.2.1.3 MARYLAND

In July 2011, Governor Martin O'Malley of Maryland appointed a 14-member commission. The Governor's Executive Order specifically states:

> Purpose. The Marcellus Shale Safe Drilling Initiative will assist State policymakers and regulators in determining whether and how gas production from the Marcellus Shale in Maryland can be accomplished without unacceptable risks of adverse impacts to public health, safety, the environment and natural resources. (O'Malley and McDonough 2011)

In summary, public health and the environment are featured in the rationale for the formation of all three committees formed to give advice on Marcellus Shale drilling.

11.2.2 COMPOSITION OF THE ADVISORY COMMITTEES.

The total number of appointments to these three advisory committees was 52, consisting of 51 individuals. One individual, Jeffrey Kupfer, an energy company executive who was the former DOE Deputy Secretary, was chosen as a member of both state advisory committees.

The SEAB Natural Gas Subcommittee consisted of seven members (DOE 2011): three in academia, three in industry, and one in an environmental group (Fred Krupp, President of the Environmental Defense Fund). The chair of the SEAB subcommittee, John Deutsch, is the former Chairman of the Department of Chemistry, Dean of Science, and Provost at Massachusetts Institute of Technology. Other members are Stephen Holditch, head of the Department of Petroleum Engineering at Texas A&M University, and D. Mark Zoback, professor of geophysics at Stanford University. Several of the subcommittee members have experience in more than one sector. For example, Deutsch is a former Deputy Secretary of Defense and the former head of the Central Intelligence Agency and is on the board of various energyrelated companies. Kathleen McGinty of Weston Corporation was previously head of the White House Council on Environmental Quality under President Bill Clinton and was Pennsylvania Secretary of the Environment under its previous governor.

The Pennsylvania Governor's Marcellus Shale Advisory Commission has 31 members: 10 from government, 1 from academia, 4 from environmental groups, 5 from civil society groups, and 11 from industry (Governor's Marcellus Shale Advisory Commission 2011a). It was chaired by the lieutenant governor. The one academic, Terry Engelder, is a professor of geoscience at the Pennsylvania State University whose research on the extent and availability of natural gas in the Marcellus Shale has been recognized as central to its current rapid development.

The Maryland Marcellus Shale Safe Drilling Initiative Advisory Commission has 14 members: 6 from government, 1 academic who chairs the commission (David Vanko, a geologist and current Dean of the College of Science and Mathematics at Towson University), 3 from environmental groups, 2 from civil society groups, and 2 from industry (Maryland Department of the Environment 2011).

11.2.3 ENVIRONMENTAL PUBLIC HEALTH OR OTHER HEALTH EXPERTISE

Our review of the background of all 51 members of the three advisory committees provided no evidence that any member had expertise in the human health aspects of environmental health or experience in health or health care. Based on the available information, we were unable to identify any public health personnel, physicians, nurses, pharmacists, dentists, or others with a health background on the three advisory committees.

11.2.4 EVALUATION OF PUBLIC CONCERNS.

Categorization of the 110 public comments at the SEAB Natural Gas Subcommittee meeting in Washington, Pennsylvania, revealed 51 speakers favorable to shale gas drilling and 59 opposed (Natural Gas Subcommittee 2011). Among the opposed, the major concerns discussed included the negative effects of Marcellus Shale drilling on the environment expressed by 46 speakers (78%), concern regarding the safety and/or regulation of the natural gas drilling industry expressed by 41 (69.5%), and concern for residents' health expressed by 37 (62.7%) (Table 1). Other concerns ranged from general effects on air, water, and ecosystems to concerns about the potential for negative health effects on humans. Of those opposed, 12 (20.3%) attributed a direct negative health impact on themselves, a family member, or a friend to Marcellus Shale drilling. Almost a quarter of the speakers opposed to drilling expressed concerns regarding the make-up of the committee, including the potential for bias toward industry interests and a lack of expertise among committee members regarding their specific concerns.

11.3 DISCUSSION

Recent technological and operational improvements in extracting natural gas resources, particularly shale gas, have increased gas drilling activities nationally and led to significantly higher natural gas production estimates for decades to come (Considine et al. 2010). The potential for adverse environmental public health consequences has been recognized by the public, who are concerned, and by members of the government, who have asked for advice. However, despite these stated concerns for potential public health impacts from Marcellus Shale activities, none of the three recently formed advisory bodies include any recognizable expertise in assessing environmental impacts on human health, nor have they invited participation from state or federal agencies with direct public health responsibilities.

This is not the only recent instance in which environmental public health expertise has not been at the table despite obvious human health implications of an environmental issue. President Obama's seven-member National Commission on the BP Deepwater Horizon Oil Spill and Offshore Drilling was co-chaired by former U.S. EPA administrator William Reilly and Senator Bob Graham and has one member, Donald Boesch, who is a professor of marine science and has extensive experience in ecosystem research. However, no members of this committee have a background in environmental public health. Other recent advisory committee reports on natural gas extraction include that of the National Petroleum Council, whose membership primarily is from the petroleum industry but also includes government personnel and university faculty. Its extensive report of impacts on wildlife habitat and on the environmental footprint of oil exploration and operations includes only an occasional reference to human health (National Petroleum Council 2011).

We can only conjecture about the reasons that environmental public health experts or organizations have not been included in advisory bodies related to Marcellus Shale activities. It is unlikely that the failure to include environmental public health expertise is due to lack of recognition that there is reason to be concerned about human health risks. Such concern is clearly stated by the two state governors and by President Obama in establishing their advisory committees. Nor can it be a lack of awareness of the well-publicized public concerns about the potential health impacts of Marcellus Shale activities, which have been forcefully described by the public in numerous hearings.

In Pennsylvania, the absence of the Department of Health or of any public health expertise in the 31-member Governor's Marcellus Shale Advisory Commission could reflect the relative weakness of that state's public health infrastructure. Although having expert health departments at the state level and in some localities, Pennsylvania ranked last nationally in a survey sponsored by the Health Resources and Services Administration on the size of the public health workforce in each state (37 per 100,000, vs. the national mean of 138 per 100,000) (Gebbie et al. 2000; Potter 2008). However, Maryland did relatively well in this regard (304 per 100,000) and ranked first in its region, which includes Pennsylvania.

University-based expertise is well represented in the federal SEAB subcommittee, with three of its seven members being respected academics, but academia is not particularly well represented in the two state advisory committees. In Pennsylvania, only one of the 31 members has a direct university affiliation. Maryland also has only one of its 14 members from academia, although in this case it is the committee chair. None of the three committees includes members with academic expertise in health science or ecosystem sciences; that is, none of the five academic members has expertise in biological systems or human health.

Ecosystem concerns are represented in these three advisory committees through the presence of leaders of environmental organizations that are particularly involved in ecosystem health, such as the Nature Conservancy, Trout Unlimited, Savage River Watershed Association, Pennsylvania Environmental Council, Western Pennsylvania Conservancy, and Chesapeake Bay Foundation. But there is no representation from organizations known primarily for their concern about human health and the environment. Some of the environmental organizations whose leaders are members of the advisory committees, such as the Environmental Defense Fund, have been involved with human health issues as well as more general environmental preservation. But none of the advisory committee members has personal expertise in human health.

The failure to choose academics that have expertise in human health issues and the environment is not because of lack of such expertise. Both Maryland and Pennsylvania have reasonably robust academic public health infrastructures, with each having two accredited schools of public health, and Maryland having three and Pennsylvania five accredited programs of public health. All four schools of public health have formal departments in the field of environmental health, and accreditation as a program requires at least sufficient faculty to teach a core course in environmental health. Both states do well in another indicator of academic expertise in environmental health sciences, that of funding from the NIEHS. In fiscal year 2010, Maryland received 4.7% of total NIEHS funding and Pennsylvania 4.3%; the states have 1.9% and 4.1% of the total U.S. population, respectively (DHHS 2011).

TABLE 1: Concerns raised by opponents (n = 59) of Marcellus Shale drilling at the Washington, Pennsylvania, public meeting with the SEAB Natural Gas Subcommittee.

Concern	*n* (%)
Environmental Concerns	46 (78.0)
Safety and Regulation of Industry	41 (69.5)
Negative Effects on Water	39 (66.1)
General Health Concerns	37 (62.7)
Negative Effects on Air	23 (39.0)
Chemicals in Water	22 (37.3)
Bias, Conflict of Interest, or Lack of Expertise in Desired Subject Area by Members of the Committee	14 (23.7)
Health Problem in Family Member Attributed to Drilling	12 (20.3)
Personal Legal Rights have been Infringed upon by Companies	8 (13.6)
Export of Domestic Natural Gas Resources	6 (10.2)
Depreciation in Property Values	4 (6.8)

Political and bureaucratic issues deserve further consideration. President Obama gave the lead on Marcellus Shale to the Secretary of Energy in consultation with the Department of Interior and the U.S. EPA. The rationale for not including the DHHS is not clear. Arguably, the DOE does have some health expertise, because it has various organizational structures dealing with health and safety related to energy or to the cleanup of atomic materials production sites. Although conceivable, we do not believe that the authorities fail to recognize the difference between environmental and public health expertise, particularly because in each case the executive orders separately specify environment and health.

One can argue that the U.S. EPA, which at its formation included components moved from the U.S. Public Health Service, does have public health responsibilities (Goldstein 1988; Johnson 2010). Evidence that the U.S. EPA does take its public health responsibilities seriously includes a recent reorganization to include an Environmental Public Health division (U.S. EPA 2012). Lisa Jackson, the Administrator of the U.S. EPA, in speaking about sustainability, often uses classic public health language by pointing out that sustainability is similar to pursuing wellness instead of treating disease (Jackson 2010).

Congress provided the U.S. EPA's Office of Research and Development with funding specifically to look at groundwater contamination from Marcellus Shale hydrofracturing (U.S. EPA 2011). This appropriation did not allow studies of the potential environmental or human toxicity of fracking compounds or of the potential for air pollution—although it is not clear why the U.S. EPA does not use other funding for such research. This at least raises the possibility that political leadership does not want research on human health because of concerns that equivocal or positive findings might inhibit economic development or offend major industries. Environmental public health experts may be seen as more likely to raise problems than to find solutions.

To explain the absence of environmental public health expertise in governmental advisory processes related to the Marcellus Shale, we are left with the distinct possibility that it is the fault of the environmental public health community. We have not worked as hard or as effectively as we could or should with local, state, or federal governmental organizations responsible for making decisions on environmental matters. These governmental organizations extend well beyond state departments of health or the federal units that are part of the DHHS. Simply being able to provide the science needed for effective environmental decision making is not sufficient if our science does not inform decisions that are made. A proactive approach to working with the broad range of federal and state agencies involved in environmental decision making, and with the public, including providing economic analysis related to potential positive and negative health impacts, is central to converting our knowledge to protection of human health and the environment (Hearne 2008; Longest and Huber 2010; Rutkow et al. 2009)

11.4 CONCLUSIONS

Environmental public health is not yet at the table in governmental advisory processes related to drilling in the Marcellus Shale. The explanation for the lack of involvement of the environmental public health community does not appear to be a failure to recognize the importance of public health to this issue by the president or the governors, nor is it a lack of public concern. Expertise in the impact of environmental factors on public health is also readily available.

Political concern that evaluation of the potential environmental public health consequences of shale gas drilling may find a problem that slows down the rush to develop the Marcellus Shale may be an important factor. However, we believe the most likely major cause is the failure of a relatively robust community of environmental public health experts to adequately project this expertise into the state and national debates about developing natural resources.

Larry Gordon (1990) has noted the lack of involvement of environmental public health in key environmental issues. He emphasizesthe importance of increasing the numbers and the training of the environmental health workforce. More than two decades later, review of the advisory processes developed to make recommendations concerning natural gas suggests that we need to go beyond standard infrastructure issues to consider how best to project environmental public health concerns at the local, state, and national levels.

REFERENCES

1. [1]Arthur JD, Bohm B, Layne M. 2009. Considerations for development of Marcellus Shale gas. World Oil, July:65–68.
2. Considine TJ, Watson R, Blumsack S. 2010. The Economic Impacts of the Pennsylvania Marcellus Shale Natural Gas Play: An Update. Available: http://marcellusco-

alition.org/wp-content/uploads/2010/05/PA-Marcellus-UpdatedEconomic-Impacts-5.24.10.3.pdf [accessed 6 October 2011].

3. Corbett T. 2011. Executive Order, Commonwealth of Pennsylvania Governor's Office: Creation of Governor's Marcellus Shale Advisory Commission. Available: http://www.portal.state.pa.us/portal/server.pt/gateway/PTARGS_0_2_785_708_0_43/http%3B/pubcontent.state.pa.us/publishedcontent/publish/global/files/executive_orders/2010___2019/2011_01.pdf [accessed 6October 2011].
4. DHHS (Department of Health and Human Services). 2011. NIH Awards by Location and Organization. Available: http://report.nih.gov/award/organizations.cfm [accessed 28 September 2011].
5. DOE (Department of Energy), Office of Fossil Energy, National Energy Technology Laboratory. 2009. Modern Shale Gas Development in the United States: A Primer. Available: http://www.netl.doe.gov/technologies/oilgas/publications/epreports/shale_gas_primer_2009.pdf [accessed 5 October 2011].
6. DOE (Department of Energy), SEAB Shale Gas Production Subcommittee. 2011. Shale Gas Production Subcommittee 90-Day Report. Available: http://www.shalegas.energy.gov/resources/081811_90_day_report_final.pdf [accessed 6 October 2011].
7. Finkel ML, Law A. 2011. The rush to drill for natural gas: a public health cautionary tale. Am J Public Health 101(5):784–785.
8. FracFocus Chemical Disclosure Registry. 2011. Groundwater Protection: Priority Number One. Available: http://fracfocus.org/ [accessed 28 September 2011].
9. FracTracker. 2011. FracTracker Homepage. Available: http://www.fractracker.org/ [accessed 5 October 2011].
10. Gebbie K, Merrill J, B'toush R, Cortazal M, Gebbie E, Gupta M, et al. 2000. The Public Health Work Force—Enumeration 2000. Document U76 AH 00001-03. New York: Center for Health Policy, Columbia University School of Nursing.
11. Goldstein BD. 1988. U.S. EPA as a public health agency. Regul Toxicol Pharmacol 8(3):328–334.
12. Goldstein BD. 2011. Potential Health Effects of Marcellus Shale Activities: The Need for Public Health Surveillance. Available: http://www.chec.pitt.edu/documents/Marcellus%20Shale/CHEC-Present-Goldstein-5.10.11-HealthMarcellus.pdf [accessed 6 October 2011].
13. Goldstein BD, Osofsky HJ, Lichtveld MY. 2011. The Gulf oil spill. N Engl J Med 364(14):1334–1348.
14. Gordon LJ. 1990. Who will manage the environment? Am J Public Health 80(8):904–905.
15. Governor's Marcellus Shale Advisory Commission. 2011a.Marcellus Shale Advisory Commission Membership List. Available: http://files.dep.state.pa.us/PublicParticipation/MarcellusShaleAdvisoryCommission/MarcellusShaleAdvisoryPortalFiles/MSAC_Members.pdf [accessed 6 October 2011].
16. Governor's Marcellus Shale Advisory Commission. 2011b. Governor's Marcellus Shale Advisory Commission Report. Available: http://www.mde.state.md.us/programs/Land/mining/marcellus/Documents/MSAC_Final_Report.pdf [accessed 5 October 2011].
17. Hearne SA. 2008. Practice-based teaching for health policy action and advocacy. Public Health Rep 123(suppl 2):65–70.

18. Jackson LP. 2010. Speeches by EPA Administrator. Administrator Lisa P. Jackson, Remarks to the National Academies of Science, as Prepared. Available: http://yosemite.epa.gov/opa/admpress.nsf/12a744ff56dbff8585257590004750b6/1c893e457b3cbb25852577ec0054048c!OpenDocument [accessed 6 October 2011].
19. Johnson BL. 2010. The U.S. EPA as an agency for public health. Hum Ecol Risk Assess 16(6):1209–1214.
20. Longest BB Jr, Huber GA. 2010. Schools of public health and the health of the public: enhancing the capabilities of faculty to be influential in policymaking. Am J Public Health 100(1):49–53.
21. Maryland Department of the Environment. 2011. Governor O'Malley Names Members of the Marcellus Shale Safe Drilling Initiative Advisory Commission. Available: http://www.mde.state.md.us/programs/PressRoom/Pages/071911.aspx [accessed 5 October 2011].
22. National Petroleum Council. 2011. Prudent Development—Realizing the Potential of North America's Abundant Natural Gas and Oil Resources. Available: http://www.npc.org/Prudent_Development.html [accessed 29 February 2012].
23. New York City Department of Environmental Protection. 2009. Final Impact Assessment Report: Impact Assessment of Natural Gas Production in the New York City Water Supply Watershed. Available: http://www.nyc.gov/html/dep/pdf/natural_gas_drilling/12_23_2009_final_assessment_report.pdf [accessed 5 October 2011].
24. Obama B. 2011. Blueprint for a Secure Energy Future. Available: http://www.whitehouse.gov/sites/default/files/blueprint_secure_energy_future.pdf [accessed 5 October 2011].
25. O'Malley M, McDonough JP. 2011. Executive Order 01.01.2011.11: The Marcellus Shale Safe Drilling Initiative. Available: http://www.governor.maryland.gov/executiveorders/01.01.2011.11.pdf [accessed 5 October 2011].
26. Pennsylvania Department of Environmental Protection. 2011. Marcellus Shale. Available: http://www.portal.state.pa.us/portal/server.pt/community/marcellus_shale/20296 [accessed 5 October 2011].
27. Potter MA. 2008. The Public Health System of the Pennsylvania Commonwealth: Where Do We Go from Here? Commonwealth J Pol Sci 14(1):1–6.
28. Range Resources. 2010. Range Resources Announced on July 14, 2010 That the Company Would Voluntarily Disclose the Composition of Each of the Hydraulic Fracturing Components for All the Wells Operated by Range Resources with the Pennsylvania Department of Environmental Protection (DEP) Completed in the Marcellus Shale. Available: http://www.rangeresources.com/getdoc/50e3bc03-3bf6-4517-a29be2b8e-f0afe4f/Well-Completion-Reports.aspx [accessed 6 October 2011].
29. Rutkow L, Levin MB, Burke TA. 2009. Meeting local needs while developing public health practice skills: a model community-academic partnership. J Public Health Manag Pract 15(5):425–431.
30. SEAB (Security of Energy Advisory Board) Natural Gas Subcommittee. 2011. Video of Public Meeting at Washington and Jefferson College. Available: http://prod-mmedia.netl.doe.gov/video/Schale_gas_Meeting1.wmv [accessed 28 September 2011].
31. U.S. EPA (Environmental Protection Agency). 2012. About the Health and Environmental Effects Research (NHEERL). Available: http://www.epa.gov/aboutepa/nheerl.html#aed [accessed 6 February 2012].

32. U.S. EPA (Environmental Protection Agency), Office of Research and Development. 2011. Hydraulic fracturing. In: Framework for an EPA Safe and Sustainable Water Resources Research Program, 89–93. Available: http://www.epa.gov/ord/priorities/docs/SSWRFramework.pdf [accessed 28 September 2011].
33. U.S. House of Representatives, Committee on Energy and Commerce. 2011. Chemicals Used in Hydraulic Fracturing. Available: http://democrats.energycommerce.house.gov/sites/default/files/documents/Hydraulic%20Fracturing%20Report%20 4.18.11.pdf [accessed 4 October 2011].
34. Witter R, Stinson K, Sackett H, Putter S, Kinney G, TeitelbaumD, et al. 2008. Potential Exposure-Related Human Health Effects of Oil and Gas Development: A White Paper. Available: http://docs.nrdc.org/health/files/hea_08091702a.pdf [accessed 14December 2011].

CHAPTER 12

MODERN NATURAL GAS DEVELOPMENT AND HARM TO HEALTH: THE NEED FOR PROACTIVE PUBLIC HEALTH POLICIES

MADELON L. FINKEL, JAKE HAYS, and ADAM LAW

12.1 INTRODUCTION

A modern form of natural gas development has become a global "game changer" in the quest for energy. Natural gas, abundant around the world, has a clean reputation compared to other fossil fuels since it burns less carbon when used. It is easy to transport, reasonably economical, and requires comparatively quick construction timelines and low capital costs.

Traditionally, natural gas was extracted using a method that bores a vertical well in single gas reservoirs close to the surface (conventional natural gas drilling). However, drilling for natural gas in shale rock was not particularly economical, primarily because shale typically has insufficient permeability to allow significant fluid flow to a well bore. With technological advances and unconventional methods (i.e., horizontal hydraulic fracturing), gas extraction from tight formations (e.g., shale) is now feasible.

This type of unconventional natural gas development relies on clustered, multi-well pads and long, horizontal laterals. Wells are drilled vertically (often thousands of feet) and horizontally in multiple directions. The method entails injecting large volumes of fluid consisting of chemicals, water, and sand into the well to fracture the shale rock that releases the natural gas. The internal pressure of the rock formation also causes a portion of the injected fracking fluids to return to the surface (flowback fluids); these fluids are often stored in a tank or pit before being pumped into trucks for transport to a disposal site. Flowback has been shown to contain a variety of formation materials, including brines, heavy metals, radionuclides, and organics, which can make wastewater treatment difficult and expensive [1]. Further, other studies found that 20% to 85% of fracturing fluids may remain in the formation, which means the fluids could continue to be a source of groundwater contamination for years to come [2]. By 2009, there were more than 493,000 active natural gas wells across 31 states, almost double the number in 1990, of which approximately 90 percent have used hydraulic fracturing to extract gas [3].

Whereas shale gas has the potential to become a significant, economical energy source, the potential for harm and the potential of giving a false sense of energy security are often dismissed by its proponents. The process is potentially polluting and damaging not only to human and animal health but also to the environment, as a result of clearing of land for well pads, drilling the wells, extracting the gas, storing the byproducts of the extraction, transporting the gas by diesel trucks, and the final capping of the well. The potential for harm to children is especially worrisome. This paper focuses on a literature review of unconventional natural gas development and its potential impact on human health.

12.2 DISCUSSION

Canaries in coal mines were used as an early-warning signal for toxic gases, primarily carbon monoxide. The birds, being more sensitive, would become sick before the miners thus providing advanced warning of a danger. Animals and children also can be viewed as sentinel species. A 1993 report, now viewed as a watershed moment for health and environmental policy, documented that children are more vulnerable and sensitive than adults to chemicals in the environment [4]. Early development (in utero and during the first few years after birth) is particularly sensitive to disruption by exposures to chemicals in the environment and to imbalanced nutrition, with potentially adverse consequences for health later in life [5]. A child's ability to metabolize toxic chemicals is different from an adult's.

Children receive proportionately larger doses of chemical toxicants than adults, and these exposures occur at a time in the life cycle when organs and tissues are rapidly growing and developing.

Endocrine disrupting chemicals (EDCs) present a particularly concerning hazard during human growth and development. EDCs can affect the reproductive system and often effect epigenetic mechanisms leading to pathology decades after exposure. Given the potential for harm, it would be a prudent course of action that chemicals used in the hydraulic fracturing process be evaluated for their EDC potential and screening assays be developed to test flowback fluids.

Landrigan et al. [6], looking at children's vulnerability to toxic chemicals, found strong evidence that toxic chemicals are important causes of disease in children. For example, benzene, being toxic to all humans, has been shown to contribute a disproportionate risk of leukemia to young children [7]. Neural tube defects, spina bifida being the most common [8], and decreased fetal growth also have been shown to occur disproportionately higher in children exposed to toxic chemicals, including benzene. Further, because of the long latency period of some diseases, toxic exposures in childhood are more likely to result in disease in adulthood compared to exposure in adulthood [10].

Trasande and Liu [11], building on Landrigan et al.'s analysis of the costs of environmental pollutants and disease [12], estimated that costs of environmentally mediated diseases in American children totaled $76.6 billion (in 2008 dollars) and called for federal policy action to limit children's exposure to known chemical hazards, including exposures to toxic chemicals. Given the staggering human and economic costs of environmentally mediated diseases, a wise course of action would be to empirically document trends in specific diseases among children living in close proximity to unconventional natural gas operations compared to those living in areas where drilling is not occurring.

We acknowledge that adverse health outcomes of medical conditions with long latency periods will not be evident for years and will depend on the exposure, duration of exposure, dose, and other factors. A higher incidence of asthma, cancer, heart disease, and the effects of endocrine disruption on developing fetuses and children, due to contaminant exposure, only become evident over time. However, baseline measurements should be recorded and updated over time. As of this writing, the extent of health risks associated with unconventional natural gas operations among children is unknown.

In an effort to assess the impact of hydraulic fracturing on children's health, we are advocating that local and state governments work together to establish a system to track the incidence and prevalence of diseases that have been shown in the literature to be causally related or exacerbated by exposure to environmental agents. In Pennsylvania, for example, there has been active, on-going unconventional natural gas development since the late 2000s. There are currently 6,773 horizontal wells drilled or under development and over 9,600 drilling permits have been issued [13]. As of May 2011, 320 daycare facilities, 67 schools, and 9 hospitals were located within two miles of natural gas wells [14]. What effect will this activity have on the development of disease among children living in counties with active drilling as compared to children living in counties with little or no drilling activity? Trends in childhood cancers, especially acute lymphoblastic leukemia, birth outcomes including birth defects, premature and low birth weight births, neurodevelopment disorders, and, respiratory disease, especially asthma, should be monitored.

The limited information available on the chemicals that are used in the drilling hampers efforts to empirically assess the potential for harm. Oil and gas companies are legally permitted to withhold information on their proprietary mixtures, and the federal government has granted oil and gas companies exemption from many environmental law restrictions such as the Clean Water Act, Clean Air Act, Safe Drinking Water Act, Resource Conservation and Recovery Act, Toxic Release Inventory under Emergency Planning and Community Right-to-Know Act, and the National Environmental Policy Act. By default, states have the responsibility for enacting regulations; however, many states have weak or no regulations. As of February 2012, only four of 31 states have significant drilling rules; of these, 5 adopted disclosure rules, although they still allow for "proprietary trade secrets" [15].

12.3 HARM TO HEALTH

The production process creates a huge pressure cooker of organics and inorganics, and even if every single compound pumped into the well is harmless by itself (which is not the case), the pressure would create hundreds if not thousands of different compounds that are highly toxic. Of the few studies that have looked at the chemical cocktails used in the process, findings have identified chemicals that are known to cause cancers, mutations, and diseases of the nervous, immune, and endocrine systems, the kidney, gastrointestinal tract and liver, heart, and skin [16]. Colburn identified almost 1,000 chemical products and nearly 650 individual chemicals used in natural gas operations, many of which have the potential to cause adverse health effects as well as to potentially cause deleterious effects on the environment [17]. Specifically, the researchers documented that the hydraulic fracturing process releases toxic and cancer-causing chemicals such as benzene, toluene, xylene (BTEX), and methylene chloride among other health-hazardous air pollutants. These health-hazardous pollutants are released from a number of sources including blowouts, flaring, condensate tanks, construction activity, engines, and venting. Methane, a powerful greenhouse gas, also is emitted throughout the oil and gas development process. Methane interacts with sunlight to produce tropospheric ozone, which is a strong respiratory irritant associated with increased respiratory morbidity and mortality [18].

Witter et al. were one of the first to present a detailed assessment of health trends in Garfield County, CO, that documented the negative impact of drilling on air, soil, water, and human health [19]. Building on those findings, McKenzie et al. estimated health risks for exposures to air emissions from a natural gas development project also in Garfield County and found that residents living less than one-half mile away from wells were at greater risk for ill health effects than those living farther away [20]. Although these studies focused on Garfield County, CO, the researchers maintain that the exposure pathways and related health risks would be similar wherever oil and gas development is occurring.

12.4 WATER CONTAMINATION

Whereas the conventional method of natural gas production utilizes about 20,000 to 80,000 gallons of fluid, the unconventional method utilizes up to 5 million gallons of fluid per hydraulic fracturing event, which includes not only water and sand, but also numerous toxic chemicals. The potential for contamination of aquifers by the residual fracking fluids that remain underground must be considered. The likelihood of spills throughout the entire lifecycle of development also must be taken into account. Blowouts (uncontrolled release of natural gas from a gas well after pressure control systems have failed) allow gas and/or highly contaminated produced waters to flow to the surface; hoses come undone, gaskets fail, pits or tanks that hold the fracking fluids leak raising the serious risk of ground and water contamination. Even small quantities of the toxic fracking fluids can contaminate shallow aquifers with hydrocarbons, toxic chemicals, heavy metals, and radioactive materials.

Further, improper wastewater disposal, specifically the handling of fracking fluids including flowback wastewater (a byproduct of the process), can lead to contamination of ground and water [21]. The flowback can be taken to sewage plants, but it is widely acknowledged that sewage plants are not equipped to handle the contaminants. There have been reports of untreated wastewater being dumped into rivers and streams and sprayed on rural roads and forests [22].

The New York Times, in its analysis of more than 30,000 pages of federal, state, and company records relating to more than 200 gas wells, found that radioactive wastewater from the process has been discharged into rivers that supply drinking water to millions of people in Pennsylvania and Maryland. At least 12 sewage treatment plants in three states have discharged waste that was only partly treated into rivers, lakes, and streams [23]. There have been well-publicized instances of water contamination in Pavillion, WY, and Dimock, PA [24, 25], and gas has seeped into underground drinking water supplies in five states (Colorado, Ohio, Pennsylvania, Texas, and West Virginia). Further, a study conducted in Northeastern Pennsylvania found that water wells near a fracking site were 17 times more likely to exhibit methane contamination than wells not near drilling sites [26].

12.5 AIR AND SOIL CONTAMINATION

Unconventional natural gas development causes air pollution from multiple sources. Many particulates and chemicals are released into the atmosphere, including sulfuric oxide, nitrogen oxides, volatile organic compounds (VOCs), benzene, toluene, diesel fuel, hydrogen sulfide, and radon gas, all of which can have serious health implications. Further, the venting or flaring of wells during drilling and production contributes to local air pollution.

The drilling sludge, which is brought to the surface during the drilling process, contains fracking fluid, drilling mud, and radioactive material from the subsurface land formation, hydrocarbons, metals, and volatile organic compounds. Sludge, often left to dry on the surface in waste pits, may be removed to waste disposal sites (but not always to hazardous waste sites) or may be tilled into the soil in "land farms." These practices raise the risk of contaminating soil, air, and surface water, as a result of the fine dust becoming airborne thus affecting local air quality and raising the risk of respiratory disease. Based on concerns about the exposure to dust containing silica sand, the US Occupational Safety and Health Administration, along with the National Institute of Occupations Safety and Health (NIOSH), released a joint hazard alert on fracking silica in June, 2013 [27].

Unconventional natural gas development requires many diesel trucks for the transportation of the products used in drilling as well as the removal of flowback fluid. Diesel emissions contain nitrogen oxides and volatile organic compounds, which can react to sunlight to produce ozone, a strong respiratory irritant associated with increased respiratory morbidity and mortality [28].

12.6 CONCLUSION

The health impacts related to unconventional natural gas development may not be evident for years, as medical conditions with long latency periods will present over time. While the potential long-term, cumulative effects will not be known for years, we argue that it would be prudent to begin to track and monitor trends in the incidence and prevalence of diseases that already have been shown to be influenced by environmental agents. Meanwhile, the natural gas industry needs to address the risks to human and animal health and take steps to limit, preferably to eliminate, the exposure pathways. We need far greater transparency and full chemical disclosure. There needs to be an end to discharging effluent into rivers, streams, and groundwater. There needs to be much more attention paid to curtailing or preferably eliminating spills and leaks of radioactive wastewater. There needs to be an end to the disposal of radioactive sludge from drilling sites in landfills. There needs to be a safer way to develop this resource to limit the exposure to silica, which can cause silicosis, chronic obstructive pulmonary disease, and lung cancer. Banning the practice of burning off the initial flow of natural gas (flaring) needs to be mandated sooner than 2015, the date when EPA ruling goes into effect. And, perhaps most importantly, there needs to be a well-designed epidemiologic study conducted to empirically assess health status among those living proximate to active development compared to those living in areas where development is not occurring.

REFERENCES

1. D. J. Soeder and W. M. Kappel, "Water resources and natural gas production from the Marcellus Shale," U.S. Department of the Interior, U.S. Geological Survey Fact Sheet 2009–3032, 2009.
2. Environmental Protection Agency (EPA), "Evaluation Impacts to Underground Sources of Drinking Water by Hydraulic Fracturing of Coalbed Methane Reservoirs," June 2004, http://www.epa.gov/ogwdw/uic/pdfs/cbmstudy_attach_uic_ch04_hyd_frac_fluids.pdf.
3. I. Urbana, Regulation Lax as Gas Wells' Tainted Water Hits Rivers, New York Times, New York, NY, USA, 2011.

4. National Research Council, Pesticides in the Diets of Infants and Children, National Academies Press, Washington, DC, USA, 1993.
5. R. Barouki, P. D. Gluckman, P. Grandjean, M. Hanson, and J. J. Heindel, "Developmental origins of non-communicable disease: implications for research and Public Health," Environmental Health, vol. 11, pp. 42–51, 2012. V
6. P. J. Landrigan, C. B. Schechter, J. M. Lipton, M. C. Fahs, and J. Schwartz, "Environmental pollutants and disease in American children: estimates of morbidity, mortality, and costs for lead poisoning, asthma, cancer, and developmental disabilities," Environmental Health Perspectives, vol. 110, no. 7, pp. 721–728, 2002.
7. K. W. Whitworth, E. Symanski, and A. L. Coker, "Childhood lymphohematopoietic cancer incidence and hazardous air pollutants in Southeast Texas, 1995–2004," Environmental Health Perspectives, vol. 116, no. 11, pp. 1576–1580, 2008.
8. P. J. Lupo, E. Symanski, D. Kim Waller et al., "Maternal exposure to ambient levels of Benzene and Neural tube defects among offspring: Texas, 1999–2004," Environmental Health Perspectives, vol. 119, no. 3, pp. 397–402, 2011.
9. R. Slama, O. Thiebaugeorges, V. Goua et al., "Maternal personal exposure to airborne benzene and intrauterine growth," Environmental Health Perspectives, vol. 117, no. 8, pp. 1313–1321, 2009.
10. R. Barouki, P. D. Gluckman, P. Grandjean, M. Hanson, and J. J. Heindel, "Developmental origins of non-communicable disease: implications for research and public health," Environmental Health, vol. 11, pp. 42–51, 2012.
11. L. Trasande and Y. Liu, "Reducing the staggering costs of environmental disease in children, estimated at $76. 6 billion in 2008," Health Affairs, vol. 30, pp. 863–870, 2011.
12. R. J. Landrigan and L. R. Goldman, "Children's vulnerability to toxic chemicals: a challenge and opportunity to strengthen health and environmental policy," Health Affairs, vol. 30, pp. 842–850, 2011.
13. February 2013, http://www.marcellusgas.org/.
14. PennEnvironment Research & Policy Center, "In the Shadow of the Marcellus Boom," Tech. Rep., May 2011, http://www.pennenvironment.org/reports/pae/shadow-marcellus-boom.
15. R. L. Kosnik, "The oil and gas industry's exclusions and exemptions to major environmental statutes. Oil and gas accountability project," Tech. Rep., 2007, http://www.ogap.org/.
16. T. Colborn, C. Kwiatkowski, K. Schultz, and M. Bachran, "Natural gas operations from a public health perspective," Human and Ecological Risk Assessment, pp. 17–15, 2011.
17. T. Colburn, Spreadsheet of Products, Chemicals and Their Health Effects, The Endocrine Disruption Exchange, Paonia, Colo, USA, 2013, http://www.endocrinedisruption.org/chemicals.multistate.php.
18. M. Jerrett, R. T. Burnett, C. Arden Pope et al., "Long-term ozone exposure and mortality," New England Journal of Medicine, vol. 360, no. 11, pp. 1085–1095, 2009.
19. R. Witter, K. Stinson, H. Sackett, et al., "Potential exposure-related human health effects of oil and gas development," A White Paper, 2008, http://docs.nrdc.org/health/files/hea_08091702A.pdf.

20. L. M. McKenzie, R. Z. Witter, L. S. Newman, and J. L. Adgate, "Human health risk assessment of air emissions from development of unconventional natural gas resources," Science of the Total Environment, vol. 424, pp. 79–87, 2012.
21. D. J. Rozell and S. J. Reaven, "Water pollution risk associated with natural gas extraction from the Marcellus Shale," Risk Analysis, vol. 32, no. 8, pp. 1382–1393, 2011.
22. R. Hammer and J. VanBriesen, "In fracking's wake: new rules are needed to protect our health and environment from contaminated wastewater," Tech. Rep., NRDC, Washington, DC, USA, 2012, http://www.nrdc.org/energy/files/Fracking-Wastewater-FullReport.pdf.
23. I. Urbana, Regulation Lax as Gas Wells' Tainted Water Hits Rivers, New York Times, New York, NY, USA, 2011.
24. EPA, "EPA Releases Draft Findings of Pavillion, Wyoming Ground Water Investigation for Public Comment and Independent Scientific Review," 2011, http://yosemite.epa.gov/opa/admpress.nsf/0/EF35BD26A80D6CE3852579600065C94E.
25. K. Sinding, "Is Dimock's Water Really Safe? One Federal Health Agency Is Not so Sure," Natural Resources Defense Council (NRDC): Switchboard. September 2012, http://switchboard.nrdc.org/blogs/ksinding/is_dimocks_water_really_safe_o.html.
26. S. G. Osborn, A. Vengosh, N. R. Warner, and R. B. Jackson, "Methane contamination of drinking water accompanying gas-well drilling and hydraulic fracturing," Proceedings of the National Academy of Sciences of the United States of America, vol. 108, no. 20, pp. 8172–8176, 2011.
27. Occupational Safety & Health Administration (OSHA) and the National Institute for Occupational Safety and Health (NIOSH), "Worker Exposure to Silica During Hydraulic Fracturing," United States Department of Labor, February 2013, http://www.osha.gov/dts/hazardalerts/hydraulic_frac_hazard_alert.html.
28. M. Jerrett, R. T. Burnett, C. Arden Pope et al., "Long-term ozone exposure and mortality," New England Journal of Medicine, vol. 360, no. 11, pp. 1085–1095, 2009.

AUTHOR NOTES

CHAPTER 2

Acknowledgments

Funding was provided by the National Council for Air and Stream Improvement (NCASI) and the Waters of East Texas (WET) Center at the Arthur Temple College of Forestry and Agriculture at Stephen F. Austin State University. The Campbell Group provided land access for this study and valuable support.

CHAPTER 3

Acknowledgments

This paper is supported by the National Basic Research Program (973) subjects (2009CB219505) and Natural Science Foundation of Shandong Province (ZR2010DM012).

CHAPTER 4

Disclaimer

The report was prepared as an account of work sponsored by an agency of the United States Government. Neither the United States Government nor any agency thereof, nor any of their employees, makes any warranty, express or implied, or assumes any legal liability or responsibility for the accuracy, completeness, or usefulness of any information, apparatus, product, or process disclosed, or represents, that its use would not infringe privately owned rights. Reference herein to any specific commercial product, process, or service by trade name, trademark, manufacturer, or otherwise does not necessarily constitute or imply its endorsement, recommendation, or favoring by the United States Government or any agency thereof. The views and opinions of authors expressed herein do not necessarily state or reflect those of the United States Government or any agency thereof.

CHAPTER 5

Notes

The authors declare no competing financial interest.

Acknowledgements

We acknowledge financial support from NSF (CBET-1039172). We thank S. Attari, J. Balch, and M. Evans-White for helpful comments.

CHAPTER 6

Acknowledgments

We thank A Bergdale, R Adams, G Adams, and L Lewis for early conversations that helped develop our interest in this topic. E D'Amico provided valuable assistance on spatial analysis of well placement and suggestions that helped to shape the manuscript. A Bergdale, W Dodds, M Drew, K Fritz, and K Larson provided comments on early drafts of the manuscript. The US Environmental Protection Agency (EPA) through its Office of Research and Development partially funded and collaborated in the research described here under contracts EP-D-06-096 and EP-D-11-073 to Dynamac Corporation. The views expressed in this article are those of the author(s) and do not necessarily reflect the views or policies of the US EPA.

CHAPTER 7

Disclaimer

This document was prepared as an account of work sponsored by an agency of the United States government. Neither the United States government nor Lawrence Livermore National Security, LLC, nor any of their employees makes any warranty, expressed or implied, or assumes any legal liability or responsibility for the accuracy, completeness, or usefulness of any information, apparatus, product, or process disclosed, or represents that its use would not infringe privately owned rights. Reference herein to any specific commercial product, process, or service by trade name, trademark, manufacturer, or otherwise does not necessarily constitute or imply its endorsement, recommendation, or favoring by the United States government or Lawrence Livermore National Security, LLC. The views and opinions of authors expressed herein do not necessarily state or reflect those of the United States government or Lawrence Livermore National

Security, LLC, and shall not be used for advertising or product endorsement purposes..

CHAPTER 8

Disclaimer

Mention of trade names or commercial products does not constitute endorsement or recommendation for use.

CHAPTER 9

Acknowledgments

P. E. Lendrum, C. R. Anderson, Jr., and R. T. Bowyer were involved with research design. Lendrum and Anderson collected field data. R. A. Long assisted with resource-selection models, and J. G. Kie helped with modeling movements of deer. All authors participated in writing and editing the paper. Our project was funded and supported by the Colorado Division of Parks and Wildlife (CPW). We thank C. Bishop, D. Freddy, and M. Michaels from CPW for helping administer the project. Additionally, we thank D. Alkire, J. Broderick, P. Damm, B. deVergie, C. Flickinger, M. Grode, C. Harty, L. Kelly, T. Knowles, J. Lewis, S. Lockwood, B. Marsh, K. Maysilles, T. Parks, B. Petch, M. Peterson, M. Reitz, T. Segal, T. Swearingen, K. Taylor, and S. Wilson for field support and coordination. We thank Quicksilver Air Inc. for assistance in capturing deer from helicopters, L. Gepfert and L. Coulter for fixed-wing aircraft support, and L. Wolfe, C. Bishop, and D. Finley of CPW for crucial assistance during capture efforts. We thank D. Freddy, K. Kaal, R. Kahn, P. Lukacs, R. Velarde and G. White for assistance in initiating this research effort. Additional funding and support came from Federal Aid in Wildlife Restoration, Colorado Mule Deer Association, Colorado Mule Deer Foundation, Colorado Oil and Gas Conservation Commission, Williams Production LMT CO., EnCana Corp., ExxonMobil Production Co., Shell Petroleum, Marathon Oil Corp., and Idaho State University. We also thank the White River Bureau of Land Management, U.S. Forest Service, and numerous private land owners for their cooperation. In addition, we thank J. Jenks, J. Thiel, K. Monteith, N. Guernsey, H. Johnson, and M. Alldredge who reviewed earlier versions of this manuscript and provided valuable comments.

CHAPTER 12

Conflict of Interests

No author has any conflict of interests or financial conflicts to declare.

Acknowledgment

The authors would like to acknowledge Nitin Kondamudi for his initial research efforts and help in the preparation of this paper.

INDEX

U

V

W

X

Z

For Product Safety Concerns and Information please contact our EU representative GPSR@taylorandfrancis.com
Taylor & Francis Verlag GmbH, Kaufingerstraße 24, 80331 München, Germany

www.ingramcontent.com/pod-product-compliance
Lightning Source LLC
LaVergne TN
LVHW022313150926
848178LV00003B/52

* 9 7 8 1 9 2 6 8 9 5 8 3 3 *